T0315155

Flexible Flat Panel Displays

Wiley– SID Series in Display Technology

Flexible Flat Panel Displays, Second Edition
Darran R. Cairns, Dirk J. Broer, and Gregory P. Crawford

Liquid Crystal Displays - Addressing Schemes and Electro-Optical Effects, Third Edition
Ernst Lueder, Peter Knoll, and Seung Hee Lee

Amorphous Oxide Semiconductors: IGZO and Related Materials for Display and Memory
Hideo Hosono, Hideya Kumomi

Introduction to Flat Panel Displays, Second Edition
Jiun-Haw Lee, I-Chun Cheng, Hong Hua, and Shin-Tson Wu

Flat Panel Display Manufacturing
Jun Souk, Shinji Morozumi, Fang-Chen Luo, and Ion Bita

Physics and Technology of Crystalline Oxide Semiconductor CAAC-IGZO: Application to Displays
Shunpei Yamazaki, Tetsuo Tsutsui

OLED Displays: Fundamentals and Applications, Second Edition
Takatoshi Tsujimura

Physics and Technology of Crystalline Oxide Semiconductor CAAC-IGZO: Fundamentals
Noboru Kimizuka, Shunpei Yamazaki

Physics and Technology of Crystalline Oxide Semiconductor CAAC-IGZO: Application to LSI
Shunpei Yamazaki, Masahiro Fujita

Interactive Displays: Natural Human-Interface Techniques
Achintya K. Bhowmik

Addressing Techniques of Liquid Crystal Displays
Temkar N. Ruckmongathan

Modeling and Optimization of LCD Optical Performance
Dmitry A. Yakovlev, Vladimir G. Chigrinov, and Hoi-Sing Kwok

Fundamentals of Liquid Crystal Devices, Second Edition
Deng-Ke Yang and Shin-Tson Wu

3D Displays
Ernst Lueder

Illumination, Color and Imaging: Evaluation and Optimization of Visual Displays
P. Bodrogi, T. Q. Khan

Flexible Flat Panel Displays

Edited by

Darran R. Cairns
University of Missouri, Kansas City, USA

Dirk J. Broer
Eindhoven Technical University, Netherlands

Gregory P. Crawford
Miami University, Florida, USA

Second Edition

Registered Offices
John Wiley & Sons, Inc., 111 River Street, Hoboken, NJ 07030, USA
John Wiley & Sons Ltd, The Atrium, Southern Gate, Chichester, West Sussex, PO19 8SQ, UK

For details of our global editorial offices, customer services, and more information about Wiley products visit us at www.wiley.com.

Wiley also publishes its books in a variety of electronic formats and by print-on-demand. Some content that appears in standard print versions of this book may not be available in other formats.

Library of Congress Cataloging-in-Publication Data
Names: Cairns, Darran R., editor. | Broer, Dirk J., editor. | Crawford, Gregory P., editor.
Title: Flexible flat panel displays / edited by Darran R. Cairns, Dirk J. Broer, Gregory P. Crawford.
Description: Second edition. | Hoboken, NJ : John Wiley & Sons, 2023. | Includes bibliographical references and index.
Identifiers: LCCN 2021052707 (print) | LCCN 2021052708 (ebook) | ISBN 9781118751114 (hardback) | ISBN 9781118751060 (pdf) | ISBN 9781118750889 (epub) | ISBN 9781118751077 (ebook)
Subjects: LCSH: Information display systems. | Liquid crystal displays. | Electroluminescent display systems.
Classification: LCC TK7882.I6 F55 2023 (print) | LCC TK7882.I6 (ebook) | DDC 621.3815/422--dc23/eng/20211116
LC record available at https://lccn.loc.gov/2021052707
LC ebook record available at https://lccn.loc.gov/2021052708

Cover Image: © metamorworks/Shutterstock
Cover Design: Wiley

Set in 9.5/12.5pt STIXTwoText by Integra Software Services Pvt. Ltd, Pondicherry, India
Printed and bound by CPI Group (UK) Ltd, Croydon, CR0 4YY

C9781118751114_060223

Contents

Series Editor's Foreword

The first edition of Professor Crawford's *Flexible Flat Panel Displays* was one of the first volumes in the Wiley-SID series of technical books to be published. Since its appearance it has achieved the distinction of becoming the single most successful work in the series, based on cumulative sales. These facts illustrate the continuing importance and the technical challenge of mass producing high-quality, reliable devices in a flexible format, as well as the comprehensive analysis of the technical issues involved, which the editor and authors brought to the topic. In 2005 when the first edition was produced, the whole field of flexible displays was immature, and the volume necessarily focused mainly on the enabling technologies and technical challenges faced by those seeking to develop flexible devices and routes to their mass manufacture.

Today, the status of these devices has been transformed, and this new and completely revised edition of the book reflects that. Foldable organic light-emitting diode (OLED) displays are widely available on mobile phones and tablet devices, albeit restricted mainly to premium devices, while flexible liquid-crystal displays (LCDs) and especially electrophoretic displays have achieved maturity. The potential of large-area conformable displays to open new fields of application and product design is rapidly expanding in areas such as automotive interiors and industrial systems, while curved computer monitors are firmly established as mainstream devices. Meanwhile, the aspiration of product engineers to exploit displays with the free flexibility of paper or fabric remains problematic. The technical challenges which flexible devices face are in many cases the same as those that could be addressed in the first edition; bending any electronic device can lead to undesirable consequences ranging from a reversible shift in semiconductor characteristics through fatigue failure of different materials to catastrophic failure of conductor tracks, encapsulation, or the thin-film transistors (TFTs) themselves. However, state-of-the-art solutions to these difficulties have in many cases advanced both in technical approach and in performance in ways which could hardly be dreamed of in 2005. Encapsulation materials, conductive layers, and advanced semiconductors are examples where the technology has been transformed.

In this new edition of *Flexible Flat Panel Displays* the editors Professor Cairns, Professor Crawford, and Professor Broer have assembled a comprehensive and thoroughly updated overview of the field, key challenges that remain to be overcome, and approaches to materials, processes, and operating modes to further improve the availability, quality, and durability of flexible display screens. The advances in the technology that have been achieved since the publication of the first edition have also impacted the application space for flexible devices and forward-looking views of flexible sensor systems are also included. The work will be a valuable resource and reference, not only for scientists and engineers concerned with flexible display and electronic devices but for all interested in current developments in display technology, in integrating displays in new products, and in such diverse areas as Internet of Things and wearable devices.

January 2023

Ian Sage
Great Malvern

List of Contributors

Karen S. Anderson
Virginia G. Piper Center for Personalized
Diagnostics, The Biodesign Institute at
Arizona State University, USA

Dirk J. Broer
Eindhoven University of Technology,
Department of Chemical Engineering &
Chemistry, Laboratory of Functional
Organic Materials & Devices, Eindhoven,
The Netherlands

Darran R. Cairns
West Virginia University, Statler College of
Engineering

University of Missouri – Kansas City, School
of Science and Engineering, USA

Marco Roberto Cavallari
Departamento de Engenharia de Sistemas
Eletrônicos, Escola Politécnica da
Universidade de São Paulo, São Paulo, Brazil

Department of Renewable Energies. UNILA,
Federal University of Latin American
Integration, Foz do Iguaçu, PR, Brazil

Progyateg Chakma
Department of Chemistry and Biochemistry,
Miami University, USA

Chi-Shun Chan
AUO Display Plus Corp., Taiwan

Yi-Hong Chen
AUO Display Plus Corp., Taiwan

Janglin Che
Industrial Technology Research Institute,
Hsinchu, Taiwan

Jennifer M. Blain Christen
School of Electrical, Computer, and Energy
Engineering, Goldwater Center #208, 650 E.
Tyler Mall, ASU Tempe Campus,
Arizona, USA

Gregory P. Crawford
President, Miami University, USA

Zachary A. Digby
Department of Chemistry and Biochemistry,
Miami University, USA

Albert I. Everaerts
3M Company St. Paul, USA

Gerwin Gelinck
Holst Centre, TNO, Eindhoven,
The Netherlands

Eindhoven University of Technology,
Eindhoven, The Netherlands

Andreas Habeck
Schott AG, Germany

Alex Henzen
Electronic Paper Display Institute, South
China Normal University, China

Norbert Hildebrand
Schott North America Inc., NY, USA

Annie Tzuyu Huang
AUO Display Plus Corp., Taiwan

Silke Knoche
Schott AG, Germany

Dominik Konkolewicz
Department of Chemistry and Biochemistry,
Miami University, USA

Anke Kruse
Schott AG, Germany

Ioannis Kymissis
Professor, Electrical Engineering Columbia
University, New York, USA

Principal Engineer, Lumiode, New York, USA

Zachary A. Lamport
Electrical Engineering Columbia University,
New York, USA

Meng-Ting Lee
AUO Display Plus Corp., Taiwan

Yves Leterrier
Laboratoire de Technologie des Composites
et Polymères (LTC), Ecole Polytechnique
Fédérale de Lausanne (EPFL), Lausanne,
Switzerland

Chun-Yu Lin
AUO Display Plus Corp., Taiwan

Johan Lub
Eindhoven University of Technology,
Department of Chemical Engineering &
Chemistry, Laboratory of Functional
Organic Materials & Devices, Eindhoven,
The Netherlands

W.A. MacDonald
DuPont Teijin Films (UK) Limited

Lorenza Moro
Vice President CTO Group, Palo Alto, USA

Uwadiae Obahiagbon
School of Electrical, Computer and Energy
Engineering, Arizona, USA

Owain Parri
Merck Chemicals Ltd., Southampton, UK

Armin Plichta
Schott AG, Germany

Dr. Grzegorz Andrzej Potoczny
OPVIUS GmbH, Nuremberg, Germany

J.W. Shiu
Industrial Technology Research Institute,
Hsinchu, Taiwan

Konstantinos A. Sierros
West Virginia University, Statler College of
Engineering

West Virginia University, Department of
Mechanical & Aerospace Engineering, USA

Jean-Pierre Simonato
Director of Research, Simonato, CEA

Jonathan HT Tao
AUO Display Plus Corp., Taiwan

Robert Jan Visser
Applied Materials, Santa Clara California,
USA

Andreas Weber
Schott AG, Germany

Anthony S. Weiss
University of Missouri – Kansas City in the
School of Science and Engineering, USA

Nicholas Winch
West Virginia University, Statler College of
Engineering

Chih-Hung Wu
AUO Display Plus Corp., Taiwan

Deng-Ke Yang
Advanced Materials and Liquid Crystal
Institute, Chemical Physics Interdisciplinary
Program and Department of Physics, Kent
State University, USA

M. H. Yang
Industrial Technology Research Institute,
Hsinchu, Taiwan

Dong Yuan
Electronic Paper Display Institute, South
China Normal University, China

Guofu Zhou
Electronic Paper Display Institute, South
China Normal University, China

1

Introduction

Darran R. Cairns[1], Gregory P. Crawford[2], and Dirk J. Broer[3]

[1] West Virginia University, Statler College of Engineering and University of Missouri – Kansas City, School of Science and Engineering
[2] President, Miami University, USA
[3] Eindhoven University of Technology, Department of Chemical Engineering & Chemistry, Laboratory of Functional Organic Materials & Devices, Eindhoven, The Netherlands

1.1   Toward Flexible Mobile Devices

Displays and how we use them have gone through some major changes already in the twenty-first century. Mobile displays have developed from displaying text and some rudimentary graphics to highly interactive, high-resolution devices capable of streaming high-definition video. In addition to the advances in the display technologies, mobile devices also have high-resolution cameras, multiple internal sensors, powerful computer processers, multiple communication chips, and large area rechargeable batteries. Against this backdrop the requirements for flexible displays to be used in many mobile device applications far exceed those near the turn of the century. However, despite these challenging requirements, there are now beginning to be commercial products with amazing capabilities. An example of the type of approach that can be used to develop commercial foldable displays is described by Meng-Ting Lee et al. in Chapter 9. The significant improvements in organic light-emitting diodes (OLEDs) have opened possibilities in the development of foldable displays.

While improvements in OLEDs have been critical to recent developments in flexible displays there are a range of other critical components that would also need to be flexible for the development of truly mobile devices. One particularly important development has been in the field of transparent conductive coatings where metallic nanowires are becoming a commercial reality with important flexible properties. This is described in detail by Jean-Pierre Simonato in Chapter 5. For some applications an outer surface of glass would be very useful as an oxygen and moisture barrier or to protect underlying layers. Some examples of flexible glass are discussed in Chapter 8 by Armin Plitchta et al. For truly flexible devices large flexible power sources such as batteries will be needed, and flexible batteries are discussed in Chapter 13 by Nicholas Winch et al.

These are certainly exciting times for the development of flexible mobile devices. We have highlighted some of the key developments we discuss in this book that could be incorporated into such a device, but we will also need to understand how flexibility impacts functionality such as the rich touch input we expect. Some important aspects of integrating touch in mobile devices is described by Darran Cairns and Anthony Weiss in Chapter 15.

1.2   Flexible Display Layers

It is likely that polymer films will be used widely in flexible displays, and this raises myriad challenges, not least of these being durability. For applications where a polymer film is part of the outermost surface of the device, cuts and abrasions of the outermost polymer layer can reduce display performance. One approach to mitigate for this is to use self-healing polymers, which is discussed by Progyateg Chakma et al. in Chapter 7. One vitally important issue is damage to inorganic layers in a flexible display which can lead to cracking this is discussed by Yves Letterier in Chapter 16.

Mechanical damage is not the only thing to be considered in flexible display components. It is also important to tune the optics of polymer for the application and to mitigate for the underlying properties of polymer substrates. This can be achieved through the design of engineered polymer films as described by Bill McDonald in Chapter 2 and through the design of optical coatings as described by Owain Parri et al. in Chapter 3. The ability to tune properties in multiple ways opens a range of ways to design display components.

Two additional layers that play critical roles in the development of flexible displays are optically clear adhesives, discussed by Albert Everaerts in Chapter 6, and the thin film encapsulation layer used to protect OLEDs, discussed by Robert Jan Visser and Lorenza Moro in Chapter 12. Optically clear adhesives allow components to be laminated with minimal optical losses and enable complex stacks to be engineered and assembled. We discussed earlier how OLEDs are enabling advances in flexible devices but for OLEDs to have reasonable lifetimes they must be encapsulated—and it is this encapsulation that had enabled OLEDs to become a display of choice in flexible applications. One additional component that is required for flexible devices is a flexible backplane and Zachary A. Lamport et al. describe flexible backplanes using organic transistors in Chapter 4.

1.3   Other Flexible Displays and Manufacturing

We discussed earlier how our expectations of mobile devices has changed with expectations for high-fidelity video and computing power necessitating flexible batteries and touch sensors with a high-resolution display. We have also highlighted how OLEDs have become widely used in large part because of these expectations. However, not all devices need to play high-resolution video or require significant computing power. For applications such as e-readers the requirements are very different with low-power consumption and high-contrast ratio being more important than speed. Two important technologies that have found important niches are cholesteric liquid crystal displays, described by Deng-Ke Yang in Chapter 10, and electronic paper, described by Guofu Zhou in Chapter 11. There are currently some commercial products that can be manufactured on flexible substrates even if they are not ultimately used in a flexible form factor.

For several years roll-to-roll manufacturing has been advanced as a justification for flexible electronics because of the ability to fabricate devices in volume. There are a number of challenges with roll-to-roll fabrication and some of these are highlighted by Greg Potoczny in Chapter 17. More recently robotic deposition and direct writing is opening new approaches to manufacturing allowing for precise deposition of coatings and circuitry. This manufacturing approach is discussed by Kostas Sierros and Darran Cairns in Chapter 18.

Finally, we also include two chapters related to applications. Uwadiae Obahiagbon et al. detail some applications of flexible displays in medical applications in Chapter 19. We expect this to be an exciting area moving forward. In Chapter 14, Gerwin Gelinck describes his work on large area flexible x-ray detectors, which we believe will be useful in incorporating additional sensing in flexible devices and displays.

2

Engineered Films for Display Technology

W.A. MacDonald

DuPont Teijin Films

2.1 Introduction

Since the early 2000s there has been an explosion in interest in printed electronics and flexible displays in terms of both exploring the potential to develop new business opportunities and developing the technology base. The opportunity to exploit flexible substrates in roll-to-roll (R2R) production has excited the interest of the plastic films and associated processing and coating industries.

To replace a rigid substrate such as glass, however, a plastic substrate needs to be able to offer some or all of the properties of glass i.e. clarity, dimensional stability, thermal stability, barrier, solvent resistance, and low coefficient of linear thermal expansion (CLTE) coupled with a smooth surface. In addition, a further functionality such as a conductive layer might be required. No plastic film offers all these properties so any plastic-based substrate replacing glass will almost certainly be a multilayer composite structure [1–3]. However, not all applications require such a demanding property set and over the past decade plastic films have found application in areas broader than the flexible organic light-emitting diode (OLED) displays initially envisaged in the early days of the technology development. These include applications such as electrophoretic displays driven by thin-film transistor (TFT) arrays printed on plastic film, printed memory, and sensors. In addition there has been a smearing of the boundary between flexible devices and the use of printed electronics and/or flexible substrates in rigid devices – an example being the use of conductive films in touchscreens incorporated into smartphones and tablets.

This chapter will review the progress made in developing plastic substrates for flexible displays over the past decade and polyester films will be used as the main examples.

2.2 Factors Influencing Film Choice

2.2.1 Application Area

The requirements for the different applications areas envisaged for printable electronics are very different and will require substrates with different property sets.

This is summarized in Figure 2.1 and this classification is divided into "simple" organic circuitry, e.g. radio frequency identification (RFID), organic-based active-matrix backplanes,

Flexible Flat Panel Displays, Second Edition. Edited by Darran R. Cairns, Dirk J. Broer, and Gregory P. Crawford.
© 2023 John Wiley & Sons Ltd. Published 2023 by John Wiley & Sons Ltd.

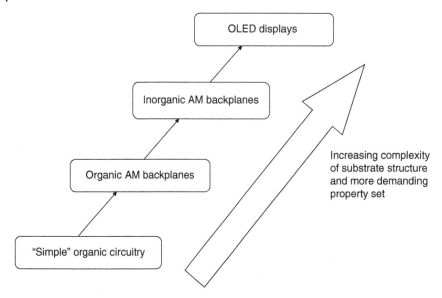

Figure 2.1   Illustration of the changing property requirements of different applications.

inorganic TFTs, and OLED displays. As one moves up the list the substrate requirements in terms of properties such as dimensional stability, surface smoothness, low coefficient of thermal stability (CLTE), conductive layers, and barrier become increasingly complex and more demanding and this will be reflected in the cost of the substrate. Substrates with low temperature stability including coated papers and oriented polypropylene film with good surface quality are adequate for some applications such as RFID and simple printed electronic devices, but a film with high temperature stability and dimensional reproducibility coupled with excellent surface quality will be required for more complex applications associated with displays such as active-matrix backplanes and OLED technology. Obviously a substrate with a property set and price commensurate with the application should be chosen and this section will focus on substrates appropriate for display application.

2.2.2   Physical Form/Manufacturing Process

The physical form of the display and whether it is

  i)   flat but exploiting light weight and ruggedness,
 ii)   conformable, one time fit to non-flat surface,
iii)   flexible and handleable, e.g. electronic newspaper, or
 iv)   rollable

will also influence film choice. Initially in the early 2000s, the vision was of flexible OLED displays and rollable displays in particular. Over the past decade, however, with the emergence of smartphones and tablets, the general public have become used to the portability and user interface of such devices and the interest in a rollable or foldable display device has waned. At the time of writing this chapter, however, serious interest in foldable and flexible displays has again emerged as the major electronic companies envisage flexibility as key to providing new innovation in the next generation of smartphones and tablets. In addition to this, flexible electronics offers robustness over glass-based rigid devices, and a further general trend on film substrates is to go thinner to reduce the bulkiness of a device.

Whether the film is manufactured by batch or R2R can influence film selection. Although R2R processing can be used for specific stages of device manufacture, for example barrier or conductive coatings, and for less complex printed electronic applications, for the most part more complex device manufacture such as active-matrix backplanes is carried out by a batch-based process on a rigid carrier. This fits with existing semiconductor manufacturing tooling equipment and the rigid carrier can also be exploited to control the dimensional stability. However, batch processing introduces new challenges such as bowing of the rigid carrier due to a mismatch in coefficient of thermal expansion and shrinkage behavior between the film and the rigid carrier. The major factors influencing this are the rigidity of the carrier (more rigid gives less bow) and the thickness of the film (less bow with thinner film). A second issue is release of the film from the carrier without damaging the printed circuitry and this remains an active area of research.

2.2.3 Film Property Set

2.2.3.1 Polymer Type

Polyethylene terephthalate (PET), e.g. DuPont Teijin (DTF) Films Melinex® polyester film, and polyethylene naphthalate (PEN), e.g. DuPont Teijin Films Teonex® polyester film, are biaxially oriented semicrystalline films [4, 5]. The difference in chemical structure between PET and PEN is shown in Figure 2.2.

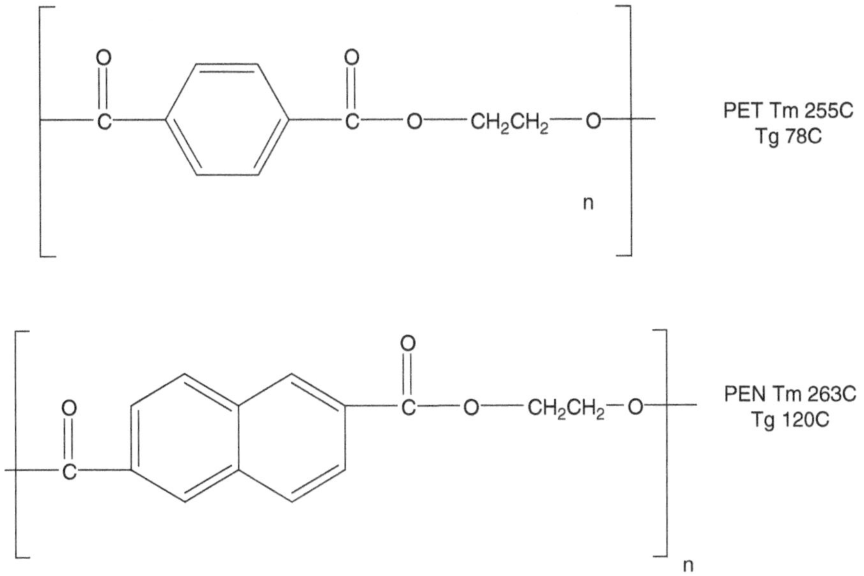

Figure 2.2   Chemical structures of PET and PEN films.

The substitution of the phenyl ring of PET by the naphthalene double ring of PEN has very little effect on the melting point (Tm), which increases by only a few degrees. There is, however, a significant effect on the glass transition temperature (Tg), the temperature at which a polymer changes from a glassy state to a rubbery state, which increases from 78°C for PET to 120 °C for PEN [2]. PET and PEN films are prepared by a process whereby the amorphous cast is drawn in both the machine direction and transverse direction. The biaxially oriented film is then heat set to crystallize the film [4, 5].

The success of polyester film in general application comes from the properties derived from the basic polymer coupled with the manufacturing process of biaxial orientation and heat setting described earlier. These properties include high mechanical strength, good resistance to a wide range of chemicals and solvents, low water absorption, excellent dielectric properties, good dimensional stability, and good thermal resistance in terms of shrinkage and degradation of the polymer chains. Fillers can be incorporated into the polymer to change the surface topography and opacity of the film. The film surface can also be altered by the use of pretreatments to give a further range of properties, including enhanced adhesion to a wide range of inks, lacquers, and adhesives. These basic properties have resulted in PET films being used in a wide range of applications, from magnetic media and photographic applications, where optical properties and excellent cleanliness are of paramount importance to electronics applications such as flexible circuitry, and touch switches, where thermal stability is key. More demanding polyester film markets, which exploit the higher performance and benefits of PEN, include magnetic media for high-density data storage and electronic circuitry for hydrolysis-resistant automotive wiring [6]. This property set provides the basis on which one can now build to meet the demands of the printed electronic market.

It is interesting to contrast these films with the other films that are currently being considered for flexible electronics applications. The main candidates are shown in Figure 2.3, which lists the substrates in terms of increasing glass transition (Tg)

The polymers can be further categorized into films that are semicrystalline (PET and PEN as mentioned earlier), amorphous and thermoplastic, and amorphous, but solvent cast. Polymers

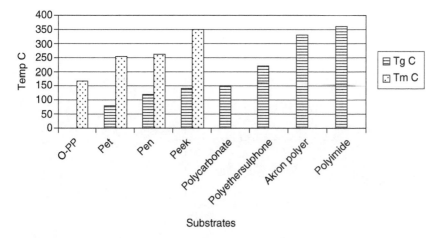

Substrates

Figure 2.3 Glass transitions of film substrates of interest for flexible electronics applications.

with a Tg higher than 150°C that are semicrystalline tend generally to have a Tm that is too high to allow the polymers to be melt processed without significant degradation – Victrex® PEEKfilm ACTIV® [7] is the highest performance semicrystalline material available in film form. The next category are polymers that are thermoplastic, but non-crystalline and these range from polycarbonate (PC), e.g. Teijin's PURE-ACE® [8] and GE's Lexan® [9], with a Tg of ~150°C to polyethersulphone (PES), e.g. Sumitomo Bakelite's Sumilite® [10], with a Tg of ~220°C. Although thermoplastic, these polymers may also be solvent cast to give high optical clarity. The third category are high-Tg materials that cannot be melt processed and include substrates based on Akron Polymer Systems (APS) resins [11] and polyimide (PI), e.g. DuPont's Kapton® [12].

PET and PEN by virtue of being semicrystalline, biaxially oriented, and heat stabilized (see Section 2.3.4) have a different property set to the amorphous polymers and for simplicity these two basic categories will be used when comparing and contrasting the properties of the film types and the importance to the property set required for flexible electronic application.

2.2.3.2 Optical Clarity

The clarity of the film is important for bottom emissive displays where one is viewing through the film and a total light transmission (TLT) of >85% over 400–800 nm coupled with a haze of less than 0.7% are typical of what is required for this application. The polymers listed earlier meet this requirement apart from polyimide, which is yellow.

Polyester films are used extensively in the light management within liquid crystal displays (LCDs). PET is the base film used in brightness enhancement films (BEF) where a prismatic coating on the surface of the film is used to recycle the scattered light and direct it through the LCD to increase the brightness.

In addition to BEF and diffuser films, highly reflective polyester films are used to reflect and recycle light from the light-guide plate in an LCD display. These films are polyester films with inorganic fillers where different levels of diffuse reflectance are achieved by generating voids during the manufacturing process in Figure 2.4. These are tailored in shape, size, and distribution by controlling filler type and film process technology to achieve different levels of performance. The total reflectance spectra of these specially designed films, e.g. Melinex® RFL1 and 2, are shown in Figure 2.5 compared to a standard white film.

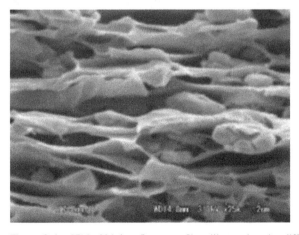

Figure 2.4 SEM of high-reflectance films illustrating the different voiding effects.

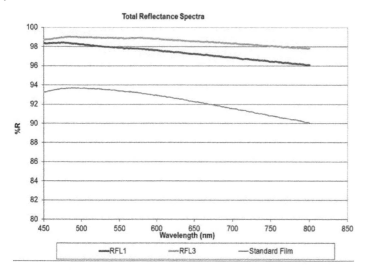

Figure 2.5   Spectra of reflectance versus wavelength for highly reflective films.

2.2.3.3   Birefringence
Biaxially oriented films such as PET and PEN are birefringent. For LCDs that depend on light of known polarization this means that birefringent films, which would change the polarization state, are unlikely to be used as base substrates. That said, polyester films are used extensively to enhance LCD performance as discussed earlier. Films based on amorphous polymer are not birefringent and are more suitable for the base substrates for LCDs. Birefringence is not an issue with OLED, electrophoretic displays, and indeed some LCDs.

2.2.3.4   The Effect of Thermal Stress on Dimensional Reproducibility
PET and PEN films (Melinex® and Teonex®) are produced using a sequential biaxial stretching technology, which is widely used for semicrystalline thermoplastics [4, 5]. The process involves stretching film in machine and transverse directions (MD and TD) and heat setting at elevated temperature. As a consequence, a complex semicrystalline microstructure develops in the material, which exhibits remarkable strength, stiffness, and thermal stability. Various studies have been made of the biaxial structure of polyester film manufactured in this way and many descriptions have been written [13, 14]. The film comprises a mosaic of crystallites or aggregated crystallites accounting for nearly 50 wt% of its material and that tend to align along the directions of stretch. Adjacent crystallites may not, however, share similar orientations. Crystallites show only a small irreversible response to temperature, which may take the form of growth or perfection. The non-crystalline region also possesses some preferred molecular orientation, which is a consequence of its connectivity to the crystalline phase. Importantly, the molecular chains residing in the non-crystalline region are on average slightly extended and therefore do not exist in their equilibrium, Gaussian conformation.

Standard PET and PEN film will shrink 1–3% at temperatures above the Tg. PET and PEN films can be further exposed to a thermal relaxation process, in which film is transported under low tension through an additional heating zone at approximately 150°C for PET and 180–200°C for PEN. Some additional shrinkage is seen, which signifies a relaxation of the molecular orientation

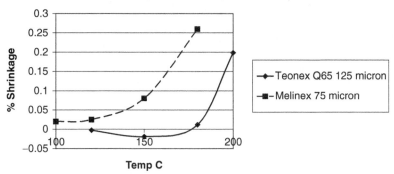

Figure 2.6 Shrinkage of heat stabilized PET and PEN film versus temperature.

in the material [1–3, 14]. Fundamental measurements of fibers and films indicate that the relaxation occurs exclusively in the non-crystalline regions [15].

Figure 2.6 illustrates how the shrinkage of heat stabilized PET (Melinex®ST504/506) and PEN (Teonex® Q65) change with temperature. For applications requiring low shrinkage above 140°C, Teonex® Q65 is the preferred option and there is continuing work within DTF to optimize the heat stabilization process without having a detrimental effect on final film properties.

In a batch-based device process it is also possible to anneal the film at temperatures around 200°C prior to processing on it. It has been shown that it is possible to achieve a level of shrinkage down to 25 ppm at 150°C with Teonex®Q65 [16].

The second factor that impacts on dimensional reproducibility is the natural expansion of the film as the temperature is cycled as measured by the CLTE. A low CLTE typically <20 ppm/°C is desirable to match the thermal expansion of the base film to the layers that are subsequently deposited. A mismatch in thermal expansion means that the deposited layers can become strained and cracked under thermal cycling. In the temperature range from room temperature up to the Tg the typical CLTE of PEN is 18–20 ppm and PET 20–25 ppm (but note that the Tg of PEN is 40°C higher than PET). Above the Tg, the natural expansion of PET and PEN films that have not been heat stabilized is dominated by the shrinkage the films undergo as the internal strains in the film relax as discussed earlier. However, the heat stabilized films discussed show only a very small increase in CLTE in the temperature range from the Tg to the temperature at which they were heat stabilized [1–3]. This contrasts favorably with the quoted coefficient of expansion of amorphous polymers, which is typically 50 ppm/°C [2, 3] below the Tg but can increase by a factor of three times above the Tg.

In addition to dimensional stability, another important factor to be considered is the upper processing temperature (T_{max}) that a film can be used at. Although, as has been outlined, the Tg does not define T_{max} with the semicrystalline polymers, it largely does with the amorphous polymers.

2.2.3.5 Low-bloom Films

Conductive coated films and their use in applications such as touch screens is outwith the scope of this chapter but an essential component of these films is that the base film in which the

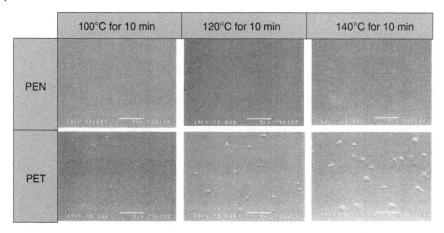

Figure 2.7 Growth of cyclic oligomers on PET and PEN film with time at 100, 120, and 140°C.

conductive coatings are deposited are dimensionally stable and do not "bloom" on heating through subsequent processing cycles. PET film contains 1.1–1.4 weight% cyclic oligomer, which can migrate to the surface of the film if held at elevated temperatures for tens of minutes [17, 18], as can occur for example in the manufacture of devices exploiting touchscreens. The presence of these oligomers on the surface gives rise to an increase in haze and this is commonly referred to as "bloom." PEN with 0.3 weight% cyclic content has significantly lower cyclic oligomer content compared with PET and there is significantly less migration to the surface of the film (Figure 2.7).

One strategy to reduce this hazing effect is to exploit planarizing coatings (see Section 2.3.8) or hard coats to act as a barrier to cyclic oligomer migration.

Control of the polymerization process and film process to reduce the cyclic oligomer content offers a further strategy to yield a "low-bloom" film. Figure 2.8 shows the scanning electron microscope (SEM) image of the surface of films aged at 150°C for 60 minutes and the reduction of cyclic oligomer on the surface of the low-bloom film can be seen compared to a "normal" film. These low-bloom films, e.g. Melinex ® TCH, typically have a haze less than or equal to 1% after aging at 150°C for 60 minutes.

As a further refinement on the low-bloom films, refractive index matching coatings have been developed that reduce the iridescence that can be observed with hardcoated films that arise because of a RI mismatch between the pretreat and the hardcoat. This is illustrated in Figure 2.9, which shows the reduction of the fringes as the RI is increased to match that of the hardcoat on a 125-micron PET film (Melinex® TCH).

2.2.3.6 Solvent and Moisture Resistance

A wide range of solvents and chemicals can potentially be used when laying down the various layers in displays depending on the processing steps involved. Amorphous polymers in general have poor solvent resistance compared to semicrystalline polymers. The solvent resistance of PET and PEN biaxially oriented films are shown in Table 2.1.

This deficiency is overcome by the application of a hardcoat to the amorphous resins that significantly improves the solvent and chemical resistance to solvents such as NMP (N-methyl-2-pyrrolidone), IPA (isopropyl alcohol), acetone, methanol, THF (tetrahydrofuran), ethyl acetate,

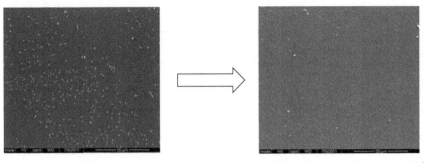

"Normal" film aged 60 mins at 150C   Low Bloom film aged 60 mins at 150C

Figure 2.8 SEM of surface of normal and low-bloom film aged at 150°C for 60 minutes illustrating the reduction in cyclic oligomer on the surface.

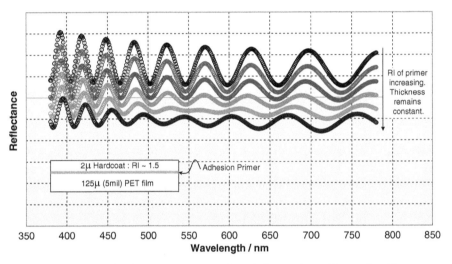

Figure 2.9 Spectra of reflectance versus wavelength for coatings of increasing RI.

98% sulfuric acid, glacial acetic acid, 30% hydrogen peroxide, and saturated bases such as sodium hydroxide [19]. With PET and PEN films a hardcoat is not required to give solvent resistance.

The residual shrinkage of Teonex®Q65 after 30 minutes at 150°C is of the order 500 ppm but can be reduced to below 100 ppm or better by careful process control – this will be discussed later. In addition to this residual shrinkage, the processing environment, in effect the prevailing humidity, must be taken into account as semicrystalline PEN will absorb up to 1500ppm of moisture under ambient conditions, yet readily loose this at higher processing temperatures.

In a previous publication [20], it has been shown that for every 100 ppm moisture absorbed the film expands by approximately 45ppm isotropically in the three film dimensions.

With the knowledge of the solubility level of moisture in PEN film and its rate of diffusion as a function of temperature, it is possible to model the impact of various environmental conditions on moisture content changes and hence volumetric changes in the different thicknesses of film. PET and PEN will typically pick up ca 1000 ppm at an RH of 40% @ 20°C and it can take up to 12 hours to reach equilibrium, depending on thickness. Although PET and PEN will reach the

Table 2.1 Table of solvent resistance.

	Unit	PEN	PET
Ketone	Acetone	Good	Good
	MEK	Good	Good
Alcohol	Methanol	Good	Good
	Ethanol	Good	Good
	Isopropanol	Good	Good
	Butanol	Good	Good
Ester	Ethyl acetate	Good	Good
Hydrocarbon	Formalin	Good	Good
	Tetrachroloethane	Good	Good
Acid	10% HCl	Good	Good
	10% HNO$_3$	Good	Good
	10% H2SO$_4$	Good	Good
	Acetic Acid	Good	Good
Alkali	10% NaOH	Good	Fair

same equilibrium level, PET picks up moisture at about three times the rate of PEN as shown in Figure 2.10.

Films can typically take over 12 hours to reach equilibrium. Figure 2.11 shows the impact of RH on this process and indicates that final (equilibrium) moisture levels will change significantly with varying RH. At an RH of 20% the film reaches an equilibrium level of ca 500 ppm but at 60% RH an equilibrium level of ca 1500 ppm (Figure 2.10).

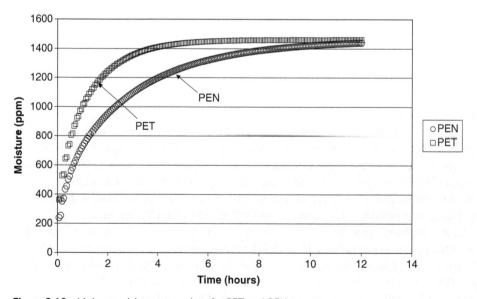

Figure 2.10 Moisture pickup versus time for PET and PEN.

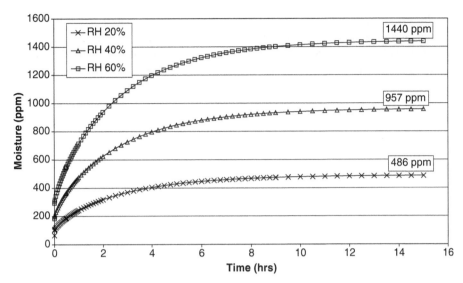

Figure 2.11   Effect of RH on moisture pickup of PEN film at 20°C.

PET and PEN films will lose moisture upon subsequent heating, but now at a much faster rate. At 150°C the film takes 6 minutes to reach an equilibrium moisture level of 5ppm, but at 90°C the film takes 30 minutes to reach an equilibrium level of 40 ppm (Figure 2.12).

Upon removal from the heated environment back into ambient conditions, the film initially picks up to 50% of its equilibrium moisture level while cooling in less than a minute or within a few seconds, but the remaining moisture to reach equilibrium is picked up at a much slower rate.

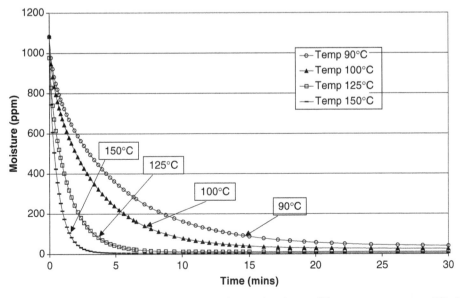

Figure 2.12   Moisture loss of PET film versus time on heating to different temperatures at RH of 40%/20°C.

Considering the variation in moisture content that can result depending upon temperature, RH, and film thickness, which can be of the order of several hundred ppm, it can be seen that uncontrolled moisture pick up will have a significant influence on dimensional reproducibility. This will be particularly significant at lower processing temperatures where there is minimal film shrinkage, but where the film takes a longer time to reach equilibrium moisture level than the time at temperature taken to carrying out the processing step. Ideally the film should be allowed to equilibrate before any registration points are established before laying down subsequent layers.

By carrying out modeling studies of the type already described, the impact of this moisture change, once understood, can be minimized.

2.2.3.7 The Effect of Mechanical Stress on Dimensional Reproducibility

One of the main drivers in moving to plastic substrates is that it opens up the possibility of R2R processing and the process and economic advantages that this brings. Under these conditions a winding tension will clearly be present and polymer film substrates with low moduli will be susceptible to internal deformation, particularly at elevated process temperatures. Figure 2.13 shows a comparison between PET and PEN films.

The storage modulus, E' is recorded using dynamic mechanical thermal analysis and as the temperature is increased, the stiffness of both materials is seen to fall. However, in the region 120–160°C PEN is significantly stiffer and stronger, with a modulus almost twice that of PET.

Insert Figure 2.13

Any small tension or load applied to film at that temperature will therefore impose strain in the material, which will be "frozen-in" upon cooling and reappear as shrinkage upon reheating. The implication of this is that if the film is constrained through a thermal processing cycle, e.g. in a R2R process or by lamination to a rigid carrier, the strain introduced into the film may result in subsequent shrinkage on reheating.

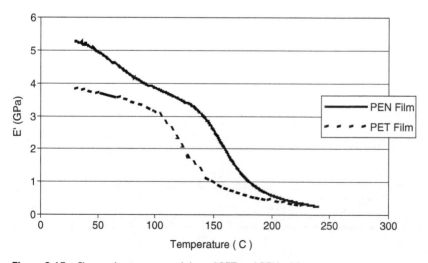

Figure 2.13 Change in storage modulus of PET and PEN with temperature.

Measures of stiffness such as Young's modulus are thickness independent and do not indicate how rigidity will change with thickness. Rigidity (D) can be defined by [20, 21]

$$D = Et^3 / 12(1 - \nu)$$

where E is the elastic modulus (6.1 GPa), t the thickness, and ν the Poisson's ratio (0.33). Assuming a Young's modulus of 2 Gpa for amorphous films and 6 Gpa for PEN film, it can be seen that (Table 2.2) 200-micron PEN film is 4 times more rigid than 125-micron PEN film and 12 times more rigid than a 125-micron amorphous film. This stiffness may prove to be an advantage in a batch-based display manufacturing process.

This extra rigidity can be exploited to reduce the stress and resulting distortion as evidenced by the difference in distortion measured through a-Si TFT process on 125-micron (5 mil) PEN compared to 200-micron (8 mil) PEN where the distortion through the process is reduced from ca 175 ppm. to 100 ppm (Figure 2.14).

Table 2.2 Comparative rigidity of amorphous and semicrystalline films.

Material	Thickness micron	Rigidity Nm x 10^4	Rigidity relative to 125 u amorphous film
Amorphous	125	5	1
Amorphous	200	20	4
PEN (Teonex®Q65)	125	15	3
PEN (Teonex®Q65)	200	61	12

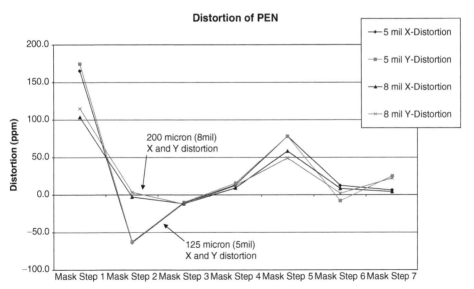

Figure 2.14 Distortion through a-Si TFT deposition. Slide courtesy of Flexible Electronic Display Center, Arizona State University.

2.2.3.8 Surface Quality

The surface quality of the film is essential for the more demanding display applications to remove any surface defects or debris that could protrude through subsequent conductive or barrier layers. Surface quality can be divided into (i) inherent surface smoothness and (ii) surface cleanliness.

2.2.3.8.1 Inherent Surface Smoothness

PET and PEN have an inherent surface smoothness of less than 1 nm, in terms of both roughness average (Ra) and root mean square roughness (Rq), as measured using white light interferometric methods [22]. However, the use of Ra and Rq can be misleading as they tend to be measured over a small sampling area and do not capture the occasional peak that occurs on the sample size of a display. The surface smoothness is largely defined by the internal cleanliness of the film substrate and typified by the presence of peaks of tens to hundreds of nm in size.

DTF have worked with a leading surface metrology tool manufacturer to develop a large area metrology (LAM) tool – the first of its type [23]. This device has the capability to defect map the surface of film fast and precisely, achieving optimum lateral resolution obtainable from white light interferometry. The equipment was designed in addition to help distinguish between both intrinsic (polymeric based) and extrinsic (external, ie air borne debris) defects.

Typical output:

1) Sampling area = 35 x 35 cm measured in around 2 hours.
2) Map of intrinsic and extrinsic defects, precise locations registered for compositional analysis.
3) Full X'Y'Z measurements of the surface defects, enabling significance testing.

Figure 2.15 shows a LAM surface analysis of a 50-micron uncoated PEN film illustrating current manufacturing quality. The graph is an exponential decay curve, expressing the intrinsic defects (area of) at a given height threshold of interest. One can imagine the height threshold as an imaginary Z plane bisecting the total X'Y field of view. The propensity of defects or opportunity to encounter one is quantified as a Six Sigma term DPMO (defects per million opportunities)

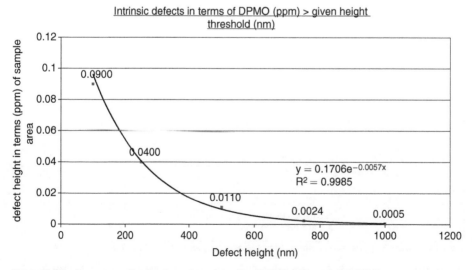

Figure 2.15 Frequency distribution of particles for PEN film (Teonex ®Q65FA).

or ppm. Since such values are effectively a fraction of the total sampled area, it can be scaled up or extrapolated for much larger and meaningful surfaces areas. For example, to put this into perspective take the example 0.01 ppm. defects of heights greater than 0.5 microns relates to of 1 mm^2 area within a total of 100 m^2 area. In other words the intrinsic defects are not the issue dominating surface quality.

2.2.3.8.2 Surface Cleanliness

In this chapter surface cleanliness is taken to mean surface defects that are dominated by dust and surface scratching. These defects can range in size up to tens of microns both laterally and vertically, and if not controlled can have a significant impact on device fabrication. A recent analysis of film entering a cleanroom carried out with CPI Printable Electronics Centre [24] and Teknek Ltd. [25] showed that the external debris on the surface was predominantly debris associated with human, packaging, and process interactions – this is largely unavoidable in film that is handled and slit in a non-cleanroom environment. There are several strategies to address this:

a) Hygiene issues associated with handling transport and packaging of films, e.g. use of plastic rather than cardboard cores, sharp blades to minimize debris on slitting, and the double bagging of the film to ensure dust is not trailed into the cleanroom.
b) Control of static as film is unwound from a reel.
c) Contact and non-contact cleaning technologies.
d) The use of protect films to protect the surface of the film from both external debris and surface scratching.
e) Planarizing coatings, coated in a cleanroom environment [2, 3] (see Section 2.4). These coatings have the dual function of providing both surface smoothness over wide area and a certain surface hardness to prevent scratching of the surface during film processing.

This subject area of minimizing the defects on the surface and at the same time being able to measure defects during actual film line processing is of considerable activity at the present. Perfection is unlikely and, given that device manufactures now claim the same yields on films that they achieve on glass (see Section 2.5), is probably not required. Understanding, however, which defects are critical and targeting elimination of them is where the focus currently is and this is an achievable target.

2.3 Summary of Key Properties of Base Substrates

The main properties of heat-stabilized PET (Melinex® ST504/506) and PEN (Teonex® Q65) relevant to flexible electronics are summarized in Figure 2.16.

Unstabilized PET and PEN films exhibit the same property set apart from the shrinkage and upper processing temperature, which for these films is largely dictated by the Tg.

It remains difficult to directly compare the property sets of different film types as the public domain information may not represent the films targeted for printed electronics applications. Table 2.3 shows the author's attempt to contrast the property set of the different substrates under consideration for flexible electronics and should only be used as a rough guide.

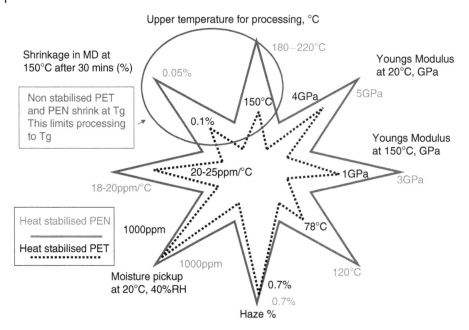

Figure 2.16 Summary of PET and PEN properties relevant to flexible electronics.

Table 2.3 Comparison of the key properties relevant to flexible electronics.

Property	Biaxially oriented heat-stabilized PET	Biaxially oriented heat-stabilized PEN	Biaxially oriented PEEK	PC	PES	Akron APS	Polyimide
CTE (-55–85 °C) ppm/°C	√√	√√	√√	√	√	√	√√
%Transmission (400–700 nm)	√√	√√	√√	√√√	√√	√√	X
Water absorption %	√√	√√	√√	√	X	Unknown	X
Young's modulus (Gpa)	√√	√√	√√	√	√	√	√
Tensile strength (Mpa)	√√	√√	√√	√	√	√	√
Solvent resistance	√√	√√	√√	X	X	√√	√√
Upper Operating Temp	√	√√	√√√	√	√√	√√√	√√√
Availability at commercial scale	√√√	√√√	√	√√√	√√√	√	√√√

The general properties have been discussed in the text and also in previous papers [2, 3, 20]. This table shows that both heat-stabilized PET (e.g. Melinex® ST504) and heat-stabilized PEN (Teonex®Q65A) have an excellent balance of the key properties required for flexible electronics and it is this property set coupled with commercial availability in grades tailored for flexible electronics that is leading to both Melinex® ST and Teonex®Q65A emerging as the leading films for the base substrate of the more demanding flexible and printed electronic applications. Victrex® PEEK film is commercially available in cast film form but to unlock the full potential of this high-performance polymer the material would have to be available in biaxially oriented form and, although development quantities are available, the cost of scale-up for what at the moment is a relatively small market remains a challenge. The Akron film developments also offer high temperature performance but face the same cost of scale-up hurdle.

The ideal substrate for high-performance display applications would have the key attributes obtained with semicrystalline and biaxially oriented polymers, i.e. low CTE, excellent solvent resistance, and the thermal stability of amorphous polymers such as polyimide, but with high clarity. Additionally, the film should be commercially available and available with the film surface quality that is only really achieved by manufacture at volume scale. Although research is being carried out on new materials such as clear polyimide and small quantities of these materials are being sampled, the development and production costs of developing a new film based on a new polymer possibly requiring novel raw materials coupled with access to both polymer and film-making facilities are prohibitively expensive, for what is a relatively small volume market. The time periods to commercialize a new film are also lengthy. In the author's opinion, this is not an attractive proposition for a substrate supplier and a more likely route forward is to push existing commercially available films to their limits and marry this with the advances that are being made in low-temperature processing of inorganic materials.

Prototype displays and backplanes have in the past been produced on flexible stainless-steel substrates but there is virtually no reported activity on this at present. Flexible glass obviously offers the advantages of glass with respect to barrier (although edge sealing remains a problem common with plastic based barrier materials), but the availability of the flexible glass and the issue of major investment to commercialize manufacture for a relatively small market remains a challenge to this industry.

2.4 Planarizing Coatings

The surface quality of the film can be controlled to an extent via recipe control, plant hygiene, web-cleaning techniques, and film line-processing conditions. The film, however, is typically not manufactured in an environment that is clean enough to give a level of extrinsic contamination suitable for the more demanding applications in flexible electronics.

While a number of cleaning techniques can reduce extrinsic contamination, it is difficult to remove them all completely. To achieve the surface quality required for most electronic devices one option taken by DTF has been to apply a planarizing coating. This coating covers all surface defects and comprises a low viscosity liquid that flows easily over the film imperfections to create a glass-smooth surface with low defects over wide area. Ideally, the planarizer coating is coated in a cleanroom to ensure the surface has low intrinsic and extrinsic contamination. The planarizer coating also has anti-scratch properties preventing scratching of the film during subsequent processing and handling.

The choices of chemistry and coating thickness of the planarizing coating have an impact on the physical properties of the planarized film. Factors to consider include maximum processing temperature during film processing and the chemistry of the layers that will be deposited on the planarizer surface. Suitable chemistry types that offer pencil hardness a >2-hour range from predominantly inorganic materials, e.g. sol gel derived coatings based on siloxane chemistry to organic materials such as polyfunctional acrylate coatings.

Figures 2.17 and 2.18 illustrate the impact of the planarizing coating. These surface peak heights in terms of propensity and height are clearly seen to reduce due to the "smoothing" effect of the planarizing coating. In Figure 2.17 it can be seen that there are no peaks above 250 nm in an area of 5 × 5 cm with the planarized sample whereas the "raw" PEN film surface in this particular study has peaks up to 600 nm.

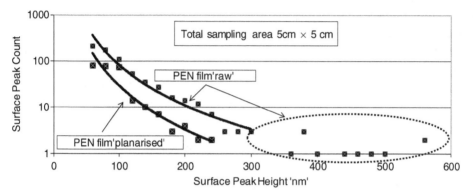

Figure 2.17 Effectiveness of planarizing coating on surface roughness of PEN film as measured by white light interferometry.

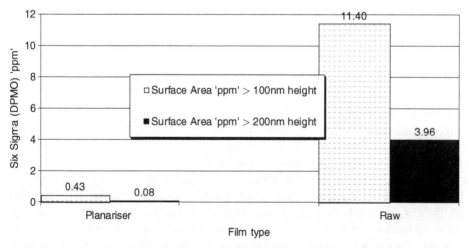

Figure 2.18 Shows that there is a factor of 50 reduction in occupied surface area of peaks greater than 200 nm in height for planarized PEN film compared to standard PEN film.

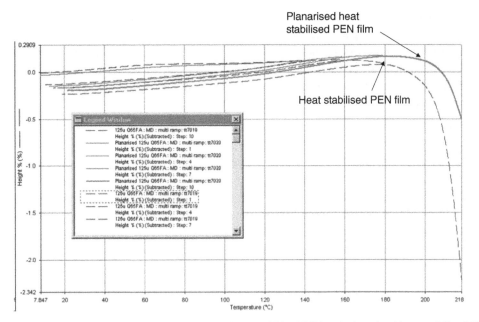

Figure 2.19 Thermal mechanical analysis of heat-stabilized PEN and planarized heat-stabilized PEN.

Another key benefit from planarizing the surface will be to reduce the surface slopes associated with such peak heights, i.e. render them less "sharp" and so less disruptive to subsequent coatings during device manufacture.

The thermal mechanical analysis carried out on heat-stabilized PEN and an inorganic–organic hybrid planarized-coated heat-stabilized PEN using a Thermomechanical Analyzer – TMA-7 (Perkin Elmer) are shown in Figure 2.19. This shows that planarized PEN film dimensional response to temperature cycling is very similar to uncoated PEN, i.e. the planarized film remains a dimensionally reproducible substrate up to 180°C. This is within the performance requirements of a flexible substrate for a display based on a-Si-H TFT.

The planarized-coated PET and PEN films show excellent chemical resistance to common solvents used in photolithography processes and solution based organic semiconductors.

2.5 Examples of Film in Use

There have been significant advances in the development of electronic paper-type applications based on electrophoretic displays driven by organic active-matrix backplanes. One of the lead technology pioneers in this area, Plastic Logic Ltd. [24], claim to achieve the same yields on polyester film as they can achieve on glass. Displays of a commercial quality are at the time of writing this chapter being trialed in different application areas such as bus information signs (exploiting low power, readability in bright light) and second screen displays for smartphones, which are always on (exploiting low power and robustness).

The key film features required for these applications are:

i) dimensional stability at processing temperatures to obtain dimensional reproducibility through multilayer processing (discussed in 2.2.3.4),

ii) excellent surface smoothness to ensure surface artifacts don't interfere with the organic TFTs (discussed in 2.2.3.8 and 2.4), and

iii) solvent resistance – the films must be able to withstand the solvents used in TFT manufacture (discussed in 2.2.3.6).

The relatively low processing temperatures of organic semiconductor materials plays to the property set of plastic films such as PET and PEN.

Similarly, significant advances are being made on reducing inorganic deposition temperatures to allow fabrication of inorganic TFTs on plastic film at temperatures below 200°C [26]. In this case the property set required of the base substrate is more demanding, in particular with respect to dimensional reproducibility at elevated temperatures, and the higher temperature performance of PEN is required. The Flexible Electronic Display Center (FEDC) [27] at Arizona State University have demonstrated the ability to manufacture mixed-oxide TFT arrays on PEN and have also increased the processing area size in the manufacture of mixed-oxide TFT arrays from 150-mm carriers to 370 mm x 470 mm scale without any yield losses due to bond failure or bow warpage. This technology has been exploited to manufacture prototype flexible X-ray detectors and flexible color OLED devices.

The aforementioned exceptional performances demonstrate that the film can be pushed beyond its data sheet specification if careful control of the key factors discussed in this chapter – thermal stress (offline stabilization and reducing processing temperatures) (Section 2.2.3.4), environment to minimize the effect of moisture (Section 2.2.3.6), and mechanical stress (exploiting rigidity through a rigid carrier) (Section 2.2.3.7) – are achieved, and by matching and optimizing device fabrication against the property set of a given film.

The plastic substrates that are under consideration as flexible substrates typically have barriers of the order 1–10 g/m^2/day for water vapor transmission rate and ca 1–10 mL/m^2/day for oxygen. OLED displays will require water vapor transmission rates of $< 10^{-6}$ g/m^2/day and oxygen transmission rates of $< 10^{-5}$ mL/m^2/day; and from the start of interest in flexible electronics achieving barrier films of this performance on flexible substrates has remained one of the key challenges. Detailed discussion of barrier technology is outside the scope of this chapter, but significant progress has been achieved over the past decade exploiting multilayer organic and inorganic coatings and barrier materials with water vapor transmission of the order 10^{-3} g/m^2/day are becoming commercially available. Higher performance barrier materials are claimed. However, the limited availability of barrier films of the order 10^{-6} g/m^2/day has limited the development of flexible OLED devices. Achieving a cost-effective high-barrier film still remains a key target for this industry and film with low surface contamination is critical to achieving this. Strategies to achieving this have been outlined in Sections 2.2.3.8 and 2.4 and this remains an area of active development.

2.6 Concluding Remarks

This chapter has only addressed the base substrate on which subsequent processing including conductive and barrier coatings in addition to electronic device fabrication will be carried out. The demanding property set required for flexible displays is pushing the existing commercially available films to the limits of their performance, but as has been discussed in this chapter, the approach of the substrate developer working closely with the display manufacturer and each understanding the limitations of their respective technologies is enabling significant progress in fabricating flexible electronics devices.

Acknowledgments

The author would like to acknowledge the significant contributions of his colleagues Robert Eveson, Duncan Mackerron, Kieran Looney, and Karl Rakos

References

1 Crawford, G. (ed.) (2005). *Flexible Flat Panel Displays*. John Wiley & Sons Ltd.
2 MacDonald, W.A. (2004). Engineered films for display application. *J. Mat. Chem.* 14: 4.
3 MacDonald, W.A., Rollins, K., MacKerron, D., Eveson, R., Rakos, K., Adam, R., Looney, M.K., and Hashimoto, K. (2007). *J. SID* 15(12): 1075–1083.
4 MacDonald, W.A., Mackerron, D.H., and Brooks, D.W. (2002). PET film and sheet. In *PET Packaging Technology* (ed. D.W. Brookes), 116–137. Sheffield Academic Press.
5 MacDonald, W.A. (2003). *Polyester Film, Encyclopedia Polymer Science & Technology*, 3rd e. John Wiley & Sons, Inc.
6 MacDonald, W.A., Mace, J.M., and Polack, N.P. (2002). New developments in polyester films for display applications, 45th Annual Technical Conference Proceedings of the Society of Vacuum Coaters, 482.
7 APTIV^{TM} Films, Victrex Technology Centre, Hillhouse International Thornton Cleveleys, Lancashire, FY5 4QD, UK.
8 PURE-ACE^{TM}, Teijin Chemicals Ltd, 1-2-2 Uchisaiwai-cho, Chiyoda-ku, Tokyo, Japan.
9 Lexan^{TM}, GE Plastics, One Plastics Ave, Pittsfield, MA 01201, USA.
10 Sumilite^{TM}, Sumitomo Bakelite Co. Ltd.,Ten-Nouzu Parkside Blgd., 5-8,2-Chome, Higashi-Shinagawa, Shinagawa-Ku, Tokyo, Japan.
11 Akron Polymer Systems, Inc., 62 N. Summit St. Akron, OH 44308, USA.
12 Kapton^{TM}, DuPont High Performance Films, P.O. Box 89, Route 23 South and DuPont, Road, Circleville, OH 43113, USA.
13 Adlen, M.J., Kuusipalo, J., MacKerron, D.H., and Savijarvi, A.-M. (2004). *J. Mat. Sci.* 39: 6909–6919.
14 MacDonald, W.A., Rollins, K., Eveson, R., Rustin, R.A., and Handa, M. (2003). Plastic displays – new developments in polyester films for plastic electronics. *Soc. Inf. Display Digest Techn. Papers* 264–267.
15 Gohil, R.M. (1994). *J. Appl. Polym. Sci.* 52: 925–944.
16 Sarma, K.R., Chanley, C., Dodd, S., Roush, J., Schmidt, J., Srdanov, G., Stevenson, M., Wessel, R., Innocenzo, J., Yu, G., O'Regan, M., Macdonald, W.A., Eveson, R., Long, K., Gleskova, H., Wagner, S., and Sturm, J.C. (2003). Active matrix OLED using 150°C a-Si TFT backplane built on flexible plastic substrate, Proceedings of the SPIE Aerosense, Techologies and Systems for Defense and Security, April 22–25.
17 Shiono, S. (1979). *J. Pol. Sci. Pol. Chem. Ed.* 17: 4123–4127.

18 Holland, B.J. and Hay, J.N. (2002). *Polymer* 43: 1797–1804.

19 Angiolini, S., Avidano, M., Bracco, R., Barlocco, C., Young, N.G., Trainor, M., and Zhao, X.-M. (2003). High performance plastic substrates for active matrix flexible FPD, *Soc. Inf. Display Digest Tech. Papers*, 1325–1327.

20 Klauk, H. (ed.) (2006). *W.A. MacDonald in Organic Electronics: Materials, Manufacturing and Applications*. Wiley-VCH. See chapter 7.

21 MacDonald, W.A., Rollins, K., MacKerron, D., Eveson, R., Rustin, R., Adam, R., Looney, M.K., and Hashimoto, K. (2005). Eurodisplay 2005. *Proceedings of the 25th International Display Research Conference*, 36–41.

22 International Standard ISO 25178-604 Geometrical product specifications (GPS) – Surface texture: Areal. Part 604 Nominal characteristics of non-contact (coherence scanning interferometry) instruments.

23 Rakos, K. (2014). WO2014/045038 A1. *Metrol. Method*.

24 CPI Printable Electronic Centre. NETPark Thomas Wright Way, Sedgefield, County Durham, TS21 3FG, UK.

25 Teknek Ltd. River Drive, Inchinnan Business Park, Renfrewshire, PA4 9RT, UK.

26 Wong, W.S., Lujan, R., Daniel, J.H., and Limb, S. (2006). Digital lithography for large-area electronics on flexible substrates. *J. Non. Cryst. Solids* 352: 1981–1985.

27 Flexible Electronic and Display Center at Arizona State University, Arizona State University Research Park, 7700 South River Parkway, Tempe, Arizona 85284-1808.

3

Liquid Crystal Optical Coatings for Flexible Displays

Owain Parri[1], Johan Lub[2], and Dirk J. Broer[2,3]

[1] *Merck Chemicals Ltd., University Parkway, Southampton, SO16 7QD, UK*
[2] *Eindhoven University of Technology, Department of Chemical Engineering and Chemistry, Laboratory of Functional Organic Materials and Devices, Den Dolech 2, 5612 AZ Eindhoven, The Netherlands*
[3] *Eindhoven University of Technology, Department of Chemical Engineering & Chemistry, Laboratory of Functional Organic Materials & Devices, Eindhoven, The Netherlands*

3.1 Introduction

In many flexible display designs optical coatings enhance the viewing experience or are an essential part of the display optics. For instance, color filters and polarizers are common in liquid crystal display (LCD) optics whereas optical retarders enhance viewing angle, contrast and color purity or create new functions like three-dimensional viewing experiences. And even non-liquid crystal-based displays like organic light-emitting diode (OLED) can be enhanced by adding liquid crystal-based optical functions such as a circular polarizer for contrast enhancement. For these functions liquid crystal network (LCN) technology based on reactive mesogens (RMs) is an interesting candidate as it can be utilized as thin-film coatings with the possibility of roll-to-roll manufacturing or patterning by lithographic procedures.

In this chapter we review the LCN technology and the various optical film technologies that can potentially be integrated in flexible display designs. Often, they are not explicitly developed for flexible display applications, but can be adapted to and employed in flexible applications. We will discuss the basic principles of LCN technology and their use for thin-film polarizers, retarders, and color filters.

3.2 LCN Technology

Crosslinked glassy LCNs are obtained from the polymerization of multifunctional liquid-crystal monomers (Figure 3.1). These materials were developed in the 1980s at Philips Research where monomer 1 (Figure 3.1a) formed the basis for the generation of a whole family of comparative monomers [1–4]. The monomers, further referred to as RMs, exhibit a liquid crystalline phase with a particular molecular orientation before polymerization and the orientation can be retained after polymerization. Preferably the polymerization is initiated by light addressing a dissolved photoinitiator, providing freedom to choose the phases independent of the polymerization temperature. Copolymerization of monoacrylates such as RM348 (**4**, Figure 3.1b) with diacrylates such as RM82 (**2**, Figure 3.1b) generate LCNs with side-chain (pendant) and main-chain mesogenic units. The advantages of this approach are numerous. Within reasonable limits the polym-

Flexible Flat Panel Displays, Second Edition. Edited by Darran R. Cairns, Dirk J. Broer, and Gregory P. Crawford.
© 2023 John Wiley & Sons Ltd. Published 2023 by John Wiley & Sons Ltd.

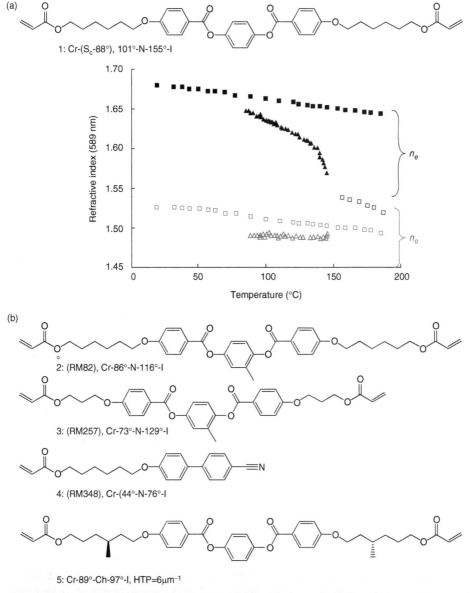

(a)

1: Cr-(S$_c$-88°), 101°-N-155°-I

(b)

2: (RM82), Cr-86°-N-116°-I

3: (RM257), Cr-73°-N-129°-I

4: (RM348), Cr-(44°-N-76°-I

5: Cr-89°-Ch-97°-I, HTP=6µm^{-1}

Figure 3.1 (a) Photopolymerization of RMs. RM monomer 1 synthesized in an early stage of RM development and its refractivive indices before and after polymerization at 110°C. (b) Examples of other RM monomers (the RM values between brackets refer to Merck notations).

erization temperature can be freely chosen, which enables the desired phase to be retained. Similar to low-molar-mass liquid crystals, the molecular alignment can be manipulated by external boundary conditions and stimuli, including surface alignment materials, surfactants, electric/magnetic/optical fields, or shear forces to prepare engineered materials with complex properties and alignments that are retained indefinitely upon polymerization. The ability to freeze-in the three-dimensional structure of the liquid crystalline phase in polymeric form has generated a number of compelling application possibilities as will be discussed in this review. The methods to produce such structures in conventional polymeric materials are limited in number and no process that we are aware of is able to generate such modularity and programmability. The interested reader can find more exhaustive reviews of the materials chemistry and processing methods to prepare LCNs in [4–6].

3.3 Thin-film Polarizers

Polarizers are key components in all LCD designs. A polarizer film consists of a thin dichroic sheet that transmits the desired polarization component and absorbs the unwanted polarization component. Currently, sheet polarizers are fabricated by stretching films, typically polyvinyl alcohol, containing rod- or needle-like molecules or crystals that align in the stretching direction. To protect from moisture and to prevent relaxation from the stretched polyvinyl alcohol, they are connected at both sides to a laminated film. Generally, cellulose triacetate films are used, although non-birefringent films are occasionally used to integrate optical retarder films into the polarizers either to reduce the overall thickness or to save costs. These polarizers are thermally stable up to 80°C and function in their ordinary mode (O-mode). Typical sheets polarizers can be on the order of hundreds of micrometers thick and in flexible displays they will inhibit bending and rolling of plastic substrates. Therefore, thin-film polarizer solutions are desired, both for flexible LCDs and for bendable OLED displays. In the latter case, circular polarizers, consisting of a dichroic polarizer and a quarter wave retarder film, improve the display on daylight contrast extinguishing light reflected from the backplane metallic electrode [7].

3.3.1 Smectic Polarizers

The performance of LCDs is strongly influenced by the quality of the polarizer. The most elegant way to define the polarization performance is the dichroic ratio in absorbance. The dichroic ratio is a materials property and therefore it is independent of the thickness of the polarizer, unlike the contrast ratio that increases with the thickness at the expense of the transmittance. The dichroic ratio (DR) in absorbance can be determined via polarized absorption spectroscopy and is defined as:

$$DR = \frac{A_{\parallel}}{A_{\perp}}$$

where $A_{//}$ and $A_{\perp}$ are defined as the absorbance parallel and perpendicular to the average orientation axis of the dye molecules, respectively.

For mobile applications of LCDs, polarizers with dichroic ratios of approximately 35 are standard in current products. For the high-end applications such as thin-film transistor (TFT) monitors and liquid crystal TVs, the requirements for the polarization performance are more demanding. Polarizers with dichroic ratios exceeding 40–50 are currently used in these applications. Numerous advantages are foreseen when the traditional sheet polarizers are replaced by ultrathin coated polarizers improving bendability, but also can be located inside the LCD cell enclosing the liquid crystal, color filter, electrodes, and TFTs (in-cell). Apart from a significant reduction in display thickness and weight, the positioning of the polarizers inside the cell is beneficial to the robustness of the display avoiding the scratch sensitive cellulose triacetate film at the exterior or the addition of a scratch resistant top layer. As the polarization optics is limited in the cell, new flexible materials can be employed to form a plastic LCD cell, such as polyesters that are normally rejected because of their birefringence.

We developed high-contrast thin coatable polarizers based on smectic guest–host reactive liquid crystals. Upon alignment of host liquid crystalline diacrylates, the guest dye molecules such as **6** (Figure 3.2), due to their designed elongated structure, co-align along the director resulting in large absorption of the dye molecules for light with the electrical field vector parallel to the high-polarization axes of the dye molecules. Upon photopolymerization the dichroic properties of the film are stabilized [8].

The optical properties of compound **1** provided with dichroic diazo dye are presented in Figure 3.2b. The film is polymerized at 120°C, in the nematic phase. From the absorbance values a dichroic ratio of ˜6 is calculated. This value, which is too low for application in LCDs, relates to an order parameter of around 0.6, a normal value for liquid crystals in their nematic phase. In order to induce higher ordered liquid crystal phases, the structure of the mesogenic group was changed. Compounds **7**, **8**, and **9** (Figure 3.3) exhibit smectic phases. Furthermore, they exhibit a nematic phase at higher temperatures, which is important to establish defect-free long-range alignment that is more difficult to establish in the highly viscous smectic phase. For fabrication, macroscopic and defect-free alignment is established in the nematic phase of the monomer. This is maintained during cooling to the smectic phase even though the molecular arrangement and packing changes. Then the alignment is fixed by polymerization to become temperature independent.

Polarizer films made from compounds **7**, **8**, and **9** modified with dichroic dye **6** (Figure 3.2) exhibit a dichroic ratio of 10, 30, and 32 respectively. The smectic A phase of **7** does not reveal a high dichroic ratio. The smectic B compounds **8** and **9** have considerable higher dichroic ratios, but still somewhat low for practical applications. It was observed that before polymerization the films prepared with **8** and **9** exhibit higher dichroic ratios while upon polymerization these values decrease by 50%. The way to create higher-ordered smectic polymers is by increasing the length of the spacer.

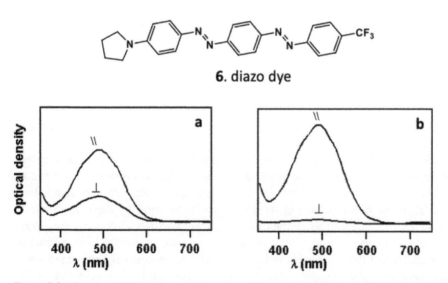

6. diazo dye

Figure 3.2 Polarized UV/V is absorption spectra of dichroic dye 6 (a) embedded in a nematic polymer network and (b) a smectic B polymer network (c).

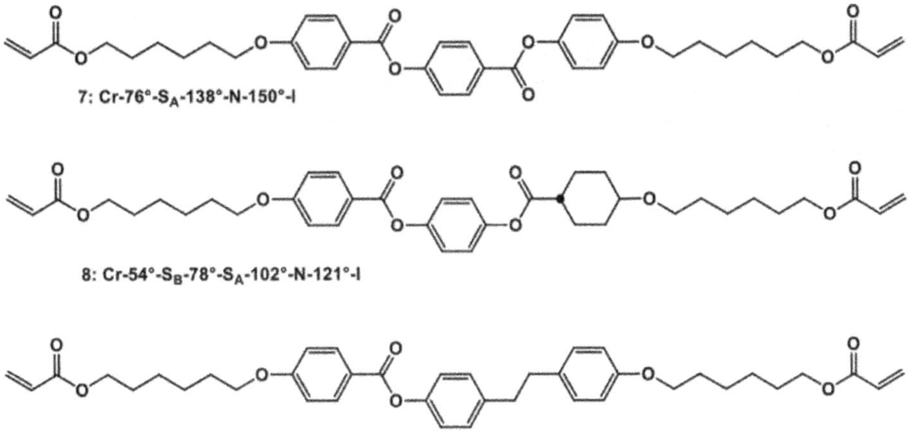

7: Cr-76°-S$_A$-138°-N-150°-I

8: Cr-54°-S$_B$-78°-S$_A$-102°-N-121°-I

9: Cr-72°-S$_B$-88°-S$_A$-105°-N-107°-I

Figure 3.3 Examples of smectic diacrylate RMs.

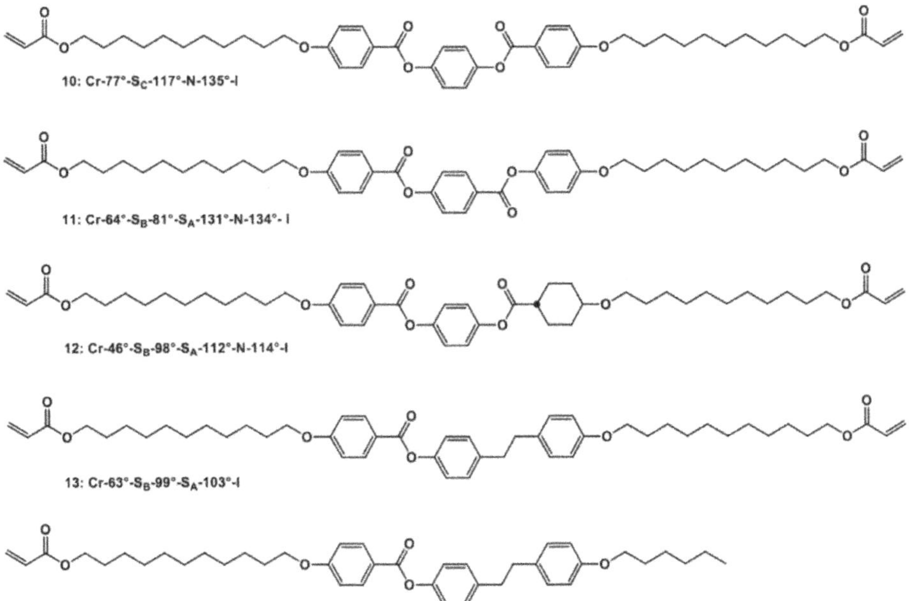

10: Cr-77°-S$_C$-117°-N-135°-I

11: Cr-64°-S$_B$-81°-S$_A$-131°-N-134°- I

12: Cr-46°-S$_B$-98°-S$_A$-112°-N-114°-I

13: Cr-63°-S$_B$-99°-S$_A$-103°-I

14: Cr-60°-S$_B$-89°-S$_A$-106°-N-112°-I

Figure 3.4 Examples of smectic diacrylate RMs with longer spacer lengths between the mesogenic center and the polymerizable acrylate moiety.

Thereto the hexamethylene spacers of the compounds 1, 7, 8, and 9 were replaced by undecamethylene spacers revealing the series of diacrylates 10, 11, 12, and 13 presented in Figure 3.4.

Compound **10** has a smectic C phase between its crystalline state and a nematic phase, which because of its tilt is not useful for the opted application. Compound **11** exhibits a smectic B phase next to a smectic A and a nematic phase and is potentially an interesting compound. The same is the case for **12**. Films made from these compounds have, after polymerization, dichroic ratios over 50, which means that the compound is very suitable for polarizer applications (Figure 3.4). Compounds **8** and **12** can be polymerized at 30°C in their supercooled state, leading to high-order parameter and related high dichroic ratios. Because the polymerization at this temperature does not reach full conversion, a post cure is needed at elevated temperatures. This can be done without affecting this high dichroic ratio. Compound **13** exhibits a broad smectic B phase. However, it lacks the nematic phase that is needed for easy alignment and is therefore difficult to use. After mixing it 1:1 with monoacrylate **14** of similar structure, the nematic phase is retained and needed for defect-free uniaxial planar alignment prior to polymerization. Also, these mixtures give polymerized films with dichroic ratios higher than 50. Films made from compound **12** show before polymerization dichroic ratios >70. The use of longer spacers has a positive effect on the order parameter when compared to compound **8** that contains a hexamethylene spacer. Furthermore, upon polymerization the decrease in dichroic ratio of the undecyl derived compounds **12** is much less dramatic than with hexamethylene derived compound **8**. The longer spacer apparently decouples the order of the central aromatic core from the steric effects of the polymeric chains formed after crosslinking. We anticipate that the interaction of the aromatic cores with the dye molecules determine the order of the latter, which explains the improved dichroic ratio. Thus, this thin-film polarizer technology based on liquid crystalline (di)acrylates exhibiting the smectic B phase is highly promising and may prove to be an attractive alternative for traditional sheet polarizers in LCD and OLED applications.

3.3.2 Cholesteric Polarizers

Nearly all LCDs visualize their images by making use of polarized light. Therefore, one of the principal components of an LCD is the polarizer. In a transmissive LCD light is emitted by a backlight, often consisting of a planar waveguide side lit by a cold cathode lamp or a light-emitting diode, toward the viewer. A sheet polarizer polarizes transmitted light by absorbing the unwanted polarization. This means a loss of the backlight intensity of at least 50%.

Cholesteric films are able to generate polarized light by transmitting one polarization direction and reflecting the other. The reflected light can be recycled in the backlight during which it can be converted into the desired polarization such that a much more efficient polarization device can be made. Disadvantages of the cholesteric films are that they generate circularly polarized light instead of linearly polarized light and they are only effective for a limited bandwidth and not for the whole visible spectrum. The first problem is solved easily by converting the circularly polarized light into linearly polarized light with the aid of a quarter wave plate, in a preferred design also consisting of a uniaxially aligned LCN with appropriate thickness and birefringence (see also Section 3.4). To solve the second problem, the bandwidth of the reflection band of the cholesteric films, which normally amount to about 60 nm, should be increased to cover the visible spectrum. In theory this can be realized by an increase in birefringence of the liquid crystals. However, these materials have disadvantages such as a low stability and absorption bands in the wavelength region of interest.

A better solution to this problem is the production of a cholesteric film in which the pitch and thus reflection wavelength has a gradient such that the blue part of the spectrum is polarized at one side of the film and the red part of the spectrum at the opposite side. Consequently, green becomes polarized halfway. To produce such a film use is made of the kinetics of the photopolymerization reaction [9, 10]. A film is made from a mixture of cholesteric diacrylate **5** [11] and nematic monoacrylate **15**, shown in Figure 3.5. The composition is chosen such that in the monomer state prior to polymerization the film reflects in the green part of the spectrum. By adding an ultra-violet (UV)

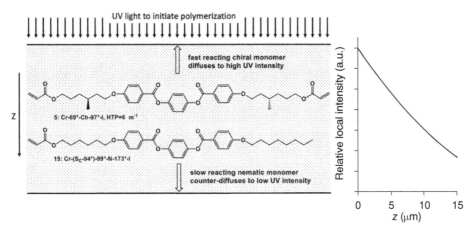

Figure 3.5 Intensity gradient for UV light in polarizing chiral-nematic film and diffusion direction. (S,S)-1,4-di-(4-(6-acryloyloxy-3-methylhexyloxy)benzoyloxy)benzene (5) is the fast reacting component and 4-(4-(6-(acryloyloxy)hexyloxy)benzoyloxy)phenyl 4-(octyloxy)benzoate (15) the slower reacting component.

absorbing dye, an intensity gradient of the UV light in the transverse direction is obtained, as shown on the right side of Figure 3.5. Due to the UV intensity gradient, the polymerization at the top proceeds faster than at the bottom of the layer. The cholesteric component **5** is a diacrylate and therefore has a twice as high capture probability as nematic monoacrylate **15** during the free-radical chain-addition polymerization. If the overall polymerization rate is tuned to the diffusion kinetics (relatively low UV intensity), depletion of the chiral diacrylate near the top of the layer generates a concentration gradient. Consequently, the diacrylate diffuses toward the top of the layer with the result that the top of the layer contains more chiral material and thus has a shorter reflection wavelength than the bottom of the layer, which contains more of the non-chiral compound.

Scanning electron microscope (SEM) results indeed show a pitch gradient over the film thickness (Figure 3.6). The effect of the pitch gradient on the width of the reflection band is apparent

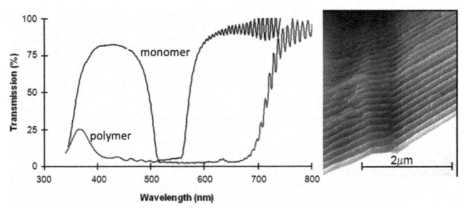

Figure 3.6 Reflection of right-handed circularly polarized light measured as transmission loss of a 1:1 mixture of compounds 5 and 15 before (monomer) and after (polymer) polymerization using a UV intensity gradient. The SEM picture shows a cross section of this polymer film obtained by freeze-fracturing, demonstrating layers of π rotation (p/2).

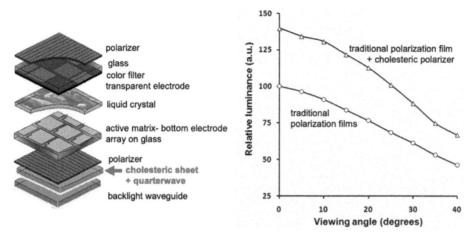

Figure 3.7 Schematic set-up of an LCD and the position of the cholesteric polarizer (left). The relative luminance measured at an LCD with and without the presence of a broadband cholesteric filter/quarter wave combination as a function of the viewing angle (right).

from the transmission measurement shown in the same figure [9]. Before polymerization the reflection of right-handed circularly polarized light is green and during polymerization it becomes metallic white. The photopolymerization reaction is not only responsible for the formation of a stable film, but it also plays a role in the formation of this architecture. It is possible to fabricate sheets of this material in a continuous coating process, which can be coated with or laminated to a quarter wave film, cut and incorporated in LCDs. When brought in a display set-up between the backlight and the a conventional sheet polarizer, the brightness of such a display is typically 1.6 times as high as with a conventional polarizer alone [10]. Figure 3.7 shows the results as a function of viewing angle. The luminance is integrated over the three colors.

3.4 Thin-film Retarders

Retarder films are used in nearly all displays. Depending on the display technology they play various roles, but in all cases they change the state of polarized light to some degree or other. Some films are designed to primarily change the on-axis state of polarized light and others primarily the off-axis properties, but in both cases, the effect the films have on both on and off axis must be considered. Different types of optical films have been developed over the years, and these are usually classified according to the optical properties [12].

There are two basic types of thin-film retarders: stretched films or coatable films. The stretched films benefit from having excellent mechanical properties, but have some limitations in their optical properties, especially their out-of-plane optical properties. The coated film has more flexibility in terms of optical properties, but requires a dedicated substrate onto which to coat the material, which has to be incorporated into the final optical layer – where the cost and the optical property of this carrier film has been considered – or the film has to be released from the carrier film, which is not an easy process. The stretched retarder films are 40–100 µm thick while the coated films are about 1–3 µm for the calamitic RM-type films and submicron for the chromonic type. Another distinguishing feature of coatable films is that both types can be patterned, a feature that was most prominently used for making filmed patterned retarders for three-dimensional TVs.

While most retarder films play a passive role, with their optical properties largely fixed, a temperature-responsive film has been developed which compensates super-twisted nematic (STN) displays. The optical properties of STN displays change with temperature giving unacceptable viewing characteristics and/or unwanted coloration. This can be overcome by optical compensation by adding another STN device with the same, but optically complementary, properties to form the so-called dSTN display. The retarder film is made from side-chain liquid crystal polymers, which are designed to change its birefringence at the same rate as the display as the temperature changes, and so provide full compensation over the operating temperature of the device [13].

3.4.1 Reactive Mesogen Retarders

The RM molecules are aligned by coating them with a solution on a plastic or other substrate that is provided with an alignment layer and evaporation of the solvent. Subsequently, the film is annealed to reduce the alignment defects and cured by UV exposure.

The alignment type is determined by a combination of surface treatment on the substrate side and formulation modifications to control the alignment on the substrate and air side. Planar-aligned achiral calamitic materials can be used to make uniaxial film with the slow (high refractive index) optical axes in the plane of the film. A film with this structure is called a positive A (+A) film. An RM polymer with homeotropically alignment (perpendicular to the film surface) can be made of similar but not identical material, which enables the production of +C films [14]. Here the film is in the isotropic plane and has the fast axis orthogonal to film. Both discotic and chiral calamitic materials can be used to make uniaxial negative C (−C), with the pitch of the cholesteric calamitic RM mixture being deep in the UV to shift the Bragg reflection peak away from the visible. Indeed, it was found that the pitch of the cholesteric mixture had to be much lower than expected because of refractive index mismatch between the top and lower surfaces leading to unwanted optical effects [15].

The fact that tight-pitch cholesteric RM films have the equivalent optics of a −C film gave rise to the idea that distorting the helix would allow the formation of a biaxial −C film. Such films are particularly useful for compensating vertically aligned LCD. It has already been demonstrated that this was possible for a broad band cholesteric film [16], and it was later shown that by introducing a novel dichroic alpha-amino liquid-crystal photoinitiator with higher-order parameter into the RMs mixture and polymerizing the film with polarized UV light, a biaxial film could be prepared with a slow axis parallel to the transmission axis of the UV polarizers. This is shown in Figure 3.8. This concept was further developed into an incell film by Kim et al. and a demonstrator was prepared. RM are particularly suitable for incell applications because they are thin. This feature also makes them particularly suitable for flexible displays.

An extension of this work was carried out by Smith et al. In this case, polymerization by polarized UV light was carried out while the coated RM film was in its non-uniform lying helix state. This process produced novel +A plate films. Unique films with radial, concentric, or tilted radial director distribution could be prepared without the need for an aligning layer [17].

RM materials, like most liquid-crystal materials are known to have achromatic behavior. For example, the birefringence of most liquid crystal materials increases with decreasing wavelength. This can give rise to unwanted optical effect. This problem has become most acute in the develop-

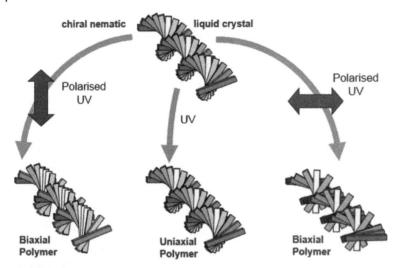

Figure 3.8 The slow axis of biaxial RM films can be controlled by adjusting the transmission axis of linearly polarized UV light.

ment of anti-reflection circular polarizers for OLED displays. Most OLED displays use a circular polarizer to control ambient light reflections, but unless the quarter wave film is achromatic, the black state can have a slight pink coloration. Several methods of producing commercial achromatic quarter wave films have been proposed. Teijin developed uniaxially oriented co-polycarbonate films containing positive and negative units by stretching a cast film. The resulting films were 27–99 um thick and had reverse birefringence dispersion. The approach of using positive and negative units has been extended to a self-organizing RM-based system. Several approaches have been developed, e.g. polymerizable T-shaped or H-shaped molecules doped into a calamitic host. Many of these systems can produce thin reverse dispersion +A, +C, and O plates. Most groups have designed molecules with strongly conjugated lateral groups and a weakly conjugated framework to hold the lateral group in the correct orientation. It's the balance of these groups that determines the efficiency of the H-shaped or T-shaped dopants. This efficiency as well as how well the dopant molecules are oriented by the host determine its concentration in the mixture. For several reasons related to processability and cost, minimizing the concentration of the dopants in the mixture is desirable. Other groups have reported that reverse-dispersion thin films can be achieved using a smectic liquid crystal-polymer composite in which a polymer is formed in-situ such that it is located at the inter-layer [18].

Yet another development in RM retarder field is in the development of thin, extremely efficient polarization gratings (Figure 3.9). These devices can be prepared using a complicated photoalignment layer in which the slow axis of the planar-aligned A-plate retarder rotates in the x–y plane. Such gratings can steer a light beam to one of two deflection angles, depending on the polarization handedness of the input light [19]. The angle of deflection is related to the pitch of the in-plane rotation of the slow axis of the retarder with shorter pitches leading to a larger deflection angle. The devices are extremely efficient (up to 99%) and have found application in projection-display devices. More elaborate versions have been made including chiral and multilayer versions. The chiral versions enable achromatic polarization gratings to be fabricated [20]. It is envisaged that such devices will find applications in head-up display units.

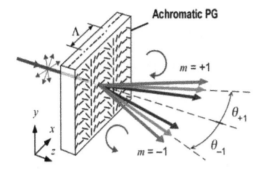

Figure 3.9 RM polarization gratings can efficiently convert unpolarized light to circularly polarized light.

It has been proposed that a polarization grating based on RMs could be used to form a very efficient polarizer for display applications, especially in the field of head-mounted displays.

3.4.2 Chromonic Liquid Crystal-based Retarders

An alternative approach to RM retarders is the chromonic liquid crystal-based materials. These materials are generally disc-like molecules that self-aggregate to form columns in polar solvents and can be coated and aligned by shear force or by using an aligning layer. These materials can also be used to form polarizers. Recent advances in this technology has demonstrated that these materials can be combined with rod-shaped polar molecules to give films that have good achromaticity. Macroscopic alignment of these materials could be achieved either by subjecting the liquid RM mixture to shear during or after the coating process or by using grooved substrate prepared using a secondary sputtering lithography [21]. It has also been demonstrated that complex patterns can be formed by coating the material onto a surface containing a photoalignment material and adding a second layer containing an RM. The function of the RM is to reduce the sensitivity of the photoalignment material to the coating solvent and, secondly, to increase the anchoring strength of the aligning layers [22].

3.4.3 Liquid Crystal Alignment and Patterned Retarders

There are two main methods of aligning liquid crystal in commercial applications: rubbing and photoalignment. The various issues concerning rubbing have been discussed in several review articles. Photoalignment has become more popular in commercial application and is used in in-plane modes such as in-plane switching (IPS)/fringe-field switching (FFS) – especially for small/medium displays – and in out-of-plane modes such as VA, in large area TVs, for example in Sharp's UV2A display technology in which photoalignment was used on a Gen10 line for the first time [23]. In the case of UV2A, one of the benefits that photoalignment offered was higher transmission relative to previous generations of vertical alignment (VA) technologies such as multi-domain VA. Other alignment methods have been described in the literature such as plasma aided alignment [24]. This approach has the advantage that the materials can be aligned directly on a glass or plastic substrate without the need for a polymer-based aligning layer. However, since this process requires vacuum chamber technology, it is more suitable to small-area devices than large-area displays. Recent development has focused on self-aligning (SA) modes such as SA-VA and SA-IPS. These new approaches enable the panel maker to fix the alignment by filling the cell using conventional filling methods such as one-drop filling equipment and exposing the cell to UV light to align the liquid crystal. In the case of SA-IPS, polarized UV light is required.

In SA-VA liquid crystal materials, additives that promote homeotropic alignment are added to the liquid crystal mixture. Also included are a small amount of RM to help stabilize the homeotropic alignment once it is formed. The additives that help to promote homeotropic alignment contain two important parts, an anchoring part and a liquid crystal-type core structure. The role of the anchoring group is to bind the additive to the substrate surface and therefore promote the homeotropic alignment. The liquid crystal-type core group is presumed to couple to the liquid crystal mixture. The SA-VA additive therefore replaces the polyimide alignment function. After the liquid crystal device is filled with this SA-VA liquid crystal mixture, the cell is treated in a manner similar to polymer-stabilized VA mode, in which a field is applied to the cell and simultaneously exposed to UV irradiation. The RM forms a thin layer on the cell surface, locking in the pretilt for the liquid crystal, which is a prerequisite for VA modes. The basic comparison between SA-VA and PS-VA is illustrated in Figure 3.10.

It is thought that the removal of the polyimide (PI) layer from LCD panels could bring several benefits. For example, it will lead to "process/design advantage." Generally, it requires high process temperature up to 200°C or above for the baking step after PI printing. The avoidance of high-temperature baking steps is clearly an advantage for the production of flexible displays. It allows the device engineers to select the substrate from a wider base.

In the case of SA-IPS or FFS, it was recently reported that *in-situ* homogeneous alignment could be achieved by adding a polymerizable RM containing a cinnamate group to the liquid crystal mixture. An IPS cell consisting of normal IPS electrodes but no polyimide was fabricated and filled with a mixture containing the modified RM. The cell was exposed to 5 mW/cm^2 of linearly polarized 365 nm UV light at 95°C for 4 minutes and homogeneous alignment was achieved. The general scheme for achieving self-alignment of in-plane liquid crystal modes is

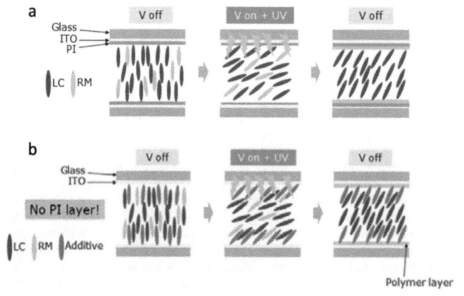

Figure 3.10 (a) RM stabilized alignment PS-VA and (b) SA-VA.

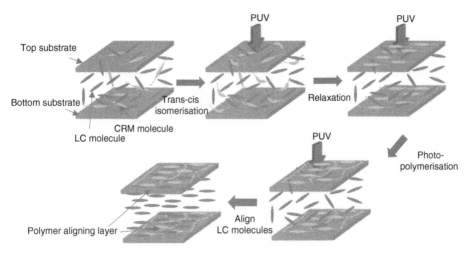

Figure 3.11 Photo-aligning RMs 16 and 17 and a general scheme for in-situ self-alignment of in-plane liquid crystal mixtures.

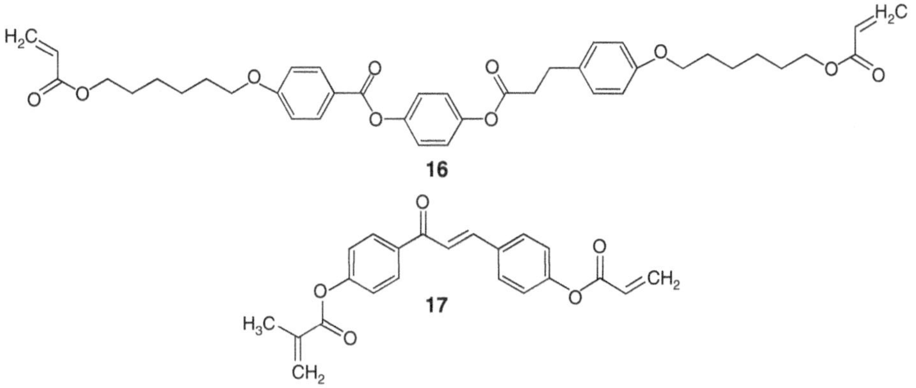

16

17

Figure 3.12 Cinnamate containing liquid crystals for photoalignment purposes.

shown in Figure 3.11. The dark state and switching characteristic of the cell was reported to be comparable to a conventional IPS cell [25]. The compound used to achieve self-alignment is diacrylate RM **16** with a cinamate group (Figure 3.12).

A similar effect was independently demonstrated by Mizusaki et al. [26, 27], however, in this case, a different, spacerless cinamate containing methacrylate RM **17** was used.

Recently it was reported that a 1.45" full color prototype has been prepared by Beijing Oriental. The Advanced Super Dimensional Switching mode prototype had almost the same level of performance as that of the same mode prepared using a conventionally rubbed polyimide aligning layer [28]. The prototype had the specifications shown in Table 3.1.

Another development in alignment technology is to coat an additional thin layer on top of the existing aligning layer. For example, it is reported that the RM enhances the surface-anchoring energy in both the out-of-plane and in-plane directions and, in addition, improves the response-time characteristics of LCDs [29, 30]. Indeed, this effect seems to be quite general and works on a variety of rubbed and photoalignment layers [31].

Table 3.1 Specification of the SA-ADS prototype

Items	Specification
Diagonal size (inch)	1.54
Resolutions	320 × 320
PPI	293
Brightness (nit)	450
Contrast	600:1
View angle ( )	89 / 89 / 89 / 89
Response time (ms)	25
Color gamut (NTSC)	53%

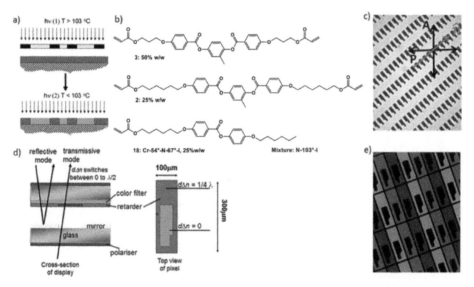

Figure 3.13 Fabrication of a patterned retarder for transflective displays. (a) The two-step photopolymerization above and below the nematic to isotropic transition temperature produces patterned quarter wave retardation area next to isotropic area. (b) An example of a mixture of RMs. (c) Patterned retarder between crossed polarizers as observed by polarization microscopy. Bright fields are birefringent, black area isotropic. (d) Application of the patterned retarder in a transflective display with a quarter wave function on the reflective mirror and isotropic area at the transmissive area. (e) Polarization microscope picture of the of the transflective display cell including the patterned retarder and the color filter.

An application that nicely demonstrates the versatility of RM-based polymer networks is their application in patterned retarders for transflective displays (Figure 3.13) [32, 33]. The objective here is to apply a retarder in the display cell that has zero retardation at the location in the pixel where light passes through the semi-transparent mirror at the backlight side of the display cell. At the location in the pixel where light is reflected at the mirror the film has a quarter wave retardation (Figure 3.13d). The light that is passing the liquid crystal layer twice (off-state/bright appearance) due to reflection is compensated by the retarder layer. Therefore, in the bright state of the film it has the same appearance as light that is transmitted through the aperture in the

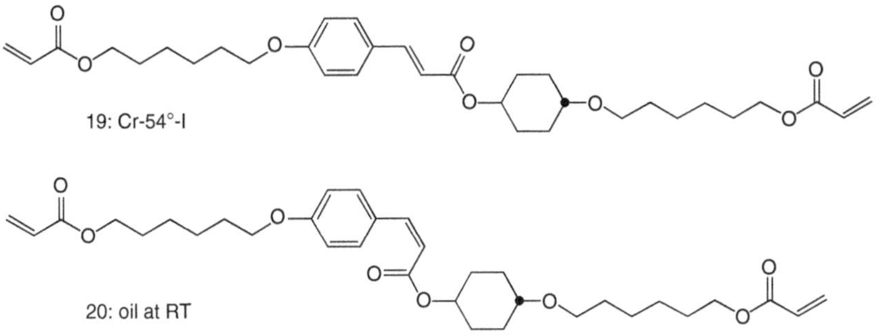

Figure 3.14 Photoisomerizable compound 19 and its isomerization product 20, used to lower the isotropic transition of its mixture with RM 3 in a pattern-wise fashion to obtain pattered retarders for use in transflective LCD.

mirror. Additionally, in the on state (black appearance) the linear polarized light is converted into circular polarized light by the quarter wave function and switches handedness upon reflection, which therefore becomes absorbed by the top polarizer film after having passed the quarter wave function for the second time. Thereby this black state matches the black state of the transmissive mode. The patterned optical retarder was conveniently made by polymerization of the RMs in a two stage process (Figure 3.13a). First, by a UV exposure through a photomask at a temperature where the RM mixture was in its nematic state to provide the quarter wave function. And, second, by UV exposure after heating the remaining non-reacted RMs to their isotropic state thus providing the non-birefringent area is located above the open area of the patterned mirror. Alternatively, a process is used that irradiates a photoactive liquid crystalline mixture through a photomask. The mixture consists of photo isomerizable compound **19** and diacrylate **3** (RM257). Upon irradiation, isomerization to compound **20** (Figure 3.14) lowers the nematic to isotropic transition of the mixture below room temperature . After photopolymerization of the whole layer, the pre-irradiated parts are non-birefringent while the non-pre-irradiated parts are birefringent with the same properties as described earlier [34]. The process is very similar to that described more extensively for the formation of cholesteric color filters discussed in Section 3.5.

3.5 Color Filters

In addition to their polarizing properties, their ability to reflect light within a well-adjusted wavelength region is an important property for the use of cholesterics in color filters, e.g. in LCDs. Thereto the reflection colors must be patterned, preferably in a single layer or, when used in transmission, in a double layer. For the manufacturing of conventional color filters, based on combinations of absorbing dyes or pigments in a polymer matrix, the three colors (R,G,B) are applied successively by e.g. lithography, in a multi-step process. This process can be simplified by the use of cholesteric materials. But there are more benefits by exchanging a conventional absorbing color filter with non-absorbing reflective cholesteric RM. For instance, they can be directly used in low-energy-consuming reflective LCDs. These displays do not require a backlight and are of relevance for mobile devices such as PDAs or mobile phones. The design of these reflective LCDs can be simplified by using the unique combination of polarization and color selection properties of cholesterics [35]. For application in transmissive LCDs, cholesteric color filters have

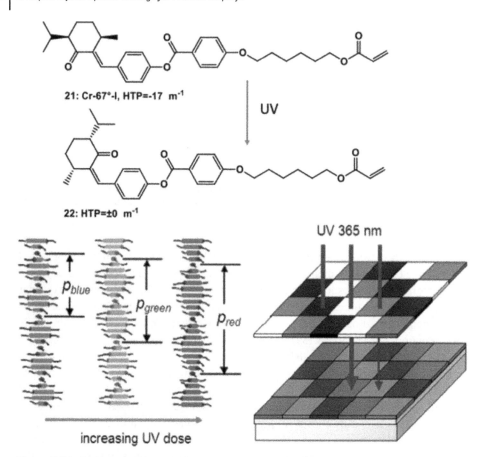

Figure 3.15 Light-responsive reactive chiral monomer that changes its helical rotation power upon light exposure. When exposing a RM mixture containing chiral monomer **21** at different dose of UV light different reflective colors are produced, which are subsequently frozen in by photopolymerization of the RM mixture.

an extra advantage. The reflective nature of the color filter offers the opportunity to recycle the two unwanted primary colors from the backlight. In conventional transmissive LCDs unwanted light (i.e. at least 66%) is absorbed by the color filter. This means that when cholesteric color filters are used in combination with the cholesteric polarizer described in Section 3.2, theoretically a six times higher light intensity can be obtained compared with the use of absorbing components [36].

Cholesteric materials offer the possibility to obtain a red–green–blue (RGB) array by changing the pitch (p) making use of the thermochromic effect. Mask-wise polymerization at different temperatures provides the different colors. However, this process has rather critical parameters because small changes in temperature change the reflection color. Therefore, another, more reliable, process based on photochemical alteration of the pitch has been developed. An array of RGB pixels are formed in a single step using a pixel-patterned gray scale mask [34]. The isomerizable menthone derivative **21** (Figure 3.15) causes a blue reflection when blended in a liquid crystalline mixture. The reflection color shifts to the red part of the spectrum depending on

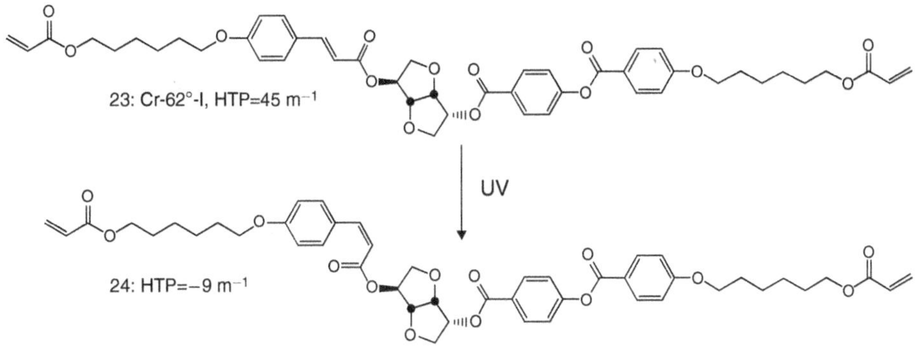

Figure 3.16 Photoisomerizable compounds with higher thermal stability compared to compounds 21 and 22 to form cholesteric color filters.

the dose. This change in pitch is a result of the fact that the helical twisting power of the cis isomer **22** is much lower than that of the initial trans isomer **21**. The three RGB colors are obtained in a single exposure step. Polymerization of the acrylate group is postponed by the presence of oxygen that inhibits radical reactions. The colored pixels are fixed by a rapid photopolymerization reaction, by UV exposure in the absence of oxygen. The fast kinetics of the polymerization reaction avoids color change upon polymerization. The remaining isomerizable compounds isomerize much slower under these conditions. After crosslinking of the layer, photoisomerization of the remaining compounds has no effect on the color because the cholesteric helix is fixed by the network formed. However, these colored layers are processed afterwards to form a display. During these subsequent process steps temperatures a high as 220°C are used. The stability of compounds such as **21** is not enough, even after incorporation in a polymeric network. The resulting reflections of the color filter become less efficient, due to blue shift and reflection loss. Fortunately, this problem could be resolved by replacing molecules derived from menthone such as **21** with derivatives of cinnamic acid **23** as the isomerizable moiety and isosorbide as the chiral moiety as depicted in Figure 3.16 [37, 38]. Alternatively, this stability can also be obtained if the isosorbide moiety in molecule **23** is replaced by a moiety derived from phenylethanediol [39].

3.6 Conclusion

The use of LCNs based on RM technology fits very well with the fabrication of flexible display based on liquid crystal technology. The optical films that they form are flexible, bendable, and adhere to plastic substrates. The polarization optics can be brought inside the cell, which enables the use of birefringent foils such as polyesters as base for the displays. The same technology can be utilized to stabilize the alignment of the liquid crystals, which becomes relevant when the displays are bent or pushed. But flexible OLED technology can also benefit from the same technology. Often OLED displays have a circular polarizer at the viewing side to improve them on daylight contrast. By constructing the circular polarizer from an RM-based dichroic polarizer in combination with an RM-based quarter wave film, the circular polarizer can be kept very thin without sacrificing the bendability of the OLED pane.

References

1 Broer, D.J., Boven, J., Mol, G.N., and Challa, G. (1989). In-situ photopolymerization of oriented liquid-crystalline acrylates, 3. Oriented polymer networks from a mesogenic diacrylate. *Makromol. Chem.* 190: 2255.

2 Broer, D.J., Hikmet, R.A.M., and Challa, G. (1989). In-situ photopolymerization of oriented liquid-crystalline acrylates, 4. Influence of a lateral methyl substituent on monomer and oriented network properties of a mesogenic diacrylate. *Makromol. Chem.* 190: 3201.

3 Broer, D.J., Mol, G.N., and Challa, G. (1991). In-situ photopolymerization of oriented liquid-crystalline acrylates, 5. Influence of the alkylene spacer on the properties of the mesogenic monomers and the formation and properties of oriented polymer networks. *Makromol. Chem.* 192: 59.

4 Broer, D.J. (1993). Photoinitiated polymerization and crosslinking of liquid crystalline systems in Radiation Curing in Polymer Science and Technology. In: *Polymerization Mechanisms*, III (ed. J.P. Fouassier and J.F. Rabek), chapter 12, 383. London and New York: Elsevier Science Publishers Ltd.

5 Broer, D.J., Crawford, G.P., and Zumer, S. (2011). *Cross-Linked Liquid Crystalline Systems: From Rigid Polymer Networks to Elastomers*. London: CRC Press.

6 White, T.J. and Broer, D.J. (2015). Programmable and adaptive mechanics with liquid crystal polymer networks and elastomers. *Nat. Mater.* 14 (11): 1087.

7 Vaenkatesan, V., Wegh, R.T., Teunissen, J.-P., Lub, J., Bastiaansen, C.W.M., and Broer, D.J. (2005). Improving the brightness and daylight contrast of organic light-emitting diodes. *Adv. Funct. Mater.* 15 (1): 138–142.

8 Peeters, E., Lub, J., Steenbakkers, J.A.M., and Broer, D.J. (2006). High-contrast thin-film polarizers by photo-crosslinking of smectic guest–host systems. *Adv. Mat.* 18 (18): 2412.

9 Broer, D.J., Lub, J., and Mol, G.N. (1995). Wide-band reflective polarizers from cholesteric polymer networks with a pitch gradient. *Nature* 378: 467.

10 Broer, D.J., Van Haaren, J.A.M.M., Mol, G.N., and Leenhouts, F. (1995). Reflective cholesteric polariser improving the light yield of back- and side-lighted flat panel liquid crystal displays, *Proceedings of the 15th International Display Research Conference, Asia Display*. California: Society for Information Displays, 735.

11 Lub, J., Broer, D.J., Hikmet, R.A.M., and Nierop, K.G.J. (1995). Synthesis and photopolymerization of cholesteric liquid crystalline diacrylates. *Liq. Cryst.* 18 (2): 319–326.

12 Maa, J., Ye, X., and Jin, B. (2011). Structure and application of polarizer film for thin-film-transistor liquid crystal displays. *Displays* 32: 49–57.

13 Bosma, M. (1998). Akzo Nobel Twistar™: LCP-retarder films with improved temperature-matched compensation. *J. Soc. Inf. Disp.* V6: 231–233.

14 Coates, D., Parri, O., Verral, M., and Slaney, K. (2000). Polymer films derived from aligned and polymerised reactive liquid crystals. *Makromolecular Symposia* 154(1): 59–72.

15 Skinnemand, K. Reactive mesogen mixtures suitable for the preparation of uniaxial and biaxial optical films; Patent Application WO 2004/013666, 2004.

16 Broer, D.J., Mol, G.N., Van Haaren, J.A.M.M., and Lub, J. (1999). Photo-induced diffusion in polymerizing chiral-nematic media. *Adv. Mater.* 11(7): 573.

17 Smith, G., Parri, O.L., Whitehouse, S., and Perrett, T., Method of preparing a birefringent polymer film; WO 2015/058832 A1. October 1, 2014.

18 Yang, S., Lee, H., and Lee, J.-H. (2015). Negative dispersion of birefringence of smectic liquid crystal-polymer composite: Dependence on the constituent molecules and temperature. *Opt. Exp.* 23(3): 2466.

19 McManamon, P.F., Bos, P.J., Escuti, M.J., Heikenfeld, J., Serati, S., Xie, H., and Watson, E.A. (2009). A review of phased array steering for narrow-band electrooptical systems. *Proceedings of the IEEE* 97(6): 1078–1096.

20 Komanduri, R.K., Lawler, K.F., and Escuti, M.J. (2013). Multi-twist retarders: Broadband retardation control using self-aligning reactive liquid crystal layers. *Opt. Exp.* 21(10): 404.

21 Nayani, K., Jeong, H.S., Jeon, H.-J., Yoo, H.-W., Lee, E.H., Park, J.O., Srinivasarao, M., Kim, J.K., and Jung, H.T. (2016). Macroscopic alignment of chromonic liquid crystals using patterned substrates. *Phys. Chem. Chem. Phys.* 18: 10362.

22 Peng, C., Guo, Y., Turiv, T., Jiang, M., Wei, Q.-H., and Baranovichi, O.D. (2017). Patterning of lyotropic chromonic liquid crystals by photoalignment with photonic metamasks. *Adv. Mater.* 29: 1606112.

23 Miyachi, K. (2014). UV^{2}A LCD panel with photo-alignment technology. In: 2014 21st International Workshop on Active-Matrix Flatpanel Displays and Devices (AM-FPD), Kyoto, 9–12.

24 Yaroshchuk, O.V., Zakrevskyy, Y., Dobrovolskyy, A., and Pavlov, S. (2001). Liquid crystal alignment on the polymer substrates irradiated by plasma beam. In: *Proceedings of the SPIE 4418, Eighth International Conference on Nonlinear Optics of Liquid and Photorefractive Crystals*, 441808, May 30.

25 He, R., Wen, P., Kang, S.-W., Lee, S.H., and Lee, M.-H. (2018). Polyimide-free homogeneous photoalignment induced by polymerisable liquid crystal containing cinnamate moiety. *Liq. Cryst.* 45(9): 1342.

26 Mizusaki, M., Tsuchiya, H., and Minoura, K. (2017). Fabrication of homogenously self-alignment fringe-field switching mode liquid crystal cell without using a conventional alignment layer. *Liq. Cryst.* 44(90): 1394.

27 Mizusaki, M., Tsuchiya, H., Itoh, T., and Minoura, K. (2018). Homogeneous self-alignment technology without forming conventional alignment layers. *Digest Tech. Papers Soc. Inf. Display Int. Sympos.* 49(1): 455.

28 Yang, R., Wu, J., Wang, R., Wu, X., Qu, L., Xu, X., You, Y., Zhao, H., Feng, Y., Yang, F., Qi, X., Qiu, Y., and Wang, D. (2019). Development of self-alignment advanced super dimensional switching technology and prototype. *Digest Tech. Papers Soc. Inf. Display Int. Sympos.* 50(1): 481.

29 Moon, Y.-K., Lee, Y.-J., Jo, S.I., Kim, Y., Heo, J.U., Baek, J.-H., Kang, S.-G., Yu, C.-J., and Kim, J.-H. (2013). Effects of surface modification with reactive mesogen on the anchoring strength of liquid crystals. *J. Appl. Phys.* 113: 234504.

30 Kim, Y., Lee, Y.-J., Baek, J.-H., Yu, C.-J., and Kim, J.-H. (2015). Dependence of planar alignment layer upon enhancement of azimuthal anchoring energy by reactive mesogens. *Jap. J. Appl. Phys.* 54: 011701.

31 Yaroshchuk, O., Kyrychenko, V., Tao, D., Chigrinov, V., Kwok, H.S., Hasebe, H., and Takatsu, H. (2009). Stabilization of liquid crystal photoaligning layers by reactive mesogens. *Appl. Phys. Lett.* 95: 021902.

32 Van Der Zande, B.M.I., Doornkamp, C., Roosendaal, S.J., Steenbakkers, J., Op't, H.A., Osenga, J.T.M., Van Glabbeek, J.J., Stofmeel, L., Lub, J., and Shibazaki, M. (2005). Technologies towards patterned optical foils applied to transflective LCDs. *J. Soc. Inf. Disp.* 13(8): 627.

33 Van Der Zande, B.M.I., Roosendaal, S.J., Doornkamp, C., Steenbakkers, J., and Lub, J. (2006). Synthesis, properties, and photopolymerization of liquid-crystalline oxetanes: Application in transflective liquid-crystal displays. *Adv. Func. Mat.* 16(6): 791.

34 Van Der Zande, B.M.I., Lub, J., Verhoef, H.J., Nijssen, W.P.M., and Lakehal, S.A. (2006). Patterned retarders prepared by photoisomerization and photopolymerization of liquid crystalline films. *Liq. Cryst.* 33(6): 72.

35 Lub, J., Van De Witte, P., Doornkamp, C., Vogels, J.P.A., and Wegh, R.T. (2003). Stable photopatterned cholesteric layers made by photoisomerization and subsequent photopolymerization for use as color filters in liquid-crystal displays. *Adv. Mat.* 15(17): 1420.

36 Lub, J., Broer, D.J., Wegh, R.T., Peeters, E., and Van Der Zande, B.M.I. (2005). Formation of optical films by photo-polymerisation of liquid crystalline acrylates and applications of these films in liquid crystal display technology. *Mol. Cryst. Liq. Cryst.* 429: 77.

37 Lub, J., Nijssen, W.P.M., Wegh, R.T., Vogels, J.P.A., and Ferrer, A. (2005). Synthesis and properties of photoisomerizable derivatives of isosorbide and their use in cholesteric filters. *Adv. Funct. Mater.* 15(12): 1961–1972. doi:10.1002/adfm.200500127. Language: English.

38 Lub, J., Nijssen, W.P.M., Wegh, R.T., De Francisco, I., Ezquerro, M.P., and Malo, B. (2005). Photoisomerizable chiral compounds derived from isosorbide and cinnamic acid. *Liq. Cryst.* 32(8): 1031–1044. doi:10.1080/02678290500284017. Language: English.

39 Lub, J., Ezquerro, M.P., and Malo, B. (2006). Photoisomerizable derivatives of phenylethanediol and cinnamic acid: Useful compounds for single-layer R, G, and B cholesteric color filters. *Mol. Cryst. Liq. Cryst.* 457: 161–180. doi:10.1080/15421400600700299. Language: English.

4

Large Area Flexible Organic Field-effect Transistor Fabrication

Zachary A. Lamport[1], Marco Roberto Cavallari[2,3], and Ioannis Kymissis[4]

[1] *Postdoctoral Research Scientist, Electrical Engineering Columbia University, New York, USA*
[2] *Departamento de Engenharia de Sistemas Eletrônicos, Escola Politécnica da Universidade de São Paulo, São Paulo, Brazil*
[3] *Department of Renewable Energies. UNILA, Federal University of Latin American Integration, PR, Brazil*
[4] *Columbia University, New York, USA*

4.1   Introduction

The development of flexible electronics is largely tied to advances in organic electronics. This is to be expected due to the well-known mechanical properties of organic materials and the vast international research effort into organic electronic devices [1–7]. The first generation of organic transistors borrowed silicon substrates and cleanroom techniques (e.g. thermal evaporation, electrochemical deposition, casting, and spin coating) to deposit semiconducting films. Dip coating, spray coating, Langmuir-Blodgett, and electrostatic self-assembly followed. Therefore, it was already possible not only to form films to integrate electronic devices but also to perform monolayer surface treatments. Altering the surface is an extremely common technique to enhanced wettability, film crystallinity, and electrical charge injection [5]. By 2000, all-polymer flexible integrated circuits were already being demonstrated by using only spin coating and photolithography [8]. Although these techniques already had the potential to be adapted to large-area processing, only later, printing, doctor blade, and slot die were developed and adapted for roll-to-roll processing of flexible electronics. The deposition of multiple organic layers requires that the subsequent films do not dissolve those underneath. That is one of several reasons why, even now, researchers opt for metal electrodes deposited from physical vapor deposition techniques such as sputtering, electron-beam assisted evaporation, and thermal evaporation. Currently, many devices also feature dielectrics deposited by sputtering, chemical vapor deposition, and atomic layer deposition [9, 10]. The major drawback associated with these techniques is that they usually require high temperatures and high vacuum conditions. Therefore, these processes can, other than slow down the fabrication, damage organic films and need to be carefully made compatible to flexible electronics. Despite these challenges, in 2014, the first flexible microprocessor was fabricated employing conventional cleanroom techniques such as thermal evaporation of pentacene and gold, atomic layer deposition of Al_2O_3, and spin coating for both photolithography and patterning of electrodes [11]. This indicates that, although new techniques are under development for improved compatibility with flexible electronics, already established cleanroom techniques will still have a role for a long time. In the following, the choice of substrate for flexible organic field-effect transistors (OFETs) is examined before addressing subtractive patterning approaches such as photolithography, as well as additive thin-film formation such as in printing processes that are compatible with the roll-to-roll fabrication promise of organic electronics.

Flexible Flat Panel Displays, Second Edition. Edited by Darran R. Cairns, Dirk J. Broer, and Gregory P. Crawford.
© 2023 John Wiley & Sons Ltd. Published 2023 by John Wiley & Sons Ltd.

4.2 Substrates

The choice of substrate for large-area flexible electronics is dictated by the constraints of the application as well as the processing steps required to fabricate devices. For the envisioned roll-to-roll processes, substrates require a higher resilience to thermal and mechanical stresses than experienced in a traditional laboratory setting, necessitating the development of robust, intrinsically flexible plastics. Examples of commonly used substrate materials include polytetra-fluoroethylene (Teflon), polyimide (Kapton), polyethylene naphthalate (PEN), Mylar, polydimethylsiloxane (PDMS), and paper [12–15]. Substrate materials must be carefully chosen in order for a material to withstand thin-film drying, polymerization, cross-linking, nanoparticle sintering, crystallization, and patterning without melting, cracking, deforming, or chemically degrading. Thermal properties are inextricably linked to other material properties such as mechanical and chemical characteristics. For these reasons among others, low-temperature processing is an essential feature of large-scale industrial printing of flexible electronics.

The manifold effects of elevated temperatures on plastic substrate materials, while usually a hindrance, can be useful when utilized carefully. Many of these polymers are thermoplastics, meaning that above the glass transition temperature the plastic becomes moldable, holding that shape when returning below glass transition. This characteristic becomes extremely useful in sophisticated techniques such as nanoimprint lithography, which will be discussed later. The glass transition temperature also functions as a soft upper limit for the available processing temperatures in roll-to-roll fabrication as well as in the laboratory setting. Many common polymers used as substrates have a glass transition temperature of around 200°C or less (e.g. PET, PEN, etc.), but Kapton is rare in that the glass transition temperature is nearly 400°C making it an attractive candidate for processes requiring elevated temperatures [16].

The mechanical flexibility of substrates is of paramount importance for device reliability, however, cracks can arise in even the most pliable materials as a result of the difference in coefficients of thermal expansion between different layers as well as the evaporation of solvents from printed components. These cracks not only reduce overall device longevity, but severely deteriorate device performance through the formation of short circuits, the separation of critical interfaces, and the reduction of π-orbital overlap, among other issues [16]. One of the most common substrates that is both easily castable and remains robust after both bending and stretching is PDMS. It is often used as the stamp in contact printing as discussed later, but also can be used for robotic touch interfaces due to its deform-ability and transparency [17]. Chemical resistance, similarly, is important for solution-based processes as well as both dry and wet etching. Careful planning of the process flow is critical to avoid the disso-lution of underlying layers upon application, which has led to the adoption of techniques such as the use of orthogonal solvents for subsequent depositions, which will be discussed later.

The surface properties also will influence the decision of which substrate to use, and in particular the surface energy and wettability must be matched to the process and liquids that will be applied. One of the myriad ways to influence the surface energy, beyond cleaning and/or plasma treatments, is the application of self-assembled monolayers (SAMs) that chemically bind to a surface. The SAM-modified surface will usually have a higher static contact angle, and pat-terning the SAM layer can provide a means to pattern further layers printed on top [18].

Biocompatibility and biodegradability are important features for medical devices. Parylene, also known as poly-(para-xylylene) (PPX), is a biocompatible electrical insulating polymer obtained by chemical vapor deposition from its dimer. The three main variations are N, C, and D.

They stand for the substitution of chlorine atoms for zero, one or two of the aromatic hydrogens, respectively. Largely applied to packaging, it is also used as a few micron-thick substrate or as a gate dielectric thin film in field-effect transistors. Paper, on the other hand, is not only biocompatible but also biodegradable, widely available, inexpensive, lightweight, and can be manipulated into seemingly endless orientations as seen in the art of origami [19]. Using paper has also allowed for older techniques to be recycled, for instance laser printing. Diemer et al. developed a laser-printing method to transfer a toner containing an organic semiconductor to a sheet of paper without the use of solvents in a pattern defined by a computer interface [20]. This method was taken further to pattern electrodes and organic–inorganic perovskites [21, 22]. Tattoo-paper transfer has also recently been used to create even edible electronics for therapeutic or diagnostic techniques from inside the human body [23]. While the mechanical properties of paper are quite favorable and well-studied through the centuries, there are major drawbacks in wet processing that stem from its high porosity, high roughness, and low thermal stability. Planarity and roughness are therefore important surface parameters for the patterning of submicron features and to prevent electrical short-circuit between different stacked layers [16]. Cellulose nanofibers (CNFs) are a potential material for paper-based electronic substrates that avoids some of the deficiencies of traditional paper [24, 25]. Finally, transparency to a predetermined photon wavelength range is often desirable depending on possible ultraviolet (UV) curing processes, the photooxidative stability of the materials, and the final application. A number of plastic films are commercially available as polycarbonate, polyethersulphone, and Kapton. As discussed before, the latter has excellent thermal properties and can be obtained from different routes, for instance through a precursor or pre-purchased, but is somewhat opaque to the visible spectrum and, as a result, finds limited use for its transparency. For instance, stainless steel foils have been demonstrated as a suitable substrate for active-matrix electronic ink and OLED displays in mobile devices (e.g. watches, phones, and tablets) [26–28]. Although planar substrates are the most commonly used device platform, there are also fibrous substrates for textile and wearable electronics [29–31]. Fabrics have the benefit of thousands of years of development from every culture, and this has resulted in a vast array of textiles with extraordinarily diverse properties. This allows modification of the surface properties, but many challenges still remain, particularly due to the roughness and complexity of the microstructures, but also because clothing and other fabrics are expected to undergo repeated stress through wear, sunlight, sweat, and washing [12, 16].

4.3 Photolithography

Borrowed from the development of silicon-based electronics, photolithography is a valuable tool used to define components with extremely high resolution and reproducibility. The process utilizes a liquid photoresist that is cast onto a substrate typically by spin coating that is then exposed to UV light through a patterned chrome on quartz mask, called a photomask. The two varieties of photoresist behave in an opposite manner, where the exposed portion of positive photoresist becomes soluble to a developer, and in negative photoresist becomes insoluble. The process flows of both types of photoresist can be seen in Figure 4.1a. The remaining photoresist can be removed by an etchant at the end of the process. The photoresist can be exposed in three common ways: contact, proximity, and projection. Naturally, many already-used photoresists were employed as dielectrics and patterned to open vias for circuits.

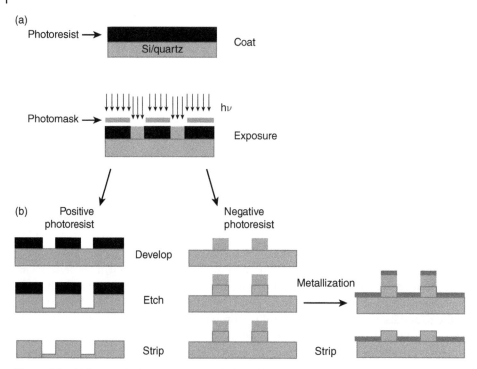

Figure 4.1 (a) Process flows of photolithography using positive and negative photoresists (b) Three main types of optical lithography. Reproduced with permission [32].

To expose either positive or negative photoresist, there are a few common methods that see significant use in both industrial and laboratory electronic development. The first and most basic is contact lithography where the hard photomask is pressed directly against the photoresist-covered substrate and exposed to a uniform UV light. This method fell out of favor due to the relatively high possibility of damage to the substrate, reducing the overall throughput and therefore raising the cost. More commonly used is the proximity lithography, where the mask is held just slightly above the substrate (10–30 μm) [32]. This reduces the damage to the wafer under test while retaining nearly the same resolution as contact printing due to the collimated UV light produced by mask aligners and the very small distance between mask and substrate.

Finally, scientists developed a fluorinated resist that would be fully compatible with processing on top of most organic films. It is called "orthogonal" as it does not dissolve conductors (usually hydrophilic, i.e. soluble or dispersed in water) or common semiconductors and dielectrics (usually oleophilic, i.e. soluble in non-polar organic solvents) [33]. Using this new chemistry, the possible fabrication methodologies are greatly expanded and complementary circuits are more straightforward to produce. Figure 4.2 shows an example of a twistable device made using this lithography system, and following complementary devices were fabricated on the same substrate in a manner that is impossible with conventional photolithography [34]. Many applications, however, may require high-speed processing with the combination of both organic and inorganic electronics. Strategies to achieve flexible interconnects in hybrid electronics are shown in Figure 4.3, in which many of the presented structures require photolithography [35].

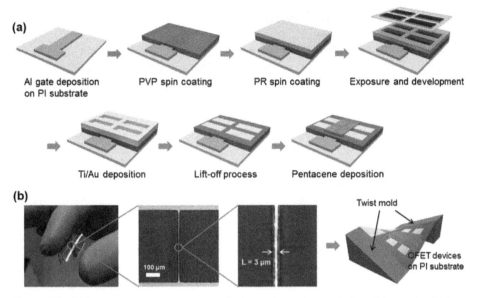

Figure 4.2 (a) Process procedure to pattern devices using orthogonal photolithography. (b) Images showing the twistable devices patterned as such [34].

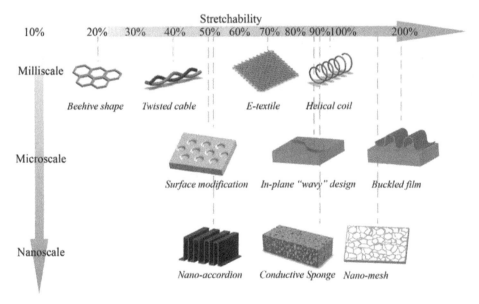

Figure 4.3 The variety of techniques used to allow stretchable electrodes and interconnects, from the nanoscale to the macroscale. Reproduced with permission [35].

4.4 Printing for Roll-to-roll Fabrication

Roll-to-roll fabrication involves myriad techniques taken from the semiconductor industry as well as the traditional printing industry to apply electronics to a flexible, moving web. Processes described in this chapter include wet coating, surface treatments (e.g. chemical, thermal, or UV), and contact printing. Conventional cleanroom processes for inorganic semiconductor electronics are also used in roll-to-roll processes, such as sputtering, thermal evaporation, chemical growth, and laser ablation. Figure 4.8 summarizes the main techniques for thin-film formation in roll-to-roll systems [36].

Printing is an additive processing technique that will generally refer to the application of some liquid "ink" in a very localized manner without the need for removing superfluous material. Printing can be achieved by direct writing such as in inkjet printing, aerosol jet printing, pneumatic meniscus dragging, by transferring an already patterned film as in contact printing, or by positioning a shadow mask in contact with the substrate as in spray coating and screen printing. The types of materials used here can vary widely, but as in many other aspects of microelectronics, the soft, organic matter allows for greater flexibility, but a lower overall resolution [16, 37]. With this lower resolution comes a new host of issues that stem from the desired flexibility and pliability of substrates when compared to traditional inorganic electronic systems, particularly the fairly high variation from component to component, but also device to device. This is true when the substrates are unflexed, but swiftly

becomes a larger issue upon bending, twisting, and other physical manipulation. One of the major effects here is the delamination of one layer from the next upon flexing, which is greatly exacerbated by any difference in coefficient of thermal expansion. Regardless of the many challenges facing the development of printed electronics, there have still been many successful developments that spur further research. As an example, fully printed radiofrequency (RF) tags had been realized as early as 2010 through the combination of inkjet and gravure printing [38]. Given the individual successes of printed organic devices, research into a vast array of flexible devices has exploded and found many new applications and processing techniques.

4.4.1 Inkjet Printing

There are two main technologies in use in contemporary inkjet printers. The first is continuous inkjet printing (CIJ), where ink is pushed through a nozzle by a high-pressure pump in a continuous stream. The nozzle is typically microscopic such that the aperture is small, and a vibrating piezoelectric breaks the stream into droplets of a desired size. The other inkjet printing technique is drop-on-demand. Through one of a variety of methods – for instance, a piezoelectric material as in CIJ – the droplets are produced as needed instead of continuously [16, 36, 39]. One of the drawbacks of using inkjet printing techniques is that they are partially constrained in the vertical direction; additional thickness can essentially only be achieved by successive prints. Increasing printing resolution can be difficult, particularly because a droplet landing on a substrate will tend to spread out, and new methodologies have been adopted to confine the spreading to a desired area. One approach shown in Figure 4.4 is to modify the wettability of the substrate through the addition of a surface energy-modifying SAM to selectively inkjet print a silver nanowire film on paper [40].

Another surface modification technique used in inkjet printing borrows from a method discussed later known as contact printing as well as imprint lithography [41]. This involves pre-patterning a PDMS stamp to UV-pattern an epoxy, serving as a template for an inkjet printing

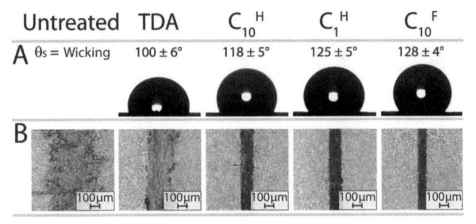

Figure 4.4 (a) Contact angle measurements on "Canson tracing paper" either unmodified or with a SAM, listed at the top, assembled on the surface. (b) the resulting silver nanowire film printed on the tracing paper modified as above. Reproduced with permission [40].

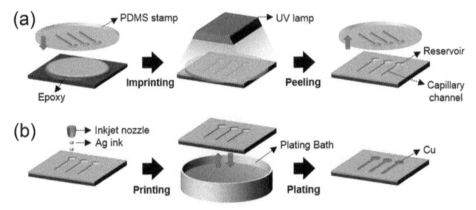

Figure 4.5 (a) Steps in imprint lithography followed by (b) an inkjet printing and plating process. Reproduced with permission [41].

process followed by a plating step. Figure 4.5a shows the process flow for creating the physical template, while Figure 4.5b shows the use of the template to build a metallized pattern.

To further improve on the resolution of inkjet printing, not only the surface must be tailored but the printing process itself. The droplet characteristics, particularly the size and spacing, have a massive effect on the resulting morphology as can be seen in Figure 4.6 [16, 42, 43]. The droplet size, then, will obviously impact the resulting dimensions as well, however, the minimum droplet size is dictated by the surface tension and can be difficult to reduce below 1 picoliter [44]. Through the clever use of electrostatics, Sekitani et al. could produce 1 μm channel lengths through inkjet printing with sub-femtoliter accuracy [44]. This result was a significant advance in inkjet-printing technology, surpassing the limit imposed by the interfacial interactions of the droplet, air, and print nozzle.

As a result of the drastic improvements in inkjet printing technologies, the field has now realized all-inkjet-printed OFET arrays, or even full integrated circuitry see Figure 4.7 [45, 46]. As

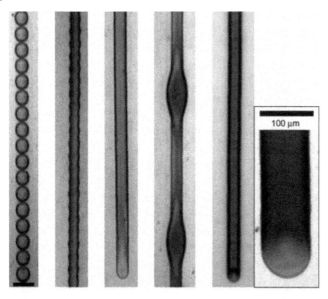

Figure 4.6 Examples of the resulting inkjet-printed morphology depending on droplet spacing. Reproduced with permission [42].

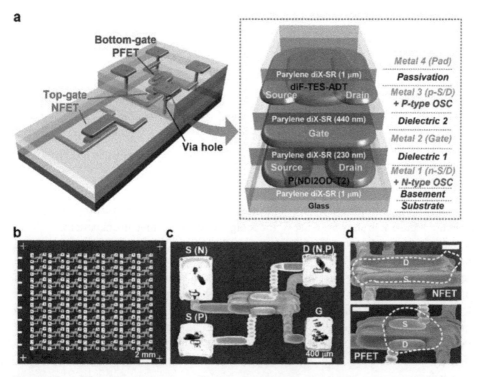

Figure 4.7 (a) Three-dimensional complementary OFET schematics with successively higher magnification shown in (b), (c), and (d). Reproduced with permission [46].

mentioned earlier, it is of the highest importance to maintain solvent orthogonality when printing successive layers to maintain component integrity and preserve critical interfaces.

4.4.2 Gravure and Flexographic Printing

Gravure printing, as seen in the top left of Figure 4.8, applies an ink to the flexible substrate through the contact of small cavities on the gravure cylinder. The ink is continuously resupplied to the gravure cylinder from a reservoir, with excess removed by a doctor blade. The substrate is sandwiched between an impression cylinder and the gravure cylinder to transfer the ink as both cylinders turn and the substrate passes through. This method can and has been used for very high-speed printing of organic semiconductors as well as dielectrics [47].

Essentially the opposite of gravure printing, flexographic printing involves the transfer of an ink from raised structures to the substrate, rather than from cavities. The transfer cylinder is typically made from a softer material such as a rubber or another polymer, and is supplied with ink from a bath, similar to the gravure printing process. This method can be seen on the top right of Figure 4.8. It finds application in the deposition of inorganic and organic conductor films as well as dielectrics [47].

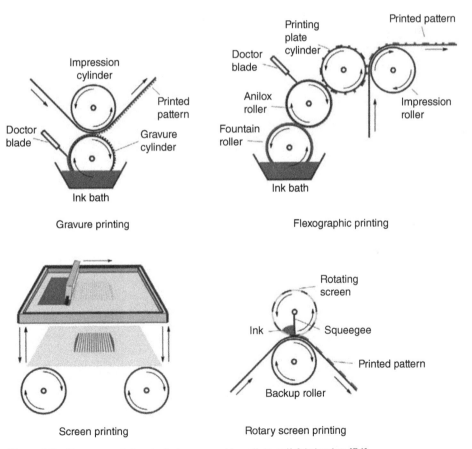

Figure 4.8 Common printing techniques used in roll-to-roll fabrication [36].

4.4.3 Screen Printing

Screen printing has been used in the textile industry for many years as an easy way to transfer patterns onto clothing or other fabrics. The process is shown in Figure 4.9a and usually follows from a metal or polymeric screen placed onto a substrate and a high-viscosity ink squeezed through the mesh using a squeegee. To print with low-viscosity inks, further considerations must be made such as the use of polymeric "banks" that restrict flow upon the use of a secondary screen, as done by Duan et al. and seen in Figure 4.9b [48]. A recent example of screen printing in this regard resulted in a fabric patch antenna that could easily be integrated into other textile electronics for RF communications [29], and there have been other extremely high-performing devices created this way as well [48]. Given the inherent size of the screen, this process is sometimes used in large-area fabrication such that patterning can be done quickly and reproducibly on roll-to-roll substrates. A further example of this technique is shown in Figure 4.11.

4.4.4 Aerosol Jet Printing

Spray coating, a precursor of aerosol jet printing, can form unpatterned films on flat and even curved substrates. In this case, the ink is pushed through a nozzle and aerosolized by an inert carrier gas toward a substrate. Drawbacks of this technique include isolated droplets (i.e. discontinuity), nonuniform (or rough) surface, and pinholes (that can induce short circuit between separate layers). Figure 4.10 provides an example of all printed metal interconnects. Spray coated metal electrodes have the potential to form connections, both pressure and strain sensors to integrate soft robotics, sensor skins, and wearable electronics [49]. Aerosol jet printing (AJP) is a microscale additive manufacturing technique that eliminates the need to mask the substrate to create a pattern. It has been referred to as an "assistive manufacturing technology" because it is often used in conjunction with other techniques to produce complex geometries [50] (Figure 4.12). An atomized ink is carried through a nozzle, surrounded by a sheath gas that directs and focuses the aerosol into a fine beam. The focusing results in feature sizes of about a few microns [50, 51]. These and any parameters are of course ink specific, however, the process can be manipulated by changing the flow rate of both aerosol and sheath gas, the print speed, ink temperature, and the size of the nozzle. As in many printing processes, optimized parameters are largely ink specific. Figure 4.5 illustrates transistors that utilized the AJP technique to formulate a part of the device [51].

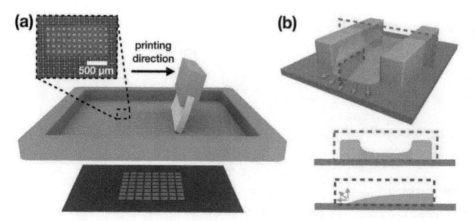

Figure 4.9 A schematic of the screen-printing process. Reproduced with permission [48].

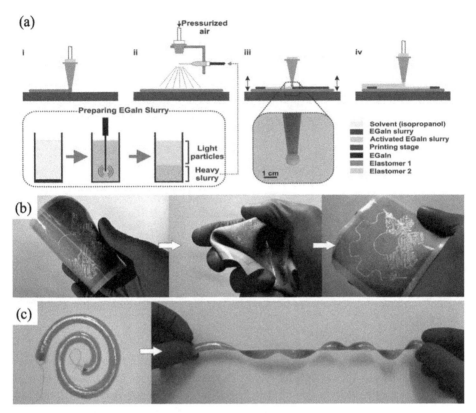

Figure 4.10 (a) Process flow for spray-printed stretchable metal contacts, (b) spray-printed complex metal patterns crumpled and flattened without damage, (c) twistable metal interconnects. Reproduced with permission [49].

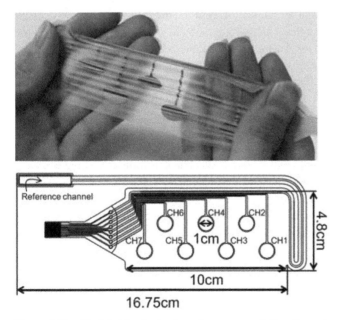

Figure 4.11 Stretchable electrodes using screen printing. Reproduced with permission [63].

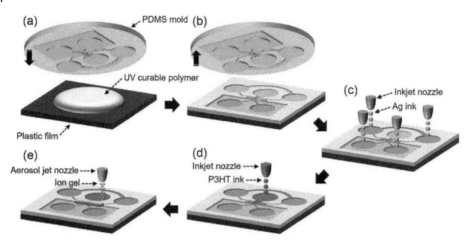

Figure 4.12 A combination printed process using imprint lithography, inkjet printing, and AJP. Reproduced with permission [64].

4.4.5 Contact Printing

Inspired by gravure and flexographic printing that have been extensively adapted and applied to roll-to-roll processing, contact printing can be interpreted as the transfer of an ink from a pre-patterned mold to a substrate through physical contact. Many techniques have arisen from that general definition. An inspired early example of a micro-contact printing by Briseno et al. utilized a PDMS stamp to transfer a commonly used SAM to silicon or plastic substrates, modifying the surface energy where contact was made [52]. Following the SAM deposition, organic single crystals of pentacene, rubrene, or C_6O were grown by physical vapor transport, nucleating only on the newly created hydrophobic regions. The process flow and images are shown in Figure 4.15.

Nanoimprint lithography (NIL) is another contact printing method; however, it can achieve resolutions of the same order of magnitude as photolithography in a roll-to-roll-compatible process. A typical thermal NIL process is shown on the left of Figure 4.16 and begins with a thin thermoplastic layer on the substrate, to which a hard mold that has been heated above the glass transition temperature is pressed. The thermoplastic layer such that an imprint is left behind as a thickness difference in the pattern of the mold. The thin layer that had been allowed to reflow out from under the hard mold pattern can then be removed by a short reactive ion etching step, defining features down to tens of nanometers [32, 53, 54]. Figure 4.14 shows a hard mold to pattern 70 nm-long transistor channels. In many cases, however, the glass transition temperature of the thermoplastic is too high to use this technique on a flexible substrate, so a new technique that is compatible with roll-to-roll processing was developed. UV-light-assisted nanoimprint lithography (UV NIL), seen on the right of Figure 4.16, uses UV light to cure in this case a photoresist, such that the temperatures necessary can stay well below the critical temperatures of the substrate [55–57]. Figure 4.13 shows devices fabricated with a combination of UV NIL, imprint lithography, and inkjet printing to realize fully flexible OTFTs [58]. Using rigid molds is very useful to achieve extremely high resolution and they are quite long lived, however, they are extremely susceptible to particulates on the surface, which renders them incapable of making direct contact across the entire substrate. On the other hand, soft lithography employs soft molds for a more conformal contact at pressures orders of magnitude lower, albeit with lower resolution (usually larger than 1 micron). In this case, a soft mold is obtained by pouring a liquid mixture of the

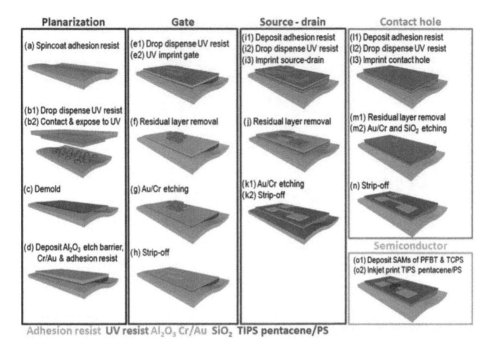

Planarization	Gate	Source - drain	Contact hole
(a) Spincoat adhesion resist	(e1) Drop dispense UV resist (e2) UV imprint gate	(i1) Deposit adhesion resist (i2) Drop dispense UV resist (i3) Imprint source-drain	(l1) Deposit adhesion resist (l2) Drop dispense UV resist (l3) Imprint contact hole
(b1) Drop dispense UV resist (b2) Contact & expose to UV	(f) Residual layer removal	(j) Residual layer removal	(m1) Residual layer removal (m2) Au/Cr and SiO₂ etching
(c) Demold	(g) Au/Cr etching	(k1) Au/Cr etching (k2) Strip-off	(n) Strip-off
(d) Deposit Al₂O₃ etch barrier, Cr/Au & adhesion resist	(h) Strip-off		Semiconductor (o1) Deposit SAMs of PFBT & TCPS (o2) Inkjet print TIPS pentacene/PS

Adhesion resist **UV resist** Al₂O₃ Cr/Au SiO₂ **TIPS pentacene/PS**

Figure 4.13 Process flow for fully printed devices using UV-NIL, inkjet printing, and imprint lithography. Reproduced with permission [58].

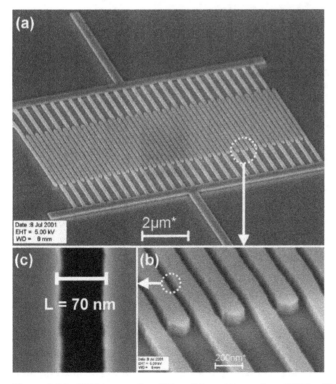

(a)

Date :8 Jul 2001
EHT = 5.00 kV
WD = 8 mm

2μm*

(c)

L = 70 nm

(b)

200nm*

Figure 4.14 SEM image of an NIL mold used to produce OFETs with a channel length of 70 nm. Reproduced with permission [65].

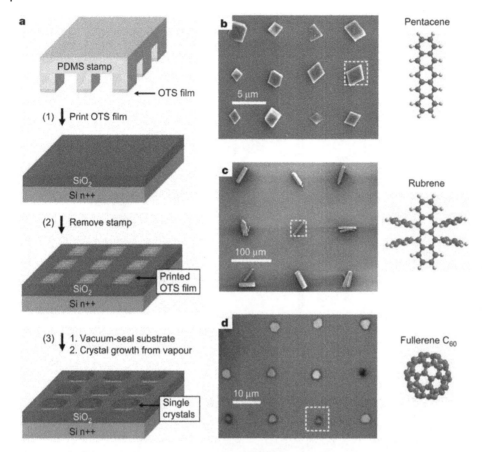

Figure 4.15 (a) Process flow for contact-printing OTS to pattern organic single crystalline semiconductors, and images showing the result from vapor deposition of (b) pentacene, (c) rubrene, and (d) C₆O. Reproduced with permission [52].

PDMS base and curing agent on a pre-patterned hard mold. After curing, the flexible mold can be detached by simply peeling it off. A new soft mold can be replicated from the obtained previous flexible one by repeating the processes described before. This is known as replica molding. It is important to treat the surfaces prior to perform the processes described in here. SAMs from fluoroalkyl or fluoroaryl silanes are a solution to prevent any deposited material to attach to the mold. At the same time, the substrate can be treated to improve adhesion and prevent detachment of the patterned films [59].

4.4.6 Meniscus Dragging

In a meniscus-dragging process, the solution is transferred due to viscous forces when the meniscus that is formed at the outlet aperture of a container is brought into contact with a substrate [6]. Film formation requires relative movement between substrate and the container of the solution. The thickness of the thin film is governed by the separation between substrate and

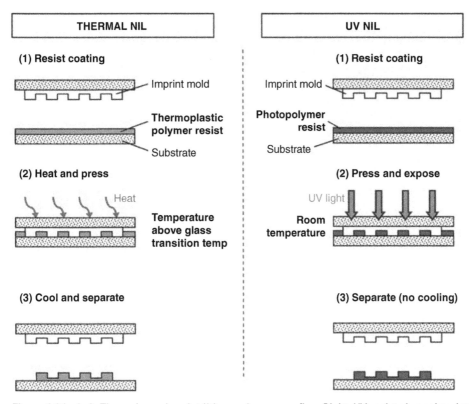

Figure 4.16 Left: Thermal nanoimprint lithography process flow. Right: UV-assisted nanoimprint lithography process flow [57].

printing head, physical properties of the ink (e.g. viscosity and surface tension), substrate temperature, and substrate-to-printing head relative speed. Differently from spin coating, where 90% of the material is discarded, meniscus dragging methods waste less than 1% and can be readily adapted to roll-to-roll processes. Figure 4.17 depicts the phenomena that occur simultaneously during thin-film formation. During meniscus displacement, a solid film is formed behind the passage of the meniscus. In the transition between solution and film, there are both temperature and concentration gradients that affect crystallinity and local flows. At this point, both capillary and Marangoni flow are at play. They differ completely from solution flow far from the solution–air interface. The resulting film highly depends on these factors, which will dictate the micro-crystal size, density, and orientation. A further modification on meniscus dragging by the introduction of micropillars on the printing "blade" allowed directionality in the meniscus to produce long crystals of a preferred orientation [60, 61].

Knife and slot die coating are other meniscus dragging techniques largely employed in roll-to-roll systems. Knife coating draws upon the doctor blading technique, except here the ink is constantly resupplied in front of the knife. Slot-die coating allows the formation of well-defined stripes along the web direction. A gear pump continuously feeds ink into the slot die, which drains down to form a meniscus that contacts the substrate as it passes by [47, 62]. Schematics of these deposition methods can be seen in Figure 4.18.

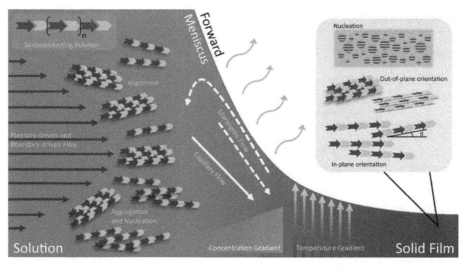

Figure 4.17 A schematic drawing of the various dynamic processes involved in meniscus-guided deposition of organic semiconductors [6].

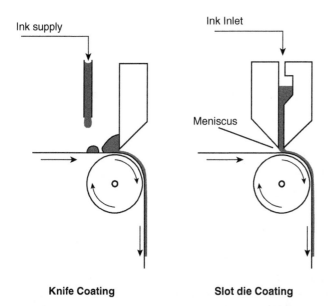

Figure 4.18 Schematics of knife coating and slot-die coating techniques. Reproduced with permission [36].

4.5 Conclusions

The development of roll-to-roll systems has heavily leaned on the ideas and innovations from the printing industry as well as silicon nanofabrication to achieve flexible electronics on a large scale. The varying techniques, used separately or in conjunction, each bring their own challenges but allow for the development of new and original structures. The clever researcher can draw upon the knowledge gained through all facets of materials science and engineering in order to bring about a cost-effective methodology for achieving massively reproducible electronics.

References

1 Horowitz, G. (1998). Organic field-effect transistors. *Adv. Mater.* 10 (5): 365–377.

2 Klauk, H. (2010). Organic thin-film transistors. *Chem. Soc. Rev.* 39 (7): 2643–2666.

3 Sirringhaus, H. (2014). 25th anniversary article: Organic field-effect transistors: The path beyond amorphous silicon. *Adv. Mater.* 26 (9): 1319–1335.

4 Ward, J.W., Lamport, Z.A., and Jurchescu, O.D. (2015). Versatile organic transistors by solution processing. *Chem. Phys. Chem.* 16 (6): 1118–1132.

5 Lamport, Z.A., Haneef, H.F., Anand, S., Waldrip, M., and Jurchescu, O.D. (2018). Tutorial: Organic field-effect transistors: Materials, structure and operation. *J. Appl. Phys.* 124 (7): 071101.

6 Xiaodan, G., Shaw, L., Kevin, G., Toney, M.F., and Bao, Z. (2018). The meniscus-guided deposition of semiconducting polymers. *Nat. Commun.* 9 (1): 1–6.

7 Matsui, H., Takeda, Y., and Tokito, S. (2019). Flexible and printed organic transistors: From materials to integrated circuits. *Org. Electron.* 75 (September): 105432.

8 Gelinck, G.H., Geuns, T.C.T., and De Leeuw, D.M. (2000). High-performance all-polymer integrated circuits. *Appl. Phys. Lett.* 77 (10): 1487–1489.

9 Ponce, O.O., Facchetti, A., and Marks, T.J. (Jan 2010). High-k organic, inorganic, and hybrid dielectrics for low-voltage organic field-effect transistors. *Chem. Rev.* 110 (1): 205–239.

10 Paterson, A.F., Mottram, A.D., Faber, H., Niazi, M.R., Fei, Z., Heeney, M., and Anthopoulos, T.D. (2019). Impact of the gate dielectric on contact resistance in high-mobility organic transistors. *Adv. Electron. Mater.* 5 (5): 1–9.

11 Myny, K., Smout, S., Rockel'e, M., Bhoolokam, A., Ke, T.H., Steudel, S., Cobb, B., Gulati, A., Rodriguez, F.G., Obata, K., Marinkovic, M., Pham, D.-V., Hoppe, A., Gelinck, G.H., Genoe, J., Dehaene, W., and Here-mans, P. (2015). A thin-film microprocessor with inkjet print-programmable memory. *Sci. Rep.* 4 (1): 7398.

12 Windmiller, J.R. and Wang, J. (2013). Wearable electrochemical sensors and biosensors: A review. *Electroanalysis* 25 (1): 29–46.

13 Liao, C., Zhang, M., Yao, M.Y., Hua, T., Li, L., and Yan, F. (2015). Flexible organic electronics in biology: Materials and devices. *Adv. Mater.* 27 (46): 7493–7527.

14 Moonen, P.F., Yakimets, I., and Huskens, J. (2012). Fabrication of transistors on flexible substrates: From mass-printing to high-resolution alternative lithography strategies. *Adv. Mater.* 24 (41): 5526–5541.

15 Facchetti, A., Yoon, M.H., and Marks, T.J. (2005). Gate dielectrics for organic field-effect transistors: New opportunities for organic electronics. *Adv. Mater.* 17(14): 1705–1725.

16 Gao, M., Lihong, L., and Song, Y. (2017). Inkjet printing wearable electronic devices. *J. Mater. Chem. C* 5 (12): 2971–2993.

17 Piacenza, P., Behrman, K., Schifferer, B., Kymissis, I., and Ciocarlie, M. (2020). A sensorized multicurved robot finger with data-driven touch sensing via overlapping light signals. *IEEE/ASME Trans. Mechatron.* 25 (5): 1–1.

18 Casalini, S., Bortolotti, C.A., Leonardi, F., and Biscarini, F. (2017). Self-assembled monolayers in organic electronics. *Chem. Soc. Rev.* 46 (1): 40–71.

19 Zschieschang, U. and Klauk, H. (2019). Organic transistors on paper: A brief review. *J. Mater. Chem. C* 7 (19): 5522–5533.

20 Diemer, P.J., Harper, A.F., Niazi, M.R., Petty, A.J., Anthony, J.E., Amassian, A., and Jurchescu, O.D. (2017). Laser-printed organic thin-film transistors. *Adv. Mater. Technol.* 2 (11): 1700167.

21 Harper, A.F., Diemer, P.J., and Jurchescu, O.D. (2019). Contact patterning by laser printing for flexible electronics on paper. *NPJ Flex. Electron.* 3 (1): 11.

22 Tyznik, C., Lamport, Z.A., Sorli, J., Becker-Koch, D., Vaynzof, Y., Loo, Y.-L., and Jurchescu, O.D. (2020). Laser printed metal halide perovskites. *J. Phys. Mater* 3 (3): 034010.

23 Bonacchini, G.E., Bossio, C., Greco, F., Mattoli, V., Kim, Y.-H., Lanzani, G., and Caironi, M. (2018). Tattoo-paper transfer as a versatile platform for all-printed organic edible electronics. *Adv. Mater.* 30 (14): 1706091.

24 Nogi, M., Iwamoto, S., Nakagaito, A.N., and Yano, H. (2009). Optically transparent nanofiber paper. *Adv. Mater.* 21 (16): 1595–1598.

25 Nogi, M., Komoda, N., Otsuka, K., and Suganuma, K. (2013). Foldable nanopaper antennas for origami electronics. *Nanoscale* 5 (10): 4395.

26 Chen, Y., Denis, K., Kazlas, P., and Drzaic, P. (2001). 12.2: A conformable electronic ink display using a foil-based a-Si TFT array. *SID Symp. Digest Tech.l Papers* 32 (1): 157.

27 Chwang, A., Hewitt, R., Urbanik, K., Silvernail, J., Rajan, K., Hack, M., Brown, J., Lu, J.P., Shih, C., Jackson, H., Street, R., Ramos, T., Moro, L., Rutherford, N., Tognoni, K., Anderson, B., and Huffman, D. (2006). 64.2: Full color 100 dpi AMOLED displays on flexible stainless steel substrates. *SID Symp. Digest Tech.l Papers* 37 (1): 1858.

28 Yun, D.-J., Lee, S., Yong, K., and Rhee, S.-W. (2012). Low-voltage bendable pentacene thin-film transistor with stainless steel substrate and polystyrene-coated hafnium silicate dielectric. *ACS Appl. Mater. Interf.* 4 (4): 2025–2032.

29 Li, Z., Sinha, S.K., Treich, G.M., Wang, Y., Yang, Q., Deshmukh, A.A., Sotzing, G.A., and Cao, Y. (2020). All-organic flexible fabric antenna for wearable electronics. *J. Mater. Chem. C* 8 (17): 5662–5667.

30 Wang, L., Fu, X., He, J., Shi, X., Chen, T., Chen, P., Wang, B., and Peng, H. (2020). Application challenges in fiber and textile electronics. *Adv. Mater.* 32 (5): 1901971.

31 Allison, L., Hoxie, S., and Andrew, T.L. (2017). Towards seamlessly-integrated textile electronics: Methods to coat fabrics and fibers with conducting polymers for electronic applications. *Chem. Commun.* 53 (53): 7182–7193.

32 Traub, M.C., Longsine, W., and Truskett, V.N. (2016). Advances in nanoimprint lithography. *Annu. Rev. Chem. Biomol. Eng.* 7 (1): 583–604.

33 Zakhidov, A.A., Lee, J.-K., DeFranco, J.A., Fong, H.H., Taylor, P.G., Chatzichristidi, M., Ober, C.K., and Malliaras, G.G. (2011). Orthogonal processing: A new strategy for organic electronics. *Chem. Sci.* 2 (6): 1178.

34 Jang, J., Song, Y., Yoo, D., Kim, T.-Y., Jung, S.-H., Hong, S., Lee, J.-K., and Lee, T. (2014). Micro-scale twistable organic field effect transistors and complementary inverters fabricated by orthogonal photolithography on flexible polyimide substrate. *Org. Electron.* 15 (11): 2822–2829.

35 Dang, W., Vinciguerra, V., Lorenzelli, L., and Dahiya, R. (2017). Printable stretchable interconnects. *Flex. Print. Electron.* 1: 013003.

36 Sondergaard, R., Hosel, M., Angmo, D., Larsen-Olsen, T.T., and Krebs, F.C. (2012). Roll-to-roll fabrication of polymer solar cells. *Mater. Today* 15 (1-2): 36–49.

37 Chang, J.S., Facchetti, A.F., and Reuss, R. (mar 2017). A circuits and systems perspective of organic/Printed electronics: Review, challenges, and contemporary and emerging design approaches. *IEEE J. Emerg. Select. Top. Circuit. Syst.* 7 (1): 7–26.

38 Jung, M., Kim, J., Noh, J., Lim, N., Lim, C., Lee, G., Kim, J., Kang, H., Jung, K., Leonard, A.D., Tour, J.M., and Cho, G. (2010). All-printed and roll-to-roll-printable 13.56-MHz-operated 1-bit RF tag on plastic foils. *IEEE Trans. Electron. Dev.* 57 (3): 571–580.

39 Driessen, T. and Jeurissen, R. (2015). Drop formation in inkjet printing. In: Stephen, D. (ed.) *Fundamentals of Inkjet Printing*, 93–116. Weinheim: Wiley-VCH Verlag GmbH & Co. KGaA.

40 Lessing, J., Glavan, A.C., Walker, S.B., Keplinger, C., Lewis, J.A., and Whitesides, G.M. (2014). Inkjet printing of conductive inks with high lateral resolution on omniphobic "R F paper" for paper-based electronics and MEMS. *Adv. Mater.* 26 (27): 4677–4682.

41 Mahajan, A., Hyun, W.J., Walker, S.B., Lewis, J.A., Francis, L.F., and Daniel Frisbie, C. (2015). High-resolution, high-aspect ratio conductive wires embedded in plastic substrates. *ACS Appl. Mater. Interf.* 7 (3): 1841–1847.

42 Soltman, D. and Subramanian, V. (2008). Inkjet-printed line morphologies and temperature control of the coffee ring effect. *Langmuir* 24 (5): 2224–2231.

43 Choi, H.W., Zhou, T., Singh, M., and Jabbour, G.E. (2015). Recent developments and directions in printed nanomaterials. *Nanoscale* 7 (8): 3338–3355.

44 Sekitani, T., Noguchi, Y., Zschieschang, U., Klauk, H., and Someya, T. (2008). Organic transistors manufactured using inkjet technology with subfemtoliter accuracy. *Proc. Natl. Acad. Sci. U. S. A.* 105 (13): 4976–4980.

45 Sowade, E., Mitra, K.Y., Ramon, E., Martinez-Domingo, C., Villani, F., Loffredo, F., Gomes, H.L., and Baumann, R.R. (2016). Up-scaling of the manufacturing of all-inkjet-printed organic thin-film transistors: Device performance and manufacturing yield of transistor arrays. *Org. Electron.* 30: 237–246.

46 Kwon, J., Takeda, Y., Fukuda, K., Cho, K., Tokito, S., and Jung, S. (2016). Three-dimensional, inkjet-printed organic transistors and integrated circuits with 100% yield, high uniformity, and long-term stability. *ACS Nano* 10(11): 10324–10330.

47 Kim, G.-E., Shin, D.-K., Lee, J.-Y., and Park, J. (2019). Effect of surface morphology of slot-die heads on roll-to-roll coatings of fine PEDOT:PSS stripes. *Org. Electron.* 66 (October): 116–125.

48 Duan, S., Gao, X., Wang, Y., Yang, F., Chen, M., Zhang, X., Ren, X., and Hu, W. (2019). Scalable fabrication of highly crystalline organic semiconductor thin film by channel-restricted screen printing toward the low-cost fabrication of high-performance transistor arrays. *Adv. Mater.* 31(16): 1807975.

49 Mohammed, M.G. and Kramer, R. (2017). All-printed flexible and stretchable electronics. *Adv. Mater.* 29(19): 1604965.

50 Wilkinson, N.J., Smith, M.A.A., Kay, R.W., and Harris, R.A. (2019). A review of aerosol jet printing—a non-traditional hybrid process for micro-manufacturing. *Int. J. Adv. Manuf. Technol.* 105(11): 4599–4619.

51 Secor, E.B. (2018). Principles of aerosol jet printing. *Flex. Print. Electron.* 3(3): 035002.

52 Briseno, A.L., Mannsfeld, S.C.B., Ling, M.M., Liu, S., Tseng, R.J., Reese, C., Roberts, M.E., Yang, Y., Wudl, F., and Bao, Z. (2006). Patterning organic single-crystal transistor arrays. *Nature* 444(7121): 913–917.

53 Chou, S.Y., Krauss, P.R., and Renstrom, P.J. 1996). Nanoimprint lithography. *J. Vacuum Sci. Technol. B Microelectron. Nanometer Struct.* 14(6): 4129.

54 Guo, L.J. (2007). Nanoimprint lithography: Methods and material requirements. *Adv. Mater.* 19(4): 495–513.

55 Leitgeb, M., Nees, D., Ruttloff, S., Palfinger, U., Götz, J., Liska, R., Belegratis, M.R., and Stadlober, B. (2016). Multilength scale patterning of functional layers by roll-to-roll ultraviolet-light-assisted nanoimprint lithography. *ACS Nano* 10(5): 4926–4941.

56 Ahn, S., Cha, J., Myung, H., Kim, S.-M., and Kang, S. (2006). Continuous ultraviolet roll nanoimprinting process for replicating large-scale nano- and micropatterns. *Appl. Phys. Lett.* 89(21): 213101.

57 Kooy, N., Mohamed, K., Pin, L., and Guan, O. (2014). A review of roll-to-roll nanoimprint lithography. *Nanoscale Res. Lett.* 9(1): 320.

58 Moonen, P.F., Vratzov, B., Smaal, W.T.T., Kjellander, B.K.C., Gelinck, G.H., Meinders, E.R., and Huskens, J. (2012). Flexible thin-film transistors using multistep UV nanoimprint lithography. *Org. Electron.* 13(12): 3004–3013.

59 Gates, B.D., Xu, Q., Love, J.C., Wolfe, D.B., and White-sides, G.M. (2004). Unconventional nanofabrication. *Annu. Rev. Mater. Res.* 34(1): 339–372.

60 Kim, J.-O., Lee, J.-C., Kim, M.-J., Noh, H., Yeom, H.-I., Ko, J.B., Lee, T.H., Park, S.-H.K., Kim, D.-P., and Park, S. (2018). Inorganic polymer micropillar-based solution shearing of large-area organic semiconductor thin films with pillar-size-dependent crystal size. *Adv. Mater.* 30(29): 1800647.

61 Diao, Y., Zhou, Y., Kurosawa, T., Shaw, L., Wang, C., Park, S., Guo, Y., Reinspach, J.A., Gu, K., Gu, X., Tee, B.C.K., Pang, C., Yan, H., Zhao, D., Toney, M.F., Mannsfeld, S.C.B., and Bao, Z. (2015). Flow-enhanced solution printing of all-polymer solar cells. *Nat. Commun.* 6 (1): 1–10.

62 Wang, B.-Y., Lee, E.-S., Lim, D.-S., Kang, H.W., and Oh, Y.-J. (2017). Roll-to-roll slot die production of 300 mm large area silver nanowire mesh films for flexible transparent electrodes. *RSC Adv.* 7(13): 7540–7546.

63 Sekitani, T., Yoshimoto, S., Araki, T., and Uemura, T. (2017). 12-2: Invited paper: A sheet-type wireless electroencephalogram (EEG) sensor system using flexible and stretchable electronics. *SID Symp. Digest Tech.l Papers* 48(1): 143–146.

64 Hyun, W.J., Bidoky, F.Z., Walker, S.B., Lewis, J.A., Francis, L.F., and Frisbie, C.D. (2016). Printed, self-aligned side-gate organic transistors with a Sub-5 um gate-channel distance on imprinted plastic substrates. *Adv. Electron. Mater.* 2(12): 1600293.

65 Austin, M.D. and Chou, S.Y. (2002). Fabrication of 70 nm channel length polymer organic thin-film transistors using nanoimprint lithography. *Appl. Phys. Lett.* 81(23): 4431–4433.

5

Metallic Nanowires, Promising Building Nanoblocks for Flexible Transparent Electrodes

Jean-Pierre Simonato

Director of Research, CEA

5.1 Introduction

The development of new transparent electrodes (TEs) has been thoroughly investigated by both academics and companies during the last decade. TEs are widely used in many functional devices such as solar cells, touchscreens, transparent heaters, and displays. The worldwide demand is very high and the requirements for TEs have evolved drastically. First of all, the economical aspect remains a major point and new low-cost technologies would be highly desirable, but only if they can give access to TEs holding at least the same performances, and with a significant fabrication gain compared to the existing technologies. Another aspect, necessary to follow the market trends, is to fabricate TEs with new mechanical properties. Indeed, beside other major challenges such as lowering energy consumption, compacting structures, using environment-friendly and lightweight materials, a major driver is the development of devices with high flexibility, and even in some cases important stretchability.

The current technologies used to fabricate TEs are essentially based on transparent conductive oxides (TCOs), in particular indium tin oxide (ITO), which is by far the most used transparent conductive material. The production cost of ITO layers is rather high, in the $5–15/m^{-2} range, and requires high capital expenditures for active material deposition systems, generally by vapor phase sputtering process. Moreover, indium is a critical metal [1]. This means that resources existing to meet future needs are limited, the abundance of indium on earth being very low. Although some indium can be recovered from ITO scrap, the recycled amount remains low and the overall extracted and recycled quantities will not be sufficient to fulfill the fast-growing demand. Corollary, its price is fluctuating with high level of uncertainty. Another foremost concern comes from the fact that the ITO-based TEs do not reach the expectations for future market applications since this material is brittle, it does not support large mechanical constraints. TCOs are thus not suitable for the fabrication of highly flexible displays, alternatives are expected for future needs.

Several technical solutions are being developed to fabricate low-cost, high-performance flexible TEs. The most advanced materials include conductive polymers, metal grids, carbon-based materials such as graphene and/or carbon nanotubes, metallic nanowires, or hybrids of those materials [2–6]. Associated fabrication techniques are well advanced and should provide relevant opportunities for future flexible TEs. Among them, the development of TEs based on metallic nanowires holds tremendous potential and hope, because excellent optoelectronic properties can be reached, at least similar to ITO-based layers, using a low-cost solution deposition technique at ambient conditions [7–10].

Specifications for flexible TEs can be divided into several categories, depending on the targeted applications. Above all, satisfying optical and electrical performances are perceived as the main criteria to evaluate the quality of TEs. Basically, transmittance in the visible spectrum and sheet resistance are the first properties to be assessed. Nevertheless, many other requirements are looked at, including spatial uniformity, stability, bendability, work function, contact resistance, and so on and so forth.

5.2   TEs Based on Metallic Nanowires

5.2.1   Metallic Nanowires, New Building Nanoblocks

The synthesis of metallic nanowires has been intensively developed since the first report of silver nanowires (AgNWs) fabrication by Sun et al. [11]. The chemical process of fabrication is remarkable since the reactive mixing of isotropic precursors (metallic precursors, polymer, salts, solvents) results in the formation of one-dimensional nanomaterials with high aspect ratio (i.e. length/diameter ratio). In general, the metallic nanowires have diameters of few tens of nm and are few µm long. The growth of the nanowires might be realized through a hydrothermal process, though the solvothermal approach is usually preferred [12, 13].

In a typical procedure, silver salts are reduced in the presence of polyvinylpyrrolidone (PVP) in hot ethylene glycol. PVP is used as a capping agent. Thanks to its adsorption onto some surfaces of the germs, and then nanowires, it passivates these surfaces and focuses the growth toward both extremities of the nanomaterials. It is adsorbed along the (100) surfaces of the growing nanowire while the (111) surfaces are free to grow [14]. A lot of work has been dedicated to this synthesis. Like for many nanomaterial syntheses, tiny modifications can induce large variations in the reaction products. Concerning the synthesis of these nanomaterials, two main aspects deserve to be discussed. First, the yield of the reaction is seldom indicated in the literature. This is a serious drawback since the cost of the process relies in part on this yield. If one intends to benchmark this technology with others, it is essential to know the actual conversion rate of silver salts into usable nanowires for the fabrication of TEs. Mostly, scanning electron microscope images of purified nanowires are shown, and in some rare instances histograms of diameter and length distributions are provided. This can be an issue, since the selection of specific images can limit the information on the overall synthesis products, and by-products. To illustrate this point, Figure 5.1 shows different images realized on the same sample of AgNW crude product obtained after synthesis, and purified nanowires. In good operating conditions, the yield can be higher than 90% in purified AgNWs, which is satisfactory from an industrial standpoint. The purification of the nanowires is also very important for several reasons. Remaining organics must be eliminated in order to enhance the electrical contacts between nanowires. Indeed, residual PVP coating along the nanowires after synthesis induces an increased contact resistance at interconnections and should be removed. The presence of small metallic nanoparticles with low form factor is also deleterious to the system because these species do not participate to the percolative system for electrical conduction, but they decrease the transparency and increase the diffusion of light. Last, the presence of large metallic particles is also unwanted due to similar optical effects and potential roughness problems at interfaces with active layers in multilayered structures.

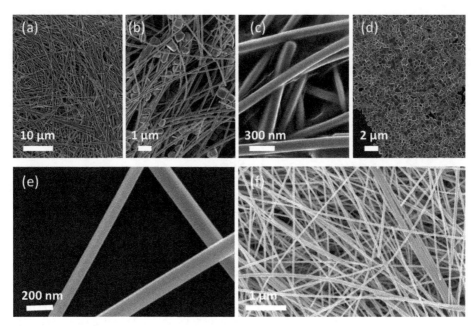

Figure 5.1 (a)(b)(c)(d) SEM images showing different parts of the same crude mixture of as-prepared AgNWs, (e)(f) SEM images of purified AgNWs.

The second point deals with the recent work dedicated to the fabrication of nanowires with controlled dimensions. Both the mean diameter and the mean length of the nanowires have tremendous impact on the TE's properties (discussed hereinafter). For instance, simply by decreasing the diameter of the nanowires from 60 nm to 30 nm, TEs' optical performances can change from inadequate to suitable for many optoelectronic applications, including flexible displays [15]. The modification of the diameter of the AgNWs can be tuned by different means. The effects of temperature or the use of additional reactants proved efficient. Probably the most common technique to reduce the diameter is the addition of KBr, which allows lessening of the diameter down to only few tens of nanometers [16].

Metals other than silver have also been studied for the fabrication of flexible TEs like copper [17–20], gold [21], Cu-Ni [22, 23], etc. Some new shapes have also been developed such as core-shell structures [24, 25]. However, so far silver remains the metal of choice, and not only because it is the most conductive metal. This can certainly be ascribed to the remarkable ease of fabrication of AgNWs. Indeed, the synthesis protocol is rather straightforward, with a good scalability and reproducibility. When compared to other metals, and in particular copper [26], AgNWs also intrinsically have a good stability [24, 25, 27, 28].

5.2.2 Random Network Fabrication

This technology relies on the use of these "building nanoblocks" in the form of random lattices. Basically, metallic nanowires are deposited randomly onto surfaces. Electrical contacts appear at nanowire–nanowire interconnections, leading to the creation of two-dimensional conductive networks. The as-made structures contain plenty of conductive pathways for electrical conduction, and at the same time plenty of empty space for light to go through (Figure 5.2).

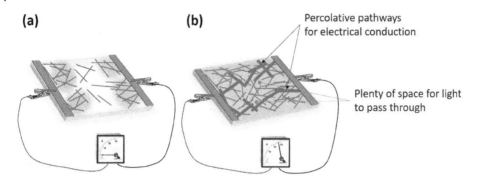

Figure 5.2 (a) AgNWs on transparent substrate without enough nanowires to reach the percolation regime. (b) Random network of AgNWs above the percolation threshold. Some AgNW-based pathways allow current to go from one side electrode to the other (thick lines), and plenty of space is still present between the nanowires, which renders the material transparent.

One of the main advantages of this technique is to allow the use of large-area solution deposition techniques. Many of them have been studied so far. Spin coating [29, 30], spray coating [31], roll-to-roll [32], drop casting [33–35], doctor-blade casting [36], dip-coating [37], Mayer rod coating [18, 38, 39], screen printing [40, 41], vacuum filtration [17, 42, 43], and brush coating [44, 45], belong to the most reported in the literature. Depending on the chosen technology, a specific ink must be developed. The choice of solvents, co-solvents, and additives has been widely studied and many formulations have been patented so far. A typical development of a water-based ink reported recently describes some of the key points that should be tackled such as stability of the dispersion, wettability, spreadability, elimination of bubbles, adjustment of the drying process, homogeneity of the performances, and so on [46]. Post-deposition treatments are also of interest since a strong improvement of the sheet conductance can be realized by several techniques, for instance, thermal annealing [47] or mechanical pressing [35].

5.2.3 Optical Characterization

The technical requirements of TEs for the fabrication of devices are manifold, and are application dependent. Concerning optical characterization, the main point for flexible display TEs is to demonstrate a good specular transmittance in the visible spectrum. The total transmittance is the percentage of light that is not reflected or absorbed by the layer. The transmittance is composed of specular transmittance and diffuse transmittance. It should be emphasized here that high transmittance does not mean transparency. For instance, frosted window glass is translucent, but not transparent. It means that it permits light to pass through, but diffusion of light is so high that any objects on the opposite side are not clearly visible. This light-scattering property of TE materials is called the haziness. The haze factor corresponds to the ratio of diffuse transmittance divided by the total transmittance. It is described and used in different protocols or norms such as ASTM D1003-13 (Standard Test Method for Haze and Luminous Transmittance of Transparent Plastics). The needed level of haziness depends on the targeted application. For instance, a high haze factor is expected for photovoltaic cells whereas light scattering should be minimized below 3%, or better 2%, for most displays. Indeed, for photovoltaic (PV) devices the scattering of light improves the efficiency of the cells by increasing the length of the photon pathways within the cell, and thus enhancing the probability of photon absorption [48–50]. On the contrary, for flexible displays a low diffusive effect is expected in order to give to the reader a fine clearness. It is commonly accepted that a haze value below 3% allows the enlightened information to pass through the TE without important fuzziness.

It must be stated that optical performances of TEs are not always adequately reported in the literature. Ideally, for each work on TE, optical measurement values should be given with and without substrate, and with a full set of optical data including at least total transmittance and diffuse transmittance. This point is too often neglected therefore it is not always possible to ensure fair comparison between different studies and to understand whether or not reported TEs would be suitable for flexible display applications.

The performances of new TEs are logically compared to ITO, the reference material. It is generally agreed that for display application, a low sheet resistance (<40 ohm.sq^{-1}) should be associated with a total transmittance of the TE layer higher than 90% in the visible spectrum, and a low haze, typically below 2% or 3%. For metallic nanowire-based TEs, the haze factor is clearly dependent on the shape of nanowires. Very long nanowires (>30 μm) allow percolation to be reached with less nanowires, which decreases the haze factor when compared to common AgNWs [51]. Lowering the diameter of AgNWs can also play a key role in the mitigation of light scattering [15, 52, 53].

5.2.4 Electrical Characterization

The electrical performance is usually given in the form of sheet resistance, expressed in ohm.sq^{-1}. It represents a resistance by unit area (a square), independently of the square size, which is often used for characterization of very thin films. One of the main goals while fabricating metallic NW-based transparent conductors is to determine the optimal trade-off between transparency and conductivity. The deposition of many nanowires induces an increase of the conductivity, but is detrimental to the optical performances. In an extreme case, the network could be very conductive (< 1 ohm.sq^{-1}) but opaque, which of course is of no interest in the TE case study.

The electrical performances of the networks depend on many effects, in particular on the morphology and the number of nanowires. The theoretical aspect can be modeled thanks to the percolation theory. This approach has been extensively studied for stick networks, and applied to nanowire networks [30, 54]. Independently of the dimensions of the nanowires, other parameters have a strong effect on the electrical performances of the networks. Many papers report methods that can lead to a dramatic decrease in sheet resistance from 104 or 105 down to about 10 Ω sq^{-1}. For instance thermal annealing [47, 55], chemical treatments [29], laser sintering [56], light-induced plasmonic nanowelding [57, 58], or mechanical pressing [35] can lead to major improvements of the network electrical conduction. The electrical optimization at the contact junctions is possible by applying increasing voltages. This is mainly ascribed to the decrease of electrical resistances at the junctions between nanowires. Indeed, local conductivity improvements at the nanoscale can generate a huge decrease of resistance at the macroscale. The most used technique to improve contacts between AgNWs is probably thermal annealing after deposition. Atomic diffusion activated by thermal energy leads to the morphological rearrangement of the junctions [55]. A TEM analysis before and after annealing shows that there is a strong modification, the nanowires are sintered as shown in Figure 5.3.

It is quite difficult to observe what happens during bias application on percolating networks. This was recently highlighted by lock-in thermography, which allows visualization in situ of the formation of active pathways with a pertinent spatial resolution [59]. Actually, it is a mapping of the heat generated locally by Joule effect across the network, which reveals how electric current is distributed through the percolating pathways. This method provided for the first time visual evidence for the irreversible activation of different pathways under various biases (Figure 5.4).

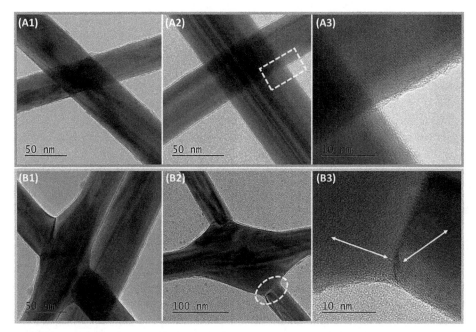

Figure 5.3 A1–A3) TEM images of "as deposited" AgNWs (without any optimization treatment). Typical junctions between AgNWs are depicted in (A1) and (A2). (A3) is a zoom in the bottom-right area of (A2) junction. B1–B3) TEM images show the morphology of junctions after isothermal annealing on a hot plate at 180°C for 1 hour. (B3) is a zoom in the grain boundary circled at the bottom of (B2) junction.

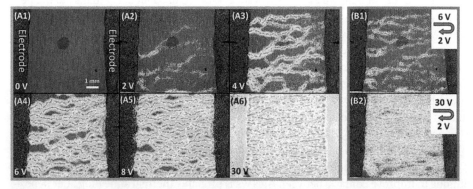

Figure 5.4 (A1–A6) Thermal lock-in images of an AgNW network close to the percolation threshold. Continuous on–off (10 Hz) voltages were applied to the network during 5-minute cycles, from 2 V up to 30 V. Each image shows the phase signal resulting from all the IR radiations collected during the 5-minute cycles at a given voltage. After (B1) 6 V and (B2) 30 V, the voltage is decreased back to 2 V: all the percolating paths previously activated are still illuminated, with a lower intensity, which confirms that the activation process is not reversible. Reprinted with permission from [59]. Copyright (2016) American Chemical Society.

5.2.5 Mechanical Aspect

As mentioned at the beginning of this chapter, current industrial technologies based on TCOs do not allow fabrication of highly flexible TEs that are needed for future applications. Mechanical flexibility is an expected characteristic of TEs for the fabrication of flexible displays, and more generally flexible or foldable electronics. Mechanical bending comparison of AgNW-based TEs with TCOs show unequivocally the advantage of using these materials. Whereas ITO or FTO (fluorine-doped tin oxide) demonstrates very high increase of the resistance at low bending radius, typically in the mm range, the AgNW networks remain very stable [7, 60]. In addition, the high conformability of AgNW networks allows deposition on surfaces with high roughness, which gives access to the fabrication of stacks onto various substrates and interfaces. The structural behavior of these networks under compression, torsion or tension stresses has been extensively studied, and in the vast majority of cases no significant alteration of the performances was observed. A simple and clear demonstration was operated through the crumpling of AgNW networks after deposition on a very thin piece of polyethylenenaphtalate (thickness 1.3 μm) without changing the electrical properties [61]. Many devices including solar cells, organic light-emitting diodes (OLEDs) and thin film heaters have been successfully fabricated on highly flexible substrates with metallic nanowire electrodes. The capability of these electrodes to withstand external mechanical stresses overtakes by far most of traditional transparent conductive materials.

5.3 Application to Flexible Displays

The key challenges related to the integration of AgNW-based TEs are presented hereinbefore. In this section, the goal is to show how these new TEs have been integrated into various flexible displays. Each application has its own peculiarities and could be subjected to various adjustments related to the intrinsic needed properties of the TEs, and the way to integrate them into functional devices. Displays taking advantage of metallic nanowire TEs and based on several technologies are presented. Emphasis is placed on touch screens, OLEDs, polymer dispersed liquid crystals, and thermochromic devices.

5.3.1 Touch Screens

Touch sensors belong to the family of tactile sensors [62]. Various properties can be measured by tactile sensors, and they especially are used in touch screens to harvest information from physical touch. The physical concepts used for detection in touch panels are essentially based on resistive or capacitive technologies. The goal is to detect touch or near proximity without need of intimate physical contact. Resistive technology has been used for a longer time than capacitive approach as this is based on simple control circuits. Basically, this kind of sensor consists of two conductive layers separated by small spacers. When a probe, a finger for instance, applies pressure on the top film, a contact is made between the two layers, inducing a voltage drop that can be detected by an X–Y coordinate controller. This system has been realized with the use of metallic nanowire-based TEs [56, 63, 64].

The second family, by far the most widely developed, is based on capacitive technology. Most smartphones, MP3 players, computer touch screens, and man–machine interfaces are based on capacitive touch sensors. Usually these sensors are separated into two categories: self-capacitance-type sensors and mutual capacitance-type sensors. The principle is close to the one of resistive sensors, but based on variation of capacitance. The capacitance of the sensor pad is measured periodically. When a conductive object such as a finger is approaching or touching the electrode the electric field is modified, corresponding to a capacitance change. This alteration is detected by the measurement circuit (by different means such as frequency or amplitude modulation, for instance) and converted into a trigger signal. A main difference with resistive sensors, is that a measurable change in capacitance is enough to transfer information, which means that a physical contact is not mandatory, the proximity between the finger and the active surface being sufficient to get a positive response of the sensor. This has been demonstrated in several papers since 2013 [65–69].

The capacitive technology using metallic nanowires for touch screens has been the subject of intense research, and few companies have claimed to develop or participate in the fabrication of industrial products (e.g. Cambrios or C3nano). The TE market is worth several billion US dollars, so even capture of a small market share would lead to important sales and would place AgNW-based TEs as a valuable incomer (Figure 5.5).

5.3.2 Light-emitting Diodes Displays

Since its invention about 30 years ago, OLED solid-state technology for use in display and lighting devices has attracted extensive attention worldwide. It represents the vast majority of flexible displays using AgNW-based TEs. Among the major challenges in producing this technology on flexible substrates, TE remains a key point. Basically, the active emitting layer is sandwiched between two electrodes, and at least one of them must be transparent. The charge carriers are driven to the active layer, which is responsible for the recombination of electron-hole pairs to generate photon emission. In order to offer appropriate solutions for the target applications, e.g. portable electronic rollable displays or conformable lighting panels, not only the active layer but also the transparent conducting layer have to be highly flexible. Large-angle bending cycles with no alteration in conductivity, transparency, or light emission properties are essential. The use of AgNW-based TEs for OLEDs or polymer light-emitting diode devices has been developed and integrated with success for several years. A main contribution was reported by Gaynor et al. in 2013 [70]. It

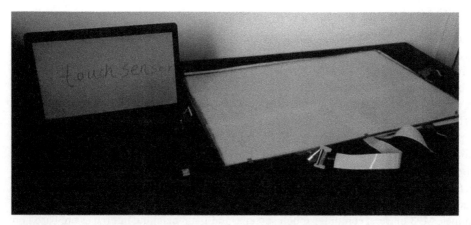

Figure 5.5 Demonstration of a 32-inch flexible touch screen fabricated from AgNW-based TE. Reprinted with permission from [46]. Copyright (2017) Royal Society of Chemistry.

was demonstrated that it is possible to fabricate very efficient ITO-free white OLEDs using an AgNW/PMMA (poly(methyl methacrylate)) composite electrode. The device showed a very high level of luminous efficacy, i.e. >30 lm.W^{-1} at a brightness of 1000 cd m^{-2}, and exhibited two other important optical characteristics: color-independent emission while increasing the viewing angle and an almost perfect Lambertian emission.

Flexible displays relying on OLED technology gather many advantages, due to the ultrathin layer constituents, the planar stacked structure (anode/hole injection and transporting layers/ light-emitting layer/electron transporting and injection layers/cathode), the low temperature fabrication process, and so on and so forth. The very high potential of this technology is commonly accepted, nevertheless, some technical hurdles still have to be overcome. Different strategies have been recently developed to fix some issues, notably the roughness of the TE layer [71, 72], the lifetime of the devices [73], the device architecture [74, 75], the charge injection [76, 77], the outcoupling structures [78–80], and the patterning of electrodes [81, 82].

The current OLED manufacturing lines are generally adapted to rigid substrates. To be able to manufacture flexible substrates, it is necessary to develop new process flows including new substrate handling. The fabrication of multilayer architectures for flexible OLED displays by solution deposition techniques still faces challenges. For instance, sequential deposition of very thin layers without damaging the structures already in place on the substrate, AgNW layer roughness, or alignment of patterns are technically difficult. A nice example of roll-to-roll process development allowing large-area uniformity led to the fabrication flexible OLEDs as shown in Figure 5.6 [32].

Beyond potential cost gain thanks to the use of large-scale solution deposition techniques, the use of flexible substrates offers new possibilities dealing with mechanical properties of OLED-based displays. High mechanical durability is expected at low bending radius. A relevant example was demonstrated by Ok et al. who prepared thin films (<10 μm thickness) of AgNW-based composite films and showed a remarkable stability under more than 100 000 iterative bendings at small bending radius (500 μm) [83]. Highly flexible blue OLEDs were successfully fabricated, with a brightness exceeding 4000 cd.m^{-2} at 12 V. High bendability and impressive stretchability were demonstrated by Liang et al., with a AgNW/PUA (poly(urethane acrylate)) composite as the transparent electrode [84]. In this case, the stretched device was found to withstand strain up to 120% while displaying fairly uniform light emission, as shown in Figure 5.7b.

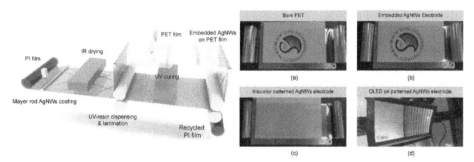

Figure 5.6 Left: schematic illustration of the roll-to-roll fabrication process for the embedded AgNWs transparent electrode on PET film. Right: photographic images of (a) bare PET, (b) embedded AgNWs on PET, (c) insulator-patterned AgNWs on PET, and (d) OLED on the embedded AgNWs/PET [32]. Copyright Elsevier (2017).

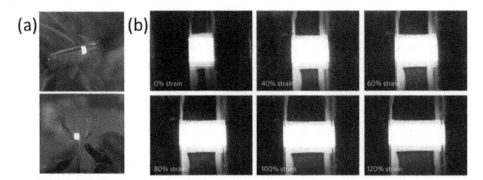

Figure 5.7 (a) A bent, lit OLED, (down) an unfolded lit OLED. Reprinted from [83]. (b) Photographs of a PLEC (original emission area, 5.0 × 4.5 mm2) biased at 14 V at specified strains. Reprinted with permission from [84]. Copyright (2013) Springer Nature.

5.3.3 Electrochromic Flexible Displays

The technology based on electrochromism has been widely developed for flexible displays application, either in transmission or reflexion modes. This is mainly due to the fact that the fabrication of these devices is rather cheap (basically electrochromic devices are composed of two electrodes separated by an electrolytic layer), and they require low power consumption. The concept of electrochromism is usually used for materials with the aptitude to modify their optical properties in a long-term and reversible way. Though initially it concerned materials, indifferently organic (e.g. polymers) or inorganic (e.g. metal oxides), that change their oxidation states when subjected to electrical fields [85], it was generalized to non-emissive displays showing "on" and "off" states while an alternating voltage potential is applied across the two electrodes.

An example of switchable display is based on the use of liquid crystals, on more precisely polymer dispersed liquid crystals (PDLC), which are liquid crystal embedded polymer films. In this case, the application of electrical fields aligns the liquid crystals droplets so that the device is in its transparent configuration. Otherwise, the random dispersion of the droplets results in a scattering of the incident light as a result of the mismatch of the refractive indices at the boundaries of the droplets, which leads to opacity. ITO is also the reference material for TEs, and recently AgNW-based systems proved efficient in replacing it [86, 87]. This is exemplified in the work published by Goldthorpe et al. who found that the transmittance of a AgNW TE based PDLC system is both higher in the on-state and lower in the off-state compared to an equivalent device fabricated using ITO [88]. As mentioned earlier, a major advantage of these new electrodes is the possibility to make devices that are flexible, and even stretchable in some cases. Stretchable and wearable WO_3 electrochromic devices on AgNW elastic conductors were reported with noticeable performances. For instance, fast coloration (1 second) and bleaching (4 seconds) times, and most importantly proper functioning at stretched state (50% strain) were achieved [89]. It was also shown that stretchable conductors based on AgNWs and polydimethylsiloxane (PDMS) could be stretched, twisted, and folded without significant loss of conductivity [90]. In an interesting approach based on PDLC devices, the AgNW networks used as TEs were also utilized to provide thermal energy to the liquid crystal embedded polymer layer via Joule heating. The flexible film changes from an opaque to a transparent state at the transition temperature as a result of the phase transition of the liquid crystal [91]. A recently published striking report showed the potential of AgNW-based TEs for the fabrication of flexible transparent electrochromic devices using $W_{18}O_{49}$ nanowires [92]. The electrochromic films were bent to a radius of 1.2 cm for more than 1000 bending cycles without obvious failure of both conductivity and electrochromic performance, and the fabrication of electrochromic pixels was convincingly demonstrated as shown in Figure 5.8.

5.3.4 Other Displays

The potential of AgNW-based TEs has been experimented on many devices, including various technologies. It is not possible to describe all of them, but hereinafter some selected further developments are presented. An example is the use of AgNW TEs as transparent film heaters, covered with thermochromic inks. When bias is applied to the TE, a temperature elevation above the switching temperature of the active material can be generated, resulting in a change of color. This is a reversible system, which could be operated with different colors and different transition temperatures controllable thanks to bias tuning [61, 93].

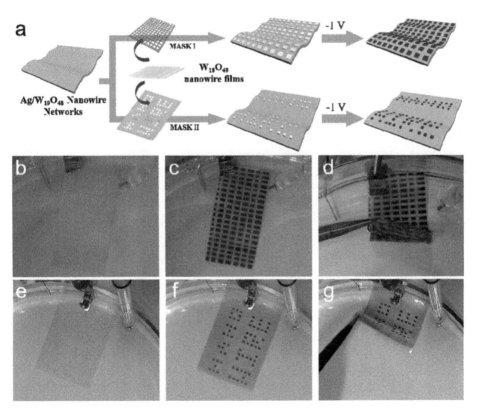

Figure 5.8 (a) Schematic illustration for the fabrication of electrochromic films with display pixels and characters of "NANO WIRE." (b–d) Photographs of the film with display pixels in the bleached state, colored state, and bent state. (e–g) Photographs of the film with characters of "NANO WIRE" that in bleached state, colored state, and while being bent manually. The film is composed of eight layers of W18O49 NWs on top of two layers of Ag/W18O49 NW networks and mAg:mW18O49 is 14:4. Reprinted with permission from [92]. Copyright (2017) American Chemical Society.

These TEs have also been integrated in purely inorganic LED based devices [94–97]. Application to flexible displays is foreseen, and stretchability is also a topic of interest. This was for instance demonstrated through highly stretchable composite fibers made of double-covered yarn, covered with AgNWs and encapsulated in PDMS. This material was used as a wiring system to integrate LEDs in series [98].

Other light-emitting systems such as electrochemical cells also proved efficient when used in combination with TEs made of AgNWs [84]. For instance, a report on a light-emitting electrochemical cell device showed that the lighting performances were similar to those of ITO-based structure, with high flexibility and stretchability [60].

5.4 Conclusions

Metallic nanowires appear as very promising building nanoblocks for the fabrication of flexible TEs for application in flexible displays. In little more than a decade, a colossal work has been accomplished, starting from the discovery of the easy, straightforward synthesis of silver nanowires in solution, up to their integration in the form of transparent and conductive networks in high-tech flexible displays. The performances of TEs depend on various parameters, and in particular on the now well-controlled dimensions of nanowires and the techniques used for deposition. A large number of functional devices has been demonstrated so far, and excellent performances were achieved. It proves that this nanomaterial-based technology is efficient, with huge potential, and can be envisaged for a wide set of applications, including flexible displays. Further developments are ongoing, like the fabrication of stretchable electrodes based on AgNWs or hybrid materials (e.g. AgNWs and graphene-based materials) for enhanced performances and stability. There is no doubt that many other stunning results dealing with AgNW-based systems will be shortly reported because it is a very active and promising field of research. However, there are still some issues to be tackled. Additional studies on stability should be carried out, both for intrinsic stability when nanowires are stacked between layers of displays structures, and for reliability in long-term operating mode. With regard to the hot topic of potential nanomaterial toxicity, it would be highly desirable to stimulate studies on the potential release of AgNWs from devices, and to foster toxicological studies, since AgNWs will probably be increasingly introduced in technological devices and consumer products in the short term.

References

1 Werner, T.T., Mudd, G.M., and Jowitt, S.M. (2015). Indium: Key issues in assessing mineral resources and long-term supply from recycling. *Appl. Earth Sci.* 124 (4): 213–226.
2 Hecht, D.S., Hu, L., and Irvin, G. (2011). Emerging transparent electrodes based on thin films of carbon nanotubes, graphene, and metallic nanostructures. *Adv. Mater.* 23 (13): 1482–1513.
3 Morales-Masis, M., De Wolf, S., Woods-Robinson, R., Ager, J.W., and Ballif, C. (2017). Transparent electrodes for efficient optoelectronics. *Adv. Electron. Mater.* 3 (5): 1600529.
4 Gueye, M.N., Carella, A., Massonnet, N., Yvenou, E., Brenet, S., Faure-Vincent, J., Pouget, S., Rieutord, F., Okuno, H., Benayad, A., Demadrille, R., and Simonato, J.-P. (2016). Structure and dopant engineering in PEDOT thin films: Practical tools for a dramatic conductivity enhancement. *Chem. Mater.* 28 (10): 3462–3468.

5 Spadafora, E.J., Saint-Aubin, K., Celle, C., Demadrille, R., Grévin, B., and Simonato, J.-P. (2012). Work function tuning for flexible transparent electrodes based on functionalized metallic single walled carbon nanotubes. *Carbon* 50 (10): 3459–3464.

6 Renault, O., Tyurnina, A., Simonato, J.-P., Rouchon, D., Mariolle, D., Chevalier, N., and Dijon, J. (2014). Doping efficiency of single and randomly stacked bilayer graphene by iodine adsorption. *Appl. Phys. Lett.* 105 (1): 011605.

7 Langley, D., Giusti, G., Mayousse, C., Celle, C., Bellet, D., and Simonato, J.-P. (2013). Flexible transparent conductive materials based on silver nanowire networks: A review. *Nanotechnology* 24(45): 452001.

8 Ye, S., Rathmell, A.R., Chen, Z., Stewart, I.E., and Wiley, B.J. (2014). Metal nanowire networks: The next generation of transparent conductors. *Adv. Mater.* 26 (39): 6670–6687.

9 Guo, C.F. and Ren, Z. (2015). Flexible transparent conductors based on metal nanowire networks. *Mater. Today* 18 (3): 143–154.

10 Sannicolo, T., Lagrange, M., Cabos, A., Celle, C., Simonato, J.-P., and Bellet, D. (2016). Metallic nanowire-based transparent electrodes for next generation flexible devices: A review. *Small* 12 (44): 6052–6075.

11 Sun, Y., Yin, Y., Mayers, B.T., Herricks, T., and Xia, Y. (2002). Uniform silver nanowires synthesis by reducing $AgNO_3$ with ethylene glycol in the presence of seeds and poly(vinyl pyrrolidone). *Chem. Mater.* 14 (11): 4736–4745.

12 Coskun, S., Aksoy, B., and Unalan, H.E. (2011). Polyol synthesis of silver nanowires: An extensive parametric study. *Cryst. Growth Des.* 11 (11): 4963–4969.

13 Zhang, P., Wyman, I., Hu, J., Lin, S., Zhong, Z., Tu, Y., Huang, Z., and Wei, Y. (2017). Silver nanowires: Synthesis technologies, growth mechanism and multifunctional applications. *Mater. Sci. Eng. B* 223: 1–23.

14 Sun, Y., Mayers, B., Herricks, T., and Xia, Y. (2003). Polyol synthesis of uniform silver nanowires: A plausible growth mechanism and the supporting evidence. *Nano Lett.* 3 (7): 955–960.

15 Lee, E.-J., Kim, Y.-H., Hwang, D.K., Choi, W.K., and Kim, J.-Y. (2016). Synthesis and optoelectronic characteristics of 20 nm diameter silver nanowires for highly transparent electrode films. *RSC Adv.* 6 (14): 11702–11710.

16 Chen, C., Wang, L., Jiang, G., Zhou, J., Chen, X., Yu, H., and Yang, Q. (2006). Study on the synthesis of silver nanowires with adjustable diameters through the polyol process. *Nanotechnology* 17 (15): 3933–3938.

17 Guo, H., Lin, N., Chen, Y., Wang, Z., Xie, Q., Zheng, T., Gao, N., Li, S., Kang, J., Cai, D., and Peng, D.-L. (2013). Copper nanowires as fully transparent conductive electrodes. *Sci. Rep.* 3: 2323.

18 Rathmell, A.R. and Wiley, B.J. (2011). The synthesis and coating of long, thin copper nanowires to make flexible, transparent conducting films on plastic substrates. *Adv. Mater.* 23 (41): 4798–4803.

19 Mayousse, C., Celle, C., Carella, A., and Simonato, J.-P. (2014). Synthesis and purification of long copper nanowires. Application to high performance flexible transparent electrodes with and without PEDOT:PSS. *Nano Res.* 7 (3): 315–324.

20 Bhanushali, S., Ghosh, P., Ganesh, A., and Cheng, W. (2015). 1D copper nanostructures: Progress, challenges and opportunities. *Small* 11 (11): 1232–1252.

21 Gong, S., Zhao, Y., Yap, L.W., Shi, Q., Wang, Y., Bay, J.A.P.B., Lai, D.T.H., Uddin, H., and Cheng, W. (2016). Fabrication of highly transparent and flexible NanoMesh electrode via self-assembly of ultrathin gold nanowires. *Adv. Electron. Mater.* 2 (7): 1600121.

22 Rathmell, A.R., Nguyen, M., Chi, M., and Wiley, B.J. (2012). Synthesis of oxidation-resistant cupronickel nanowires for transparent conducting nanowire networks. *Nano Lett.* 12 (6): 3193–3199.

23 Song, J., Li, J., Xu, J., and Zeng, H. (2014). Superstable transparent conductive Cu@Cu$_4$ Ni nanowire elastomer composites against oxidation, bending, stretching, and twisting for flexible and stretchable optoelectronics. *Nano Lett.* 14 (11): 6298–6305.

24 Stewart, I.E., Ye, S., Chen, Z., Flowers, P.F., and Wiley, B.J. (2015). Synthesis of Cu–Ag, Cu–Au, and Cu–Pt Core-Shell nanowires and their use in transparent conducting films. *Chem. Mater.* 27 (22): 7788–7794.

25 Chen, J., Chen, J., Li, Y., Zhou, W., Feng, X., Huang, Q., Zheng, J.-G., Liu, R., Ma, Y., and Huang, W. (2015). Enhanced oxidation-resistant Cu–Ni core–shell nanowires: Controllable one-pot synthesis and solution processing to transparent flexible heaters. *Nanoscale* 7 (40): 16874–16879.

26 Celle, C., Cabos, A., Fontecave, T., Laguitton, B., Benayad, A., Guettaz, L., Pélissier, N., Nguyen, V.H., Bellet, D., Muñoz-Rojas, D., and Simonato, J.-P. (2018). Oxidation of copper nanowire based transparent electrodes in ambient conditions and their stabilization by encapsulation: Application to transparent film heaters. *Nanotechnology* 29 (8): 085701.

27 Mayousse, C., Celle, C., Fraczkiewicz, A., and Simonato, J.-P. (2015). Stability of silver nanowire based electrodes under environmental and electrical stresses. *Nanoscale* 7 (5): 2107–2115.

28 Deignan, G. and Goldthorpe, I.A. (2017). The dependence of silver nanowire stability on network composition and processing parameters. *RSC Adv.* 7 (57): 35590–35597.

29 Stewart, I.E., Rathmell, A.R., Yan, L., Ye, S., Flowers, P.F., You, W., and Wiley, B.J. (2014). Solution-processed copper–nickel nanowire anodes for organic solar cells. *Nanoscale* 6 (11): 5980–5988.

30 Lagrange, M., Langley, D.P., Giusti, G., Jiménez, C., Bréchet, Y., and Bellet, D. (2015). Optimization of silver nanowire-based transparent electrodes: Effects of density, size and thermal annealing. *Nanoscale* 7 (41): 17410–17423.

31 Scardaci, V., Coull, R., Lyons, P.E., Rickard, D., and Coleman, J.N. (2011). Spray deposition of highly transparent, low-resistance networks of silver nanowires over large areas. *Small* 7 (18): 2621–2628.

32 Jung, E., Kim, C., Kim, M., Chae, H., Cho, J.H., and Cho, S.M. (2017). Roll-to-roll preparation of silver-nanowire transparent electrode and its application to large-area organic light-emitting diodes. *Org. Electron.* 41: 190–197.

33 Reinhard, M., Eckstein, R., Slobodskyy, A., Lemmer, U., and Colsmann, A. (2013). Solution-processed polymer–silver nanowire top electrodes for inverted semi-transparent solar cells. *Org. Electron.* 14 (1): 273–277.

34 Gaynor, W., Burkhard, G.F., McGehee, M.D., and Peumans, P. (2011). Smooth nanowire/polymer composite transparent electrodes. *Adv. Mater.* 23 (26): 2905–2910.

35 Tokuno, T., Nogi, M., Karakawa, M., Jiu, J., Nge, T.T., Aso, Y., and Suganuma, K. (2011). Fabrication of silver nanowire transparent electrodes at room temperature. *Nano Res.* 4 (12): 1215–1222.

36 Krantz, J., Richter, M., Spallek, S., Spiecker, E., and Brabec, C.J. (2011). Solution-processed metallic nanowire electrodes as indium tin oxide replacement for thin-film solar cells. *Adv. Funct. Mater.* 21 (24): 4784–4787.

37 Sepulveda-Mora, S.B. and Cloutier, S.G. (2012). Figures of merit for high-performance transparent electrodes using dip-coated silver nanowire networks. *J. Nanomater.* 2012: 7.

38 Hu, L., Kim, H.S., Lee, J.-Y., Peumans, P., and Cui, Y. (2010). Scalable coating and properties of transparent, flexible, silver nanowire electrodes. *ACS Nano* 4 (5): 2955–2963.

39 Liu, C.-H. and Yu, X. (2011). Silver nanowire-based transparent, flexible, and conductive thin film. *Nanoscale Res. Lett.* 6 (1): 75–82.

40 Guo, F., Zhu, X., Forberich, K., Krantz, J., Stubhan, T., Salinas, M., Halik, M., Spallek, S., Butz, B., Spiecker, E., Ameri, T., Li, N., Kubis, P., Guldi, D.M., Matt, G.J., and Brabec, C.J. (2013). ITO-free and fully solution-processed semitransparent organic solar cells with high fill factors. *Adv. Energy Mater.* 3 (8): 1062–1067.

41 Stubhan, T., Krantz, J., Li, N., Guo, F., Litzov, I., Steidl, M., Richter, M., Matt, G.J., and Brabec, C.J. (2012). High fill factor polymer solar cells comprising a transparent, low temperature solution processed doped metal oxide/metal nanowire composite electrode. *Sol. Energy Mater. Sol. Cells* 107: 248–251.

42 De, S., Higgins, T.M., Lyons, P.E., Doherty, E.M., Nirmalraj, P.N., Blau, W.J., Boland, J.J., and Coleman, J.N. (2009). Silver nanowire networks as flexible, transparent, conducting films: Extremely high DC to optical conductivity ratios. *ACS Nano* 3 (7): 1767–1774.

43 Lee, P., Lee, J., Lee, H., Yeo, J., Hong, S., Nam, K.H., Lee, D., Lee, S.S., and Ko, S.H. (2012). Highly stretchable and highly conductive metal electrode by very long metal nanowire percolation network. *Adv. Mater.* 24 (25): 3326–3332.

44 Lim, J.-W., Cho, D.-Y., Jihoon-Kim, Na, S.-I., and Kim, H.-K. (2012). Simple brush-painting of flexible and transparent Ag nanowire network electrodes as an alternative ITO anode for cost-efficient flexible organic solar cells. *Sol. Energy Mater. Sol. Cells* 107: 348–354.

45 Kang, S.-B., Noh, Y.-J., Na, S.-I., and Kim, H.-K. (2014). Brush-painted flexible organic solar cells using highly transparent and flexible Ag nanowire network electrodes. *Sol. Energy Mater. Sol. Cells* 122: 152–157.

46 Chen, S., Guan, Y., Li, Y., Yan, X., Ni, H., and Li, L. (2017). A water-based silver nanowire ink for large-scale flexible transparent conductive films and touch screens. *J. Mater. Chem. C* 5 (9): 2404–2414.

47 Langley, D.P., Lagrange, M., Giusti, G., Jiménez, C., Bréchet, Y., Nguyen, N.D., and Bellet, D. (2014). Metallic nanowire networks: Effects of thermal annealing on electrical resistance. *Nanoscale* 6: 13535–13543.

48 Lee, J.-Y., Connor, S.T., Cui, Y., and Peumans, P. (2008). Solution-processed metal nanowire mesh transparent electrodes. *Nano Lett.* 8 (2): 689–692.

49 Chiba, Y., Islam, A., Komiya, R., Koide, N., and Han, L. (2006). Conversion efficiency of 10.8% by a dye-sensitized solar cell using a TiO2 electrode with high haze. *Appl. Phys. Lett.* 88 (22): 223505.

50 Tang, Z., Tress, W., and Inganäs, O. (2014). Light trapping in thin film organic solar cells. *Mater. Today* 17 (8): 389–396.

51 Yu, X., Yu, X., Zhang, J., Chen, L., Long, Y., and Zhang, D. (2017). Optical properties of conductive silver-nanowire films with different nanowire lengths. *Nano Res.* 10 (11): 3706–3714.

52 Liu, Y., Chen, Y., Shi, R., Cao, L., Wang, Z., Sun, T., Lin, J., Liu, J., and Huang, W. (2017). High-yield and rapid synthesis of ultrathin silver nanowires for low-haze transparent conductors. *RSC Adv.* 7 (9): 4891–4895.

53 Jang, H.-W., Kim, Y.-H., Lee, K.-W., Kim, Y.-M., and Kim, J.-Y. (2017). Research Update: Synthesis of sub-15-nm diameter silver nanowires through a water-based hydrothermal method: Fabrication of low-haze 2D conductive films. *APL Mater.* 5 (8): 080701.

54 Li, J. and Zhang, S.-L. (2009). Finite-size scaling in stick percolation. *Phys. Rev. E* 80: 4.

55 Coskun, S., Selen Ates, E., and Emrah Unalan, H. (2013). Optimization of silver nanowire networks for polymer light emitting diode electrodes. *Nanotechnology* 24 (12): 125202.

56 Lee, J., Lee, P., Lee, H., Lee, D., Lee, S.S., and Ko, S.H. (2012). Very long Ag nanowire synthesis and its application in a highly transparent, conductive and flexible metal electrode touch panel. *Nanoscale* 4 (20): 6408–6414.

57 Han, S., Hong, S., Ham, J., Yeo, J., Lee, J., Kang, B., Lee, P., Kwon, J., Lee, S.S., Yang, M.-Y., and Ko, S.H. (2014). Fast plasmonic laser nanowelding for a Cu-Nanowire percolation network for flexible transparent conductors and stretchable electronics. *Adv. Mater.* 26 (33): 5808–5814.

58 Garnett, E.C., Cai, W., Cha, J.J., Mahmood, F., Connor, S.T., Greyson Christoforo, M., Cui, Y., McGehee, M.D., and Brongersma, M.L. (2012). Self-limited plasmonic welding of silver nanowire junctions. *Nat. Mater.* 11 (3): 241–249.

59 Sannicolo, T., Muñoz-Rojas, D., Nguyen, N.D., Moreau, S., Celle, C., Simonato, J.-P., Bréchet, Y., and Bellet, D. (2016). Direct imaging of the onset of electrical conduction in silver nanowire networks by infrared thermography: Evidence of geometrical quantized percolation. *Nano Lett.* 16 (11): 6052–6075.

60 Miller, M.S., O'Kane, J.C., Niec, A., Carmichael, R.S., and Carmichael, T.B. (2013). Silver nanowire/optical adhesive coatings as transparent electrodes for flexible electronics. *ACS Appl. Mater. Interfaces* 5 (20): 10165–10172.

61 Celle, C., Mayousse, C., Moreau, E., Basti, H., Carella, A., and Simonato, J.-P. (2012). Highly flexible transparent film heaters based on random networks of silver nanowires. *Nano Res.* 5 (6): 427–433.

62 Tiwana, M.I., Redmond, S.J., and Lovell, N.H. (2012). A review of tactile sensing technologies with applications in biomedical engineering. *Sens. Actuators Phys.* 179: 17–31.

63 Madaria, A.R., Kumar, A., and Zhou, C. (2011). Large scale, highly conductive and patterned transparent films of silver nanowires on arbitrary substrates and their application in touch screens. *Nanotechnology* 22 (24): 245201.

64 Chou, S.-Y., Ma, R., Li, Y., Zhao, F., Tong, K., Yu, Z., and Pei, Q. (2017). Transparent perovskite light-emitting touch-responsive device. *ACS Nano* 11 (11): 11368–11375.

65 Mayousse, C., Celle, C., Moreau, E., Mainguet, J.-F., Carella, A., and Simonato, J.-P. (2013). Improvements in purification of silver nanowires by decantation and fabrication of flexible transparent electrodes. Application to capacitive touch sensors. *Nanotechnology* 24 (21): 215501.

66 Shuai, X., Zhu, P., Zeng, W., Hu, Y., Liang, X., Zhang, Y., Sun, R., and Wong, C. (2017). Highly sensitive flexible pressure sensor based on silver nanowires-embedded polydimethylsiloxane electrode with microarray structure. *ACS Appl. Mater. Interfaces* 9 (31): 26314–26324.

67 Liu, S., Li, J., Shi, X., Gao, E., Xu, Z., Tang, H., Tong, K., Pei, Q., Liang, J., and Chen, Y. (2017). Rollerball-pen-drawing technology for extremely foldable paper-based electronics. *Adv. Electron. Mater.* 3 (7): 1700098.

68 Yang, B.-R., Liu, G.-S., Han, S.-J., Cao, W., Xu, D.-H., Huang, J.-F., Qiu, J.-S., Liu, C., and Chen, H.-J. (2016). Coating, patterning, and transferring processes of silver nanowire for flexible display and sensing applications: Process of AgNW for flexible display and sensors. *J. Soc. Inf. Disp.* 24 (4): 234–240.

69 Wang, J., Jiu, J., Araki, T., Nogi, M., Sugahara, T., Nagao, S., Koga, H., He, P., and Suganuma, K. (2015). Silver nanowire electrodes: Conductivity improvement without post-treatment and application in capacitive pressure sensors. *Nano-Micro Lett.* 7 (1): 51–58.

70 Gaynor, W., Hofmann, S., Christoforo, M.G., Sachse, C., Mehra, S., Salleo, A., McGehee, M.D., Gather, M.C., Lüssem, B., Müller-Meskamp, L., Peumans, P., and Leo, K. (2013). Color in the corners: ITO-free white OLEDs with angular color stability. *Adv. Mater.* 25 (29): 4006–4013.

71 Song, C.-H., Ok, K.-H., Lee, C.-J., Kim, Y., Kwak, M.-G., Han, C.J., Kim, N., Ju, B.-K., and Kim, J.-W. (2015). Intense-pulsed-light irradiation of Ag nanowire-based transparent electrodes for use in flexible organic light emitting diodes. *Org. Electron.* 17: 208–215.

72 Lee, J., An, K., Won, P., Ka, Y., Hwang, H., Moon, H., Kwon, Y., Hong, S., Kim, C., Lee, C., and Ko, S.H. (2017). A dual-scale metal nanowire network transparent conductor for highly efficient and flexible organic light emitting diodes. *Nanoscale* 9 (5): 1978–1985.

73 Chen, D., Zhao, F., Tong, K., Saldanha, G., Liu, C., and Pei, Q. (2016). Mitigation of electrical failure of silver nanowires under current flow and the application for long lifetime organic light-emitting diodes. *Adv. Electron. Mater.* 2 (8): 1600167.

74 Zhang, M., Höfle, S., Czolk, J., Mertens, A., and Colsmann, A. (2015). All-solution processed transparent organic light emitting diodes. *Nanoscale* 7 (47): 20009–20014.

75 Guo, F., Karl, A., Xue, Q.-F., Tam, K.C., Forberich, K., and Brabec, C.J. (2017). The fabrication of color-tunable organic light-emitting diode displays via solution processing. *Light Sci. Appl.* 6 (11): e17094.

76 Kim, J.-H., Triambulo, R.E., and Park, J.-W. (2017). Effects of the interfacial charge injection properties of silver nanowire transparent conductive electrodes on the performance of organic light-emitting diodes. *J. Appl. Phys.* 121 (10): 105304.

77 Lee, H., Lee, D., Ahn, Y., Lee, E.-W., Park, L.S., and Lee, Y. (2014). Highly efficient and low voltage silver nanowire-based OLEDs employing a n-type hole injection layer. *Nanoscale* 6 (15): 8565–8570.

78 Lee, K.M., Fardel, R., Zhao, L., Arnold, C.B., and Rand, B.P. (2017). Enhanced outcoupling in flexible organic light-emitting diodes on scattering polyimide substrates. *Org. Electron.* 51: 471–476.

79 Li, L., Liang, J., Chou, S.-Y., Zhu, X., Niu, X., Yu, Z., and Pei, Q. (2015). A solution processed flexible nanocomposite electrode with efficient light extraction for organic light emitting diodes. *Sci. Rep.* 4 (1).

80 Kim, D.W., Han, J.W., Lim, K.T., and Kim, Y.H. (2018). Highly enhanced light-outcoupling efficiency in ITO-free organic light-emitting diodes using surface nanostructure embedded high-refractive index polymers. *ACS Appl. Mater. Interfaces* 10 (1): 985–991.

81 Kang, H., Kang, I., Han, J., Kim, J.B., Lee, D.Y., Cho, S.M., and Cho, J.H. (2016). Flexible and mechanically robust organic light-emitting diodes based on photopatternable silver nanowire electrodes. *J. Phys. Chem. C* 120 (38): 22012–22018.

82 Liu, S., Ho, S., and So, F. (2016). Novel patterning method for silver nanowire electrodes for thermal-evaporated organic light emitting diodes. *ACS Appl. Mater. Interfaces* 8 (14): 9268–9274.

83 Ok, K.-H., Kim, J., Park, S.-R., Kim, Y., Lee, C.-J., Hong, S.-J., Kwak, M.-G., Kim, N., Han, C.J., and Kim, J.-W. (2015). Ultra-thin and smooth transparent electrode for flexible and leakage-free organic light-emitting diodes. *Sci. Rep.* 5: 1.

84 Liang, J., Li, L., Niu, X., Yu, Z., and Pei, Q. (2013). Elastomeric polymer light-emitting devices and displays. *Nat. Photonics* 7 (10): 817–824.

85 Thakur, V.K., Ding, G., Ma, J., Lee, P.S., and Lu, X. (2012). Hybrid materials and polymer electrolytes for electrochromic device applications. *Adv. Mater.* 24 (30): 4071–4096.

86 Choi, D., Lee, M., Kim, H., Chu, W., Chun, D., Ahn, S.-H., and Lee, C.S. (2017). Fabrication of transparent conductive tri-composite film for electrochromic application. *Appl. Surf. Sci.* 425: 1006–1013.

87 Yuksel, R., Ataoglu, E., Turan, J., Alpugan, E., Ozdemir Hacioglu, S., Toppare, L., Cirpan, A., Emrah Unalan, H., and Gunbas, G. (2017). A new high-performance blue to transmissive electrochromic material and use of silver nanowire network electrodes as substrates. *J. Polym. Sci. Part Polym. Chem.* 55 (10): 1680–1686.

88 Hosseinzadeh Khaligh, H., Liew, K., Han, Y., Abukhdeir, N.M., and Goldthorpe, I.A. (2015). Silver nanowire transparent electrodes for liquid crystal-based smart windows. *Sol. Energy Mater. Sol. Cells* 132: 337–341.

89 Yan, C., Kang, W., Wang, J., Cui, M., Wang, X., Foo, C.Y., Chee, K.J., and Lee, P.S. (2014). Stretchable and wearable electrochromic devices. *ACS Nano* 8 (1): 316–322.

90 Liu, H.-S., Pan, B.-C., and Liou, G.-S. (2017). Highly transparent AgNW/PDMS stretchable electrodes for elastomeric electrochromic devices. *Nanoscale* 9 (7): 2633–2639.

91 Wee, D., Song, Y.S., and Youn, J.R. (2016). Thermal-induced optical modulation of liquid crystal embedded polymer film on silver nanowire heater. *Sol. Energy Mater. Sol. Cells* 147: 150–156.

92 Wang, J.-L., Lu, Y.-R., Li, -H.-H., Liu, J.-W., and Yu, S.-H. (2017). Large area co-assembly of nanowires for flexible transparent smart windows. *J. Am. Chem. Soc.* 139 (29): 9921–9926.

93 Huang, Q., Shen, W., Fang, X., Chen, G., Guo, J., Xu, W., Tan, R., and Song, W. (2015). Highly flexible and transparent film heaters based on polyimide films embedded with silver nanowires. *RSC Adv.* 5 (57): 45836–45842.

94 Liu, B., Li, C., Liu, Q.-L., Dong, J., Guo, C.-W., Wu, H., Zhou, H.-Y., Fan, X.-J., Guo, X., Wang, C., Sun, X.-M., Jin, Y.-H., Li, -Q.-Q., and Fan, -S.-S. (2015). Hybrid film of silver nanowires and carbon nanotubes as a transparent conductive layer in light-emitting diodes. *Appl. Phys. Lett.* 106 (3): 033101.

95 Jeong, G.-J., Lee, J.-H., Han, S.-H., Jin, W.-Y., Kang, J.-W., and Lee, S.-N. (2015). Silver nanowires for transparent conductive electrode to GaN-based light-emitting diodes. *Appl. Phys. Lett.* 106 (3): 031118.

96 Park, J.-S., Kim, J.-H., Kim, J.-Y., Kim, D.-H., Kang, D., Sung, J.-S., and Seong, T.-Y. (2016). Hybrid indium tin oxide/Ag nanowire electrodes for improving the light output power of near ultraviolet AlGaN-based light-emitting diode. *Curr. Appl. Phys.* 16 (5): 545–548.

97 Kim, J.-W. and Kim, J. (2017). Flexible InP based quantum dot light-emitting diodes using Ag nanowire-colorless polyimide composite electrode. *J. Vac. Sci. Technol. B Nanotechnol. Microelectron. Mater. Process. Meas. Phenom.* 35 (4): 04E101.

98 Cheng, Y., Wang, R., Sun, J., and Gao, L. (2015). Highly conductive and ultrastretchable electric circuits from covered yarns and silver nanowires. *ACS Nano* 9 (4): 3887–3895.

6

Optically Clear Adhesives for Display Assembly

Albert I. Everaerts

3M Company St. Paul, Minnesota, USA

6.1   Introduction

Pressure-sensitive adhesives and liquid-curable adhesives have been used for decades in a wide range of industrial, electronics, aerospace, and medical applications. The majority of their performance attributes have been focused on reliable adhesion against a broad range of substrates under a wide set of environmental conditions. In some cases, more specialized requirements like being non-irritant to skin, non-corrosive to metals, damping for impact resistance, ultra-violet (UV) stability for outdoor type applications, etc., have also been required. While these same types of adhesives have been used in electronic applications for a long time, it is only recently that they have also made their entry into the actual viewing area of the electronic display components. Indeed, while adhesives for perimeter bonding, for example front lens attachment to a housing, have been very prevalent for years; the application for full coverage bonding in the viewing area has been limited, only to rapidly accelerate in the last 10 years or so. Prior to the advent of smartphones, optically clear adhesives (OCAs) were mainly used in the assembly of resistive touch panels where they had to be thin (~25–50 microns) and adhere well to the conductive films that made up the sensor. However, with the launch of smartphones and tablets, not only did the demand for adhesive bonding increase but their performance requirements also shifted toward much higher levels. Driven by the success of the mobile handheld devices, adhesive bonding of the viewing area has now also penetrated into other devices, such as desktop and all-in-one computer displays, notebooks, and navigation panels for automotive, marine, and aerospace use. With the emergence of smart watches, flexible displays, and wearable displays, additional requirements for adhesive bonding continue to drive the industry. In the coming years the corresponding increases in volume and revenue for both the film format (abbreviated as OCA for optically clear adhesive) and liquid format (abbreviated as LOCA for liquid optically clear adhesive) materials is expected to continue; however, the pace of revenue growth is expected to slow down, mainly because of the continued cost-down pressure in the electronics industry. Despite this trend, a number of adhesive suppliers have entered the market, making it quite competitive. However, we still see constantly changing display technology, moving from standard resolution liquid crystal displays (LCDs) to high resolution, or from LCDs to organic light-emitting displays (OLEDs); and also shifting form factors, going from rigid, planar to curved or bending, and now to flexible. Because of these changes, we expect that the requirements on the adhesives will continue to increase and further advances in formulation, performance, and fit with new assembly requirements will have to be met.

Flexible Flat Panel Displays, Second Edition. Edited by Darran R. Cairns, Dirk J. Broer, and Gregory P. Crawford.
© 2023 John Wiley & Sons Ltd. Published 2023 by John Wiley & Sons Ltd.

6.2 OCA Definition and General Performance Specifications

Unlike adhesives for application outside of the viewing area of a display, OCAs have a unique set of performance requirements. From a high-level viewpoint, the OCAs in a display have at least three main functions: optical, mechanical, and electrical. The most common optical function is to enhance contrast and brightness of a display, while mechanically the adhesive holds the display together under a broad range of use environments. Electrically, the OCA may simply act as an insulator, but it may also be used to control the capacitance of the display stack.

The optical benefits of using an OCA in display assembly can be directly correlated to reflection losses at interfaces between solid surfaces and air. Figure 6.1 exemplifies the refractive index of a few common materials and the significant difference in reflectance between an air and glass lens interface, for example, versus an OCA and glass interface.

The net result of filling the air gap with an OCA that is a close match to the substrates is to enhance contrast and brightness of the display. This is demonstrated in Figure 6.2.

By filling the air gap with an OCA, the refection losses from sunlight are reduced from about 12% to about 4%. If an anti-reflective coating is applied on the surface of the lens, most of the remaining 4% can be eliminated as well. From the formula, it can be derived that the higher the ambient light reflection off the display, the lower the contrast ratio and the more likely the display will look washed out. Likewise, while not shown in Figure 6.2, light emitted from the display may also get reflected back internally without passing toward the viewer and thus brightness is reduced. Reduced brightness will frequently be compensated for with higher power driving the display, which in turn compromises battery life of the device.

Because of the close proximity of the display to the viewer, the demand on the quality of the OCA materials is also much higher than for industrial grade adhesives, even those that are clear. Indeed, it is no longer sufficient for the adhesive just to be clear and transparent, as it may be for a window film application, but it also has to meet a number of additional requirements in terms of cleanliness, acceptable level of coating defects, and compatibility with the rest of the electronic device. For example, as with most electronic-grade adhesives the presence of outgassing volatiles or halogens needs to be avoided, or at least significantly minimized. Likewise, particles, small air

Vacuum	:	1.00
Air	:	1.0028
Water	:	1.33
Silicone	:	1.41
Glass	:	1.50(1.43–1.74)
PMMA	:	1.49
Polycarbonate	:	1.58
PET	:	1.60
Titanium Oxide	:	2.40
OCA	:	1.47

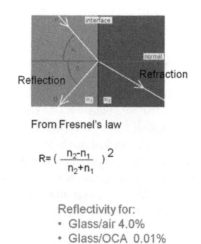

From Fresnel's law

$$R = \left(\frac{n_2 - n_1}{n_2 + n_1} \right)^2$$

Reflectivity for:
- Glass/air 4.0%
- Glass/OCA 0.01%

Figure 6.1 Refractive index of different substrates and Fresnel's law.

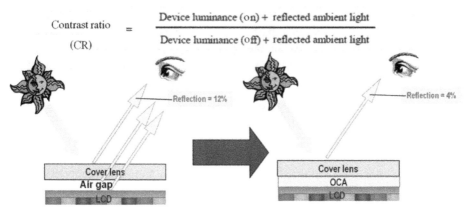

$$\text{Contrast ratio (CR)} = \frac{\text{Device luminance (on)} + \text{reflected ambient light}}{\text{Device luminance (off)} + \text{reflected ambient light}}$$

Figure 6.2 Effect of display reflections on contrast ratio.

bubbles, gels, and even coating unevenness needs to be well-controlled. Today, the vast majority of the OCA applications are in front of the LCD, electrophoretic display, or OLED; leaving little or no opportunity to hide the defects with diffusers or other film layers positioned between the light source and the viewer. With the adhesive in the direct path of the light, contaminants or coating defects as small as 50 microns may become very visible or they might create optical distortions of the light as it passes through the adhesive. In addition, some LCD modules can be very sensitive to uneven pressure or stress, causing a localized change in the display's cell gap and thus the light that passes through. The resulting non-uniform luminance of the display is often referred to as Mura. While there are several other root causes for Mura that are unrelated to the OCA, when direct bonding to the LCD panel is required, close attention needs to be paid to the adhesive caliper uniformity, its stiffness, and if used during assembly, any curing shrinkage related stresses on the panel.

Since the adhesive is in the direct path of the light and display color is a critical quality attribute, the ideal OCA is color neutral or at least close to it when measured, for example, on the CIE Lab color scale. It also has to remain color stable even when exposed to hundreds of hours of weathering, such as light irradiation from a Xenon lamp or in a QUV test. As a result of these requirements, acrylate copolymers and silicones are dominating this adhesives market. Beyond color neutrality, a high percent of visible light transmission and a low level of haze are part of the basic properties of the material. When corrected for reflectance, visible light transmission of about 98% or more is expected and the haze levels are typically less than 0.5%, even if the adhesive thickness gets to be on the order of 200 microns or even more. While technically not really optically clear, in some applications it may be required that the adhesive provides some level of diffusivity. This is commonly achieved by dispersing light-scattering particles into an OCA. Light scattering will be dependent on the concentration, size, shape, and polydispersity of the particles, and also the refractive index difference between adhesive matrix and scattering material [1, 2].

Mechanically, OCA performance is typically measured in the form of the durability of the display. While adhesion and cohesive strength matter, there is far more emphasis on reliability testing of the display and the device. For example, on the mobile handheld or portable device level, this may mean exposure to a number of tumble cycles or drop testing, where failure in the form of display delamination, display cracking, etc. may be the main criteria used to check performance. In some cases, these tests may also be performed over a broad range of temperatures, sometimes as low a −40°C and as high as 85°C. The high impact frequency of the test combined with the temperature effect can impart very high demands on the adhesive. With the narrowing or elimination of bezels, the load on the OCA is even further increasing.

Besides the device-level testing, there is also a very significant number of display-level testing to be passed. Examples of these exposure tests include the following:

- Hundreds and sometimes more than a thousand hours of continuous exposure to 85°C dry heat. For automotive type applications, that temperature may even exceed 100°C.
- Hundreds and sometimes more than 1000 hours of continuous exposure to heat and humidity, for example 85% relative humidity and 85°C, or 90% relative humidity and 65°C.
- Temperature cycling from sub-freezing to high heat, with different ramping or dropping rates, and different hold times in between those changes.

Because of the changes in temperature and in some cases the absorption or outgassing of moisture from the substrate, additional stress is being put on the adhesive. For example, changes in temperature can result in warping of extruded plastic panels or significant shear stress on the adhesive due to the thermal expansion coefficient mismatch between substrates.

Finally, when the adhesive is used in a three-layer display panel – cover lens/capacitive touch panel/LCD – the OCA may also have some electrical function. In some cases, the OCA may directly contact the conductive traces of the touch sensor, the integration circuit, or both, requiring that the adhesive be non-corrosive to these metals or metallic oxides like indium tin oxide (ITO). In such case, the adhesive will also have to be non-conductive to prevent electrical shorting. A lesser-known OCA property that may be required is control and stability of the dielectric constant. Since the touch sensor is closely positioned to the LCD, the constantly switching LCD causes electronic noise that can be picked up by the sensor. As a result the signal to noise ratio of the sensor may become compromised and a more expensive control chip may be needed to maintain the functionality of the sensor. An alternative way may be to drop the dielectric constant of the OCA that fills the gap between the sensor and the LCD. While between an LCD and a touch panel a lower dielectric OCA may be desired, the opposite may be true between the lens and the touch panel. A schematic of how the different capacitances interplay in a touch device is shown in Figure 6.3.

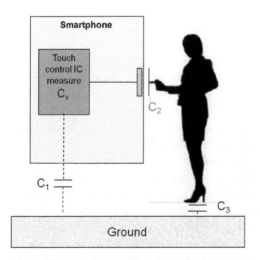

$$\frac{1}{C_x} = \frac{1}{C_1} + \frac{1}{C_2} + \frac{1}{C_3}$$

C1, C3 = Several hundreds pF

C2 = Several decades pF, when touched

Touch control chip measures Cx.

Figure 6.3 Relative capacitances in display sensor.

When the equation gets solved for Cx, as Cx = C1C2C3/C1 + C2 + C3, one finds that with C1 and C3 being large that Cx ~ C2. The capacitance can be calculated if the geometry of the conductors and the dielectric properties of the insulator between the conductors are known. For example, the capacitance of a parallel-plate capacitor constructed of two parallel plates both of area S separated by a distance d is approximately equal to the following:

$$C = \varepsilon_r \varepsilon_o S / d$$

where C is the capacitance; S is the area of overlap of the two plates; ε_r is the relative static permittivity (sometimes called the dielectric constant) of the material between the plates (for air $\varepsilon_r = 1$); ε_o is the electric constant ($\varepsilon_o \approx 8.854 \times 10^{-12}$ F m^{-1}); and d is the separation between the plates. It follows that if one wants a high capacitance C2, an ideal material may have a high dielectric constant or relative static permittivity and low thickness. If the dielectric constant of the gap-filling material is low, one would have to drop the thickness of the material to maintain the same capacitance. Unfortunately, for display design reasons this may not always be possible.

In summary, while an OCA has to adhere well to different substrates and provide mechanical integrity to the display module, the demands on the optical quality, the electrical characteristics, the environmental durability, and the coating quality are among the most severe in the adhesive industry.

6.3 Application Examples and Challenges

Dependent on the design of the display panel, OCAs may be applied in many different thicknesses and in many different positions in the display stack. They may also be used in film format or in the liquid state, the latter to be fully polymerized once the laminate is assembled. The touch and display industry is very dynamic, with constantly changing design and substrates being used to make the ever thinner, lighter, and brighter devices. We will focus this part of the chapter on only a few, key technical challenges and their solutions, while the reader is referred to the general literature for details of the many display module designs.

In order to exemplify the main OCA applications in a typical stack, Figure 6.4 shows the cross sections of the more common projective capacitance-enabled touch displays. The adhesive thicknesses shown are only representative and can be quite different in some modules.

The cover lens is typically selected from chemically hardened glass or a hard-coated plastic material like polymethylmethacrylate (PMMA), polycarbonate (PC), or a multilayer composite thereof. The touch panel substrates are commonly chosen from polyester (PET), cyclic olefin copolymer (COP), or glass for some more rigid sensors.

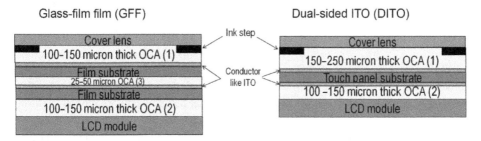

Figure 6.4 Typical cross section of a touch enabled display module stack.

The vast majority of cover lenses have a printed ink border around the viewing area of the display. The OCA is typically oversized versus the actual viewing area, and thus at least partially overlaps with the ink step. This adhesive layer (1) is sometimes just called pressure-sensitive adhesive (PSA) 1. On the bottom side between the touch panel and the actual display module an air gap may be used, but increasingly the touch module will be bonded directly to the display. This OCA layer (2) is commonly called the direct bonding OCA or just PSA 2.

The film sensor may have the conductive ITO traces on both sides of the substrate (dual ITO or DITO), or it may be assembled by laminating two single-sided ITO (SITO) film layers together. When a film sensor is assembled from multiple, single-sided conductive films, they are laminated with thinner OCAs (3) that have less demanding properties in terms of compliance and flow. Indeed, if not already protected with a barrier coating, the conductive traces are very low profile and easy to cover with adhesive, while the individual films can also be easily laminated to each other with the assistance of a roller. This roller lamination is commonly carried out in a roll-to-roll format, with the individual sensors being cut after lamination.

In contrast, the assembly of the more rigid components such as attaching the lens to the LCD is much more challenging, and indeed vacuum lamination is commonly employed. To finish the process, an autoclave step is used for film-based adhesives, or a UV curing step for liquid-based adhesives. Actual process parameters will depend on the exact construction of the display component, the choice of OCA materials, and the equipment that is available for assembly. The different OCA suppliers will supply recommendations for adhesive selection and process parameter guidelines to make successful assembly possible.

6.3.1 Outgassing Tolerant Adhesives

One of the challenges in the assembly of electronic displays is the need for the OCAs to resist environmental cycling conditions. When one of the substrates is PMMA or polycarbonate, the presence of water in the substrate can lead to severe outgassing when the assembly is exposed to high heat. In particular when the plastic substrates are thick and large, a lot of that outgassing vapor may end up moving toward the OCA layer without having a good escape route, potentially causing it to delaminate or bubble. One way the industry has tried to address this issue is to bake the plastics prior to assembly. While this process eliminates the initial moisture, over time the plastic panel may re-equilibrate with the atmosphere to become loaded with water once again. Thus, a safer way is to utilize an OCA that is tolerant to the outgassing of the substrates, for example, by using a polymer that is very breathable. One way to prevent the adhesive from becoming plasticized by the water or the interfacial adhesion from becoming compromised is to use an adhesive with high moisture–vapor transmission rate and low moisture retention, such as a silicone. Another way is to provide an adhesive with high adhesion to the substrate, both at room temperature and at elevated temperature, while also making it less susceptible to moisture. Because PMMA and polycarbonate are quite polar, the industry has been using the more polar acrylic adhesives to provide good adhesion. In order to further promote adhesion and enhance the high temperature reliability, additives like tackifiers, high glass transition (Tg) acrylic oligomers, and optically clear polymer blends have been used. An example of the latter takes advantage of the acid–base interaction between a commercially available acrylic acid containing PSA and a high Tg acrylic copolymer [3]. While most polymer blends will phase-separate to form hazy compositions, the acid–base interaction of the components can be tailored to reduce the degree of phase separation to the point where a stable OCA can be obtained. The pictures in Figure 6.5 demonstrate how the weight ratio in a high Tg copolymer of methylmethacrylate (MMA) and a weak,

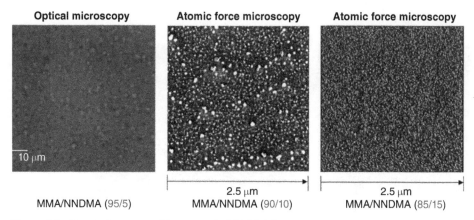

Figure 6.5 Morphology control in an acrylic PSA/high Tg acrylic additive blend.

basic comonomer N,N-dimethyl acrylamide (NNDMA) can be used to change the degree of phase separation in a blend with a typical PSA copolymer of isooctylacrylate and acrylic acid.

The amount of high Tg polymer was based on the dry adhesive weight using a target MMA content in the final adhesive composition of 19.5%. Thus, the lower the MMA amount in the high Tg copolymer, the higher its concentration in the final adhesive. Despite using a higher concentration of the 85/15 MMA/NNDMA copolymer in the blend, a smaller domain size was observed for the phase-separated high Tg material, resulting in an adhesive composition that is also optically clear. Thus, by increasing the amount of favorable interaction between the NNDMA in the polymer additive and the acrylic acid in the PSA component one can offset the large degree of entropy driven immiscibility of the two components. The Tg of the polymer additive is chosen to be sufficiently high to prevent premature softening under the durability testing conditions for the display assembly. As a result, the final adhesive composition is reinforced with high Tg nanodomains, which make the original PSA much more tolerant to outgassing even at the elevated temperatures commonly used for durability evaluation.

6.3.2 Anti-whitening Adhesives

Display assemblies are often subjected to heat and humidity both in actual use and as part of the durability testing. Exposures to 65°C/90% relative humidity or 85°C/85% relative humidity are common in the industry. Unless the panel is sealed, some of the moisture may penetrate into the adhesive to eventually saturate it. More polar adhesives such as acrylates, polyesters, and polyurethanes may be more susceptible to this than non-polar adhesives, like silicones that absorb very little water. As long as the display is warm, the moisture will stay solubilized in the polar adhesive. However, upon rapid cooling the water may exceed its solubility limit in the adhesive matrix and form little droplets that can scatter the light. Figure 6.6 shows this whitening phenomenon for a couple of regular OCAs (left and right images), and a non-whitening version (center) as a function of exposure time to 65°C/90% relative humidity condition. The test samples were constructed as follows: microscope glass/OCA/125-micron thick polyester film. In this case both the glass and the thick polyester are not breathable, causing the moisture to penetrate from the edges first and fogging up the edges once the panels cool to room temperature. Over time these edges may dry out and optical clarity is restored. When exposed to the heat and humidity,

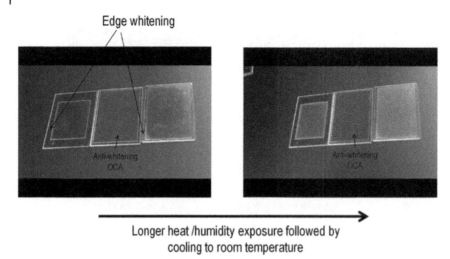

Edge whitening

Longer heat /humidity exposure followed by
cooling to room temperature

Figure 6.6 Whitening phenomena in OCA bonded panels occlusive to moisture.

for longer times, the moisture eventually penetrates throughout the whole panel and turns it completely hazy when cooled. When kept under drier condition the panel will eventually clear up with the center being last to do so, as it requires a longer diffusion path for water to leave the adhesive again through the exposed edges.

The OCA is not the only component in the display stack that can pick up moisture, indeed plastic lenses made out of PMMA or polycarbonate, or polarizers made with triacetyl cellulose, will do the same. In some cases, these substrates may not only provide an additional pathway for moisture to get into the adhesive but at the same time they may provide an escape route as well; and depending on the relative rates of these two processes whitening of the adhesive may be avoided.

However, if the substrates are occlusive, the OCA will have to accommodate the extra moisture content as the adhesive returns to room temperature and solubility in the matrix diminishes. Several methods have been proposed to address this issue with the more common ones increasing the water solubility in the adhesive by incorporating hydrophilic monomers in the acrylate copolymer, reducing the water uptake, enhancing the water vapor transmission rate, or incorporation of hydrophilic additives that help minimize the formation of light-scattering water droplets in the adhesive [4–6].

6.3.3 Non-corrosive OCAs

As shown in Figure 6.7, the cross-section of a resistive touch sensor is quite different from a capacitive touch sensor, represented here by a so-called GFF configuration, which stands for glass/film/film. Indeed, while in a resistive touch panel the ITO coatings face the inside (and thus the air gap, instead of the OCA) of the sensor, in the case of a capacitive sensor the ITO may face the OCA directly and indeed make interfacial contact with the adhesive. ITO can be quite sensitive to corrosion from acids, especially if the ITO is coated on a film where lower processing temperatures need to be used than what is possible on glass. Even when deposited on glass, direct contact between ITO and acid containing adhesives needs to be avoided. To address this issue, sensor makers have applied either organic or inorganic barrier coatings to protect the sensitive ITO. While on glass higher temperature processed inorganic barrier materials are acceptable, the industry has to rely more on solvent-cast organic barriers for the heat-sensitive film substrates.

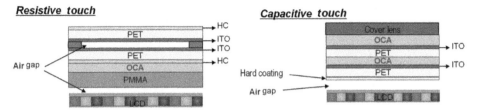

Figure 6.7 Cross section of a resistive and capacitive touch sensor (note HC = hard coating).

While these barrier coatings can be very effective, additional processing is required to apply them and concerns remain about possible cracking of the barrier with an associated risk of ITO deterioration and loss of trace conductivity.

In order to evaluate the sensitivity of the ITO coating to direct contact with the OCA, a few different ways to measure electrical resistance change as a function of contact time have been proposed. Perhaps one of the more sensitive techniques is to apply the adhesive as a strip to a simple, patterned ITO panel, and expose the sandwich to a controlled environment such as 65°C/90% relative humidity to monitor the change in resistance over the individual traces. Figure 6.8 shows a typical set-up for testing, while Figure 6.9 shows a typical result for the resistance change as a function of exposure time to the OCA.

As can be seen in the graph of Figure 6.9, an acid-containing OCA will cause a very significant change in resistance and eventually lead to failure of the trace. In contrast, the hydroxy-functional acrylic adhesive of this example (labeled as "no-acid" in the graph) shows pretty much the same negligible resistance change as the bare ITO itself (where only the air in the environmental chamber makes contact with the trace). Typical specifications for bare ITO compatibility change from customer to customer, but a resistance change of less than 10–20% in three weeks of exposure is quite common.

While ITO corrosion is perhaps the most common concern for direct contact with OCA, other materials such as silver nanowires or copper-based metal meshes can be sensitive to other forms of deterioration that can lead to color changes, excessive resistance increase, and even failure, and new OCAs are being developed to address those issues.

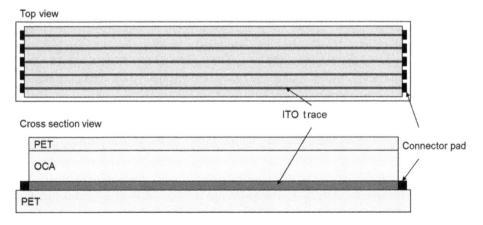

Figure 6.8 Test configuration for ITO compatibility testing of an OCA.

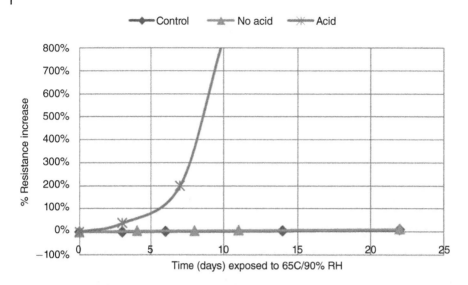

Figure 6.9 Resistance of the ITO trace as function of time for an acid-containing OCA, an acid-free OCA, and the control of bare ITO.

6.3.4 Compliant OCAs for High Ink-step Coverage and Mura-free Assembly of LCD Panels

As shown in the cross sections in Figure 6.4, most of the OCAs will have to accommodate an ink step, allow for direct bonding to an LCD without causing Mura, or do both. Compliance to the ink step can be quite challenging, especially as the step height increases (e.g. to about 50–60 microns) and is printed with a single step instead of a staircase configuration. Since the OCAs are in essence incompressible and in some cases have significant elastic character, some level of stress can be trapped in the OCA layer as a result of the ink step trying to deform the OCA. In Figure 6.10, the stress around the ink step is predicted from mechanical modeling for a more elastic adhesive and a softer adhesive with more viscous character. The adhesives are 250 microns thick and the ink step was chosen to be 25 microns. The stress was plotted as a function of distance along the glass lens OCA interface.

It is clear that the more elastic adhesive (labeled CEF8180) shows significantly more principal stress near the ink step than an OCA (labeled CEF2210) with more viscous character and lower stiffness. The shear storage modulus G' versus temperature plot for these two adhesives is shown in Figure 6.11. Not only can the OCA be put under a tensile stress near the ink step, but this value also peaks a short distance into the viewing area of the display panel. In addition, as expected there can also be a significant compressive stress under the ink itself.

It is well-known that soft materials like these pressure-sensitive OCAs can be subject to cavitation and formation of bubbles once a critical tensile stress is reached [7, 8]. The more elastic material is not only subject to a higher tensile load but the stress may also be trapped for a longer time, thus making the material more vulnerable to cavitation bubbles. It is also known that the presence of defects such as local surface roughness or the presence of a particle or fiber can facilitate the formation of these bubbles by acting as nucleating sites. It is to be noted that the formation of these bubbles can start well below the peak tensile stress experienced by the OCA, and they may indeed grow from sub-visible size to a visible defect in the display as long as the stress is present, thus causing so-called delayed bubbles. These bubbles are quite different from

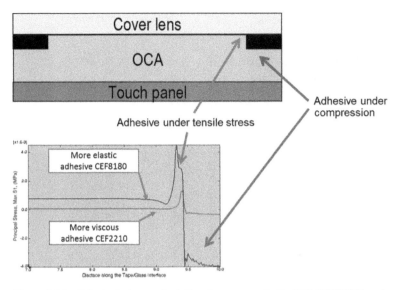

Figure 6.10 Principal stress calculation for a more elastic OCA (CEF8180) and a more viscous OCA (CEF2210) covering an ink step.

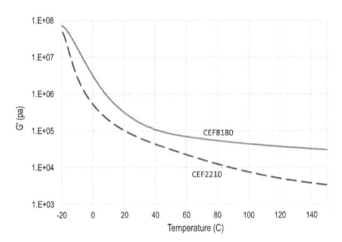

Figure 6.11 Shear storage modulus of CEF8180 and CEF2210 (both available from 3 M Company, St. Paul, MN).

those bubbles that may be present right after lamination but prior to the autoclave assembly step. Indeed, lamination bubbles are the result of trapped air that can be quite readily removed from the display during autoclaving. In contrast, the delayed bubbles can result from trapped assembly stress, especially if defects are present in the bond line. To minimize the risk for delayed bubbles, one has to carefully select the OCA and take significant care during the assembly process to insure that nucleating sites like tiny glass or metal shards, fibers, or oily stains are avoided from being trapped between the adhesive and the substrates.

A typical tensile load curve is shown in Figure 6.12. In this test, the OCA is used to bond two heavy glass coupons. The coupons are placed in the jaws of a tensile tester and pulled apart in

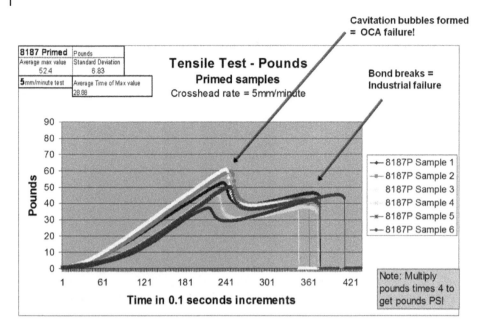

Figure 6.12 Tensile load testing of five 8187 OCA samples (available from 3 M. St. Paul, MN) applied between two primed glass plates.

so-called pluck mode, where the load is normal to the glass coupons and thus also the OCA layer. As the displacement increases the tensile force goes up to reach a peak value. At this value, naked-eye visible bubbles are already present, now growing in size and eventually causing adhesive fibrillation as displacement increases. Eventually, the adhesive fails cohesively (or adhesively dependent on the choice of adhesive and/or substrate) and the bond breaks. Note that the cavitation process may start slightly ahead of reaching peak tensile force and indeed can be the onset of an OCA bubble failure once the bubbles grow to a critical size where they are self-sustaining and clearly visible. In contrast, while cavitation may also happen in an industrial PSA application, actual breaking of the bond is considered failure; thus making OCAs far more sensitive to significant loading. Since bubbles are most likely to form near an ink edge in the viewing area of the display, it is essential to avoid this type of defect.

In order to address this risk for delayed bubble formation, the OCA suppliers have developed a number of alternative OCAs that are far less susceptible to this type of failure. Earlier generations relied on the manipulation of the overall rheology of the adhesive by dropping the shear storage modulus G', controlling its slope between room temperature and some elevated reference temperature, or by changing the visco-elastic balance of the adhesive as can be reflected in the ratio of shear loss modulus G" over shear storage modulus G', or thus tan delta [9, 10]. Each of these earlier products provided significant improvements in their ink-step coverage capacity, approaching something like 30–40% of their thickness. In other words, to reliably cover a 30–40-micron ink step an adhesive with a thickness of about 100 microns would be recommended.

However, as the trend toward thinner displays continues, increasing demands for better ink-step coverage persists. In addition, some of the applications now require direct bonding to an LCD, and dependent on the type of module, they can become very sensitive to stress, in particular localized stress as may result near an ink step. To further enhance the adhesive compliance and stress relief near the ink step, additional viscous character was incorporated in the adhesives. To some extent this was achieved by fine tuning of the rheology, but in addition the suppliers have tried to take advantage of OCAs with lower crosslinking density and modulus. However, a reduced crosslinking level may also lower the cohesive strength of the material and durability during environmental testing may be compromised. Fortunately, the assembly industry was willing to accept an additional process step to help address this issue. While normal assembly processes consist of a first lamination step (commonly flexible to rigid substrate), followed by a second lamination step (typically rigid to rigid substrate) and a finishing autoclave step, in the modified assembly process a UV curing step was introduced after autoclaving. Using longer wavelength, high intensity UV most display assemblies can be irradiated from at least one side, and if needed also from the edge. As a result, assembly induced stresses can be relieved during the autoclave step, where heat and time can be leveraged to enhance this process. Once the display assembly leaves this step, UV induced crosslinking is used to increase the cohesive strength of the OCA, resulting in a much more durable assembly without significant levels of trapped stress. Figure 6.13 shows a typical response of such an OCA. Prior to UV curing the tan delta value is quite high, indicating a significant contribution from the shear loss modulus G" and thus viscous behavior. Upon curing, this tan delta value drops; indicative of a growing elastic contribution from the increasing network content in the adhesive.

These types of adhesives may be used as single layer or multilayer constructions to further balance the properties of the material. One significant advantage of these types of materials is the low to moderate shrinkage upon curing. Indeed, in most cases a significant polymer fraction is already present in the adhesive, leaving little or no small molecules to react that may result in a volume reduction of the material. This can be particularly important in direct bonding applications to an LCD, where shrinkage-induced stress can also deform the cell gap in the display and cause Mura.

While optimization of the adhesive rheology has been used to eliminate the vast majority of assembly defects in a display, some issues with larger panel bonding remain. Vacuum lamination is commonly used for rigid-to-rigid panel lamination and total assembly cycle times on the order of 10–15 seconds are not uncommon. However, as the panel size increases, the vacuum-lamination hardware becomes more expensive and the time to evacuate the chamber may also increase, thus adding to the total assembly cycle time (TACT) and cost. Several alternatives to vacuum lamination

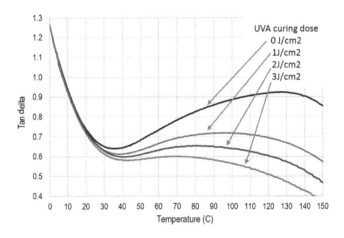

Figure 6.13 OCA UV curing response as measured in the form of tan delta versus temperature.

have been proposed, including lamination with liquid assistance or the use of air-bleed features. One example of the latter is the use of micro-structured channels in the adhesive that not only enable the evacuation of trapped air between the adhesive and the substrate, but may also facilitate positioning and early reworkability of the assembly due to the reduced initial contact between the OCA and the substrate. The micro-structure can be generated using a tool or structured release liner to emboss a smooth adhesive surface [11, 12]. As can be seen in Figure 6.14, the time to heal the micro-structure can be controlled by changing the degree of elastic behavior in the OCA, with more elastic character speeding up the recovery of the original flat surface of the adhesive. However, as discussed earlier, one also has to consider the possible consequences of using a more elastic adhesive.

A very different approach to address the need for flow and stress relief in the display assembly process is the use of liquid optically clear adhesives (LOCAs). While the liquids are not very compressible either, they offer the significant benefit of excellent flow over film-based OCAs. Thus, they are readily wetting ink steps, even as the ink-step height approaches the thickness of the wet adhesive layer. While dominated by acrylate-based materials that are UV cured, both UV-cured and thermally cured silicones are also available in the market. Some of these materials are actually not real OCAs, instead acting more as soft gels that optically couple the display subcomponents to achieve enhanced brightness and contrast, without providing any major mechanical properties other than damping.

While the excellent flow of some LOCAs can be an asset, it may also create some significant challenges in controlling that flow. Indeed, especially when bonding to an LCD module, one has to be very careful to prevent overflow of the liquid as the lens or touch panel is attached to the LCD. Due to the compression of the liquid as the panels come together, the liquid adhesive has no choice but to expand its footprint in order to preserve its volume. This can result in LOCA moving beyond its intended boundaries to penetrate in areas designated for camera windows, microphones, or flooding of the backlight unit of the LCD module. One way to prevent this from

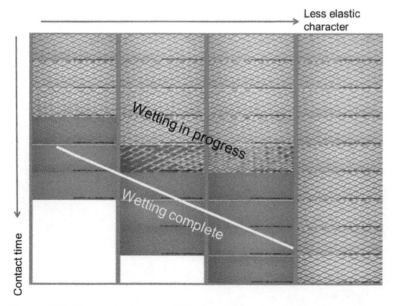

Figure 6.14 Time-lapse images of OCA wetting on glass as a function of the elastic properties of the adhesive.

happening is to carefully meter the amount of LOCA and use a higher viscosity dam material that sets the limit of flow for the less viscous fluid used to fill the balance of the display gap [13]. To prevent optical distortions between the dam material and lower-viscosity LOCA, one typically tries to match chemistry and refractive index of the cured materials. An alternative method used for direct bonding to an LCD has been the so-called open-cell assembly process, where the backlight unit is only applied after the touch panel or lens is bonded with the cured LOCA.

A very different challenge results from the change in volume and the increasing modulus of the LOCA as it is being cured. Indeed, due to the high fraction of relative low molecular weight components in the formulation, some level of volume shrinkage is expected. This is especially true for free-radically curing acrylate-based LOCAs that have high levels of monomeric units as the reagents. Shrinkage levels can be several percent, where levels below a few percent are most desirable. Managing this shrinkage level becomes even more important as the modulus of the cured adhesive increases. Indeed, it has been found that the stress on a display is very much related to the adhesive modulus and its shrinkage level [14]. If both are high, high adhesion strength may be obtained but the odds for trapped curing stress, and as a result perhaps also Mura in the direct bonded LCD module, increases. Likewise, if both the shrinkage and modulus are low, little or no curing stress will be generated, but generally the bond strength will also be low and some level of creep may result if the panel components are subject to a steady load. The modulus of the cured adhesive can readily be measured with conventional techniques like dynamic mechanical analysis in a rheometer. Volumetric shrinkage can be more challenging to measure. Density can be measured using ASTM Test Method D792 using buoyancy. However, helium gas pycnometry according to ASTM B923 seems to be the more accurate and reliable method to measure density because of the inertness of the gas toward the soft material. Knowing the mass of the sample, one can then derive the volume change.

Typical LOCA curing will be triggered by UV light, followed by polymerization of the material. In some cases, the polymerization catalyst is released by UV exposure and the reaction may proceed in the absence of light, often promoted with heat to drive the addition or ring-opening polymerization process [15]. When the light gets obstructed by an ink border, an inserted flex connector, or the like, it can be challenging to get rapid and complete curing in the shaded areas. To address this issue the assemblers rely on edge curing with the light being directed by mirrors or the use of focused LED light sources. Another approach has been to use a combination of UV and thermal initiators, so the LOCA can be cured with light in the exposed areas and shadow cured by heating the panel. With each of these approaches care has to be taken to ensure full conversion of the reagents because residual monomers may migrate and foul or damage other components in the device. Likewise, one needs to make sure that non-reactive materials such as plasticizers and oils that may be present in the LOCA will not cause any issues upon aging of the device. Thus, it may be advantageous to formulate LOCA with monomers and reactive oligomers that can co-react to form a polymer free of outgassing or migratory species.

As briefly discussed, in addition to the curing challenges one also has to pay close attention to the LOCA dispensing. Fortunately, there are a lot of hardware makers out there that can assist with this process. For smaller displays, the so-called dam-and-fill process is quite common. The dam material can set the boundaries for the LOCA, while the lower viscosity material needs to be metered quite precisely to avoid either underfilling and possibly leave air pockets, or overfilling so liquid spills over the dam. To help the spreading of the fluid without trapping air, controlled patterns are being used to dispense the material from one or more needles. However, as the display panels get larger, it becomes increasingly complex to dispense the right amount of LOCA and maintain good yields. To help address this challenge, screen-printing [16] and stencil printing [17] have been proposed. The latter utilizes a stencil to set the adhesive boundaries and applies the liquid in the stencil gap to precisely place the LOCA on the first substrate. Both the adhesive thickness and the adhesive pattern are set by the stencil, while the adhesive itself can be metered with a squeegee or simply dispensed from a slot die. To further enable this process a thixotropic adhesive was designed, so adhesive sagging and oozing could be avoided when the stencil is removed. Figure 6.15 demonstrates this difference in material creep between a regular dispensing LOCA and a thixotropic LOCA.

More recently, slot die coating hardware has become one of the main tools to apply LOCA on larger surfaces. The process is fast and quite precise, but the rheological properties of the LOCA are still important to ensure success with this coating process. One way to try to manage the flow of the LOCA right after coating is to partially cure the material and increase its viscosity [18]. In this hybrid or pre-gelation approach, the adhesive is only reacted to a level where it shows sufficient shape retention to allow for successful lamination to the second substrate, while also avoiding overcuring that would compromise flow. The pre-gelation process also allows for some shrinkage to happen prior to final assembly. However, one has to be careful to strike the right balance between increasing the viscosity of the LOCA, while not increasing the modulus and elasticity of the material too much, which otherwise may start to influence lamination of the second substrate and adhesion of the fully cured LOCA. A thixotropic LOCA offers an alternative approach to manage de-wetting and shape retention of the coated adhesive. Indeed, this material is able to maintain the printed shape and profile of the coating without any pre-curing step. This not only eliminates the need for this extra pre-curing process step, but it also allows the adhesive to retain its viscous character as formulated. As shown in Figure 6.16, the thixotropic LOCA has a higher viscosity at lamination where shear is typically low and squeeze out or ooze need to be managed, while at higher shear rates such as those encountered during printing or die coating the viscosity is much reduced, thus facilitating wetting of the substrate.

The process of shear thinning, and thickening in the absence of shear, is fully reversible. So, right after coating the material will become very viscous to retain the shape of the coating, eliminating both the need for pre-curing and dam material.

Another advantage for thixotropic LOCA is the possibility to provide a surface pattern or structure that allows for easy lamination and air evacuation from between the coated adhesive and the second substrate. Much like the micro-structured adhesive discussed earlier, the structured

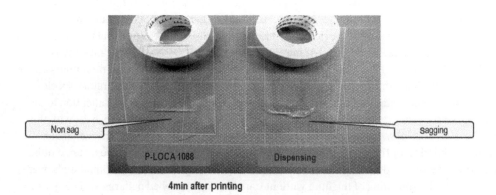

Figure 6.15 Thixotropic, printable LOCA (P-LOCA) vs. dispensing LOCA, both coated on a tilted glass slide.

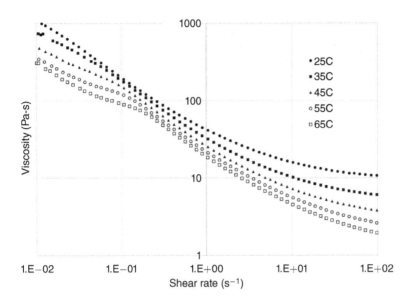

Figure 6.16 Shear-thinning behavior of a typical thixotropic LOCA as a function of shear rate and temperature.

or patterned adhesive can make initial contact with the substrate in only a few locations, while allowing the air to escape as the material continues to wet. In contrast, with a perfectly smooth adhesive coating full contact with the second substrate is desired but in practice very difficult to control, so the industry relies a lot on vacuum bonding to reduce the risk of air bubble entrapment. However, vacuum bonding requires extra capital equipment and the process typically increases TACT, so vacuumless bonding should bring significant advantages to the industry.

Taking advantage of the shape retention in the absence of shear, one can apply a thixotropic LOCA patch that is substantially flat, but add some raised features to initially offset the coating from the second substrate [19]. By using these local thickness differences, one allows the adhesive to make contact in a very controlled fashion and in essence dictate what that contact line looks like. By doing so, not only can one design a path for air to escape but also control the amount of fluid that needs to be displaced to fully wet out the substrate. This is in contrast with the dam-and-fill approach discussed earlier, where large volumes of LOCA need to be moved in a lateral fashion as the material gets compressed, and the dispensing pattern needs to be very well designed and managed to avoid air entrapment. Indeed, with the thixotropic LOCA, one can in essence put a full sheet of adhesive down, while only needing a minimum extra amount to create a topographical pattern that gets fully embedded with the rest of the material as the liquid-coated substrate comes together with the second substrate at assembly. This small additional volume also ensures minimal displacement of the adhesive patch, so registration on the substrates can be readily maintained and adhesive squeeze out is well-controlled. One can consider topographies that include individually raised points, raised lines, or even slightly crowned adhesive patches. An example of such a raised line pattern is shown in Figure 6.17. The line can be applied from a simple needle dispenser using the exact same thixotropic adhesive as the rest of the patch. This ensures seamless blending of the two, so no visible knitting lines exist once the panel is assembled and cured.

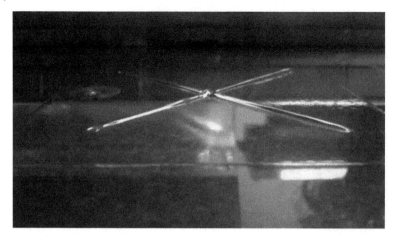

Figure 6.17 LOCA adhesive patch with raised line pattern.

6.3.5 Reworkable OCAs

As the number of displays keeps growing and their competitive pricing drives the assemblers toward ever-increasing yields, the need for reworkability has also become more important. This rework may be needed at the time of assembly, where the recovery of even a couple tenths of percent of the devices can result in significant cost savings and reduction of waste. Likewise, there may be a need to rework the display at a refurbishing center, or also at end of life of the device where disposal of electronics needs to be managed.

While flexible components may be peelable from a substrate, removing a bonded rigid panel from a rigid panel is nearly impossible. Prior to curing, LOCA assembled displays are fairly easy to rework. Indeed, panels can be shifted relative to each other to take them apart and the remaining liquid can be cleaned up by conventional means. In some cases, the manufacturer may do some spot curing to prevent shifting of the panels prior to inspection and final curing, but this shows minimal interference with the rework process. However, once the whole LOCA surface is fully cured, it is not much different from film-OCA-bonded displays with perhaps only the lower bond strength allowing for an occasional easier cleavage of the panel. Unless the adhesive is cohesively weak, once the panels are bonded with a fully cured adhesive, rework can be quite challenging because cleavage or shearing of a bonded pair of substrates with an interposed visco-elastic material is almost impossible and likely to cause damage to one or both of the panels. Especially the more costly LCD or OLED components need to be recovered, which unfortunately are also the most prone to mechanical damage.

A few common ways to rework a display include freeze-fracture and wire cutting. In the case of freeze-fracture, the assembly is chilled to a temperature at, or ideally below, the glass transition temperature of the adhesive. Once in the glassy state and using the right fixtures and tools, the adhesive may be fractured away from the substrate, ideally leaving little or no residue on that panel. At the same time the second substrate, now still coated with adhesive, can be brought back to room temperature and if needed the adhesive can be peeled from the panel, either by itself or with the assistance of a flexible backing or tape.

An alternative method to gain access to the individual substrates is to use wire cutting. Figure 6.18 provides an image of one such device used in the industry.

The process involves careful insertion of a thin metallic wire at the edge of the panel. Once inserted, the wire is pulled by hand or mechanically between the display panels, to eventually

Figure 6.18 Automated wire-cutting tool.

separate them. If all works well, no mechanical damage, like scratching, is caused to the panel to be recovered, and the adhesive residue, if present, can be picked or peeled off now that it is readily accessible.

A very different and quite elegant way leverages the stretch removable adhesive technology applied for tape strips used for mounting applications. Indeed, OCA sheets were developed that yield durable display assemblies, yet remain readily reworkable by simply pulling on a tab to stretch the adhesive from between the substrates [20]. Unlike the much narrower mounting strips, these adhesive sheets have very high aspect ratios (thickness to width) increasing the risk for premature breakage of the sheet during removal, so they have to be very carefully designed in order to maintain optical clarity and provide sufficient mechanical strength to be fail proof. To rework the display, the sheets are stretched at a near zero angle relative to the substrates and after an initial elongation of the adhesive/reinforcing carrier film construction, the detachment of the adhesive follows by deforming a very small volume of material and creating a sharp crack propagation front. During the actual debonding the stress maintains a controlled plateau value with minimal normal to the substrate component, facilitating the detachment without mechanical damage to the fragile LCD.

6.3.6 Barrier Adhesives

While LCDs are quite insensitive to moisture, others like electro-phoretic displays used in E-readers and OLEDs need better protection. Electro-phoretic displays may have to be sealed with assistance of an optically clear adhesive where the water-vapor transmittance rate needs to be minimized to a level not exceeding single $g/m^2/day$. In such cases, simple elastomer-based adhesives may be used. Not only do OLEDs need to be protected from moisture but oxygen penetration into the device must also be managed. Penetration levels on the order of 10^{-3}–10^{-5}g/m^2/day are only tolerated for such devices. While primary encapsulants are used during the manufacturing of the OLED, there is still a need for adhesives to provide secondary barrier properties. Especially as displays move to true flexible format, the industry may have to increasingly rely on these adhesives to complement the barrier films, which mainly perform in the z-direction of the device but do little in the plane of the assembly. Thus, barrier adhesives may become more critical in the future. To date, most of the barrier adhesive technology has been reliant on the use of UV-curable glassy adhesives where their barrier characteristics are typically enhanced with so-called getters that preferentially react with oxygen or moisture as they try to diffuse through the bulk of the adhesive.

6.4 Summary and Remaining Challenges

Over the last decade or so, the importance of OCAs as a bonding solution for display assembly has rapidly grown from mainly being a thin-polarizer attachment adhesive to a material that not only bonds but also provides an optical and electrical function to the display stack. Especially with the rapid growth of capacitive touch enabled devices and the ever-increasing resolution of the displays the use of thicker gap filing materials has penetrated this industry very quickly. As designs continue to change, we expect the need for newer gap-filling materials to continue. For example, as displays get thinner and lighter, the OCAs will also have to get thinner, increasing the demand on both the dielectric management and the mechanical performance. Indeed, to eliminate thickness and weight, newer devices are moving away from the use of bezels, thus increasingly relying on the OCA to provide mechanical integrity to the device. Unfortunately, as the devices get thinner they also become more fragile, driving a trend toward gentler assembly processes and better damping properties of the adhesive to enhance the drop durability of the device. Now that the users have become very familiar with touch-enabled devices, we expect further adoption of capacitive and other touch technology into automotive displays and larger devices, up to and including TVs and advertising panels in kiosks and malls. Together with touch functionality, these newer applications will also require more and more direct bonding to enhance brightness and contrast, and to ruggedize the display panels. As a result, the assembly industry and the adhesive suppliers will have to develop technologies that facilitate the bonding of these newer panels that are not only larger in size but also may have increasingly complex shapes. Finally, with curved displays already commercially available and the imminent emergence of bendable and truly flexible or rollable displays, one can anticipate that the demands on the mechanical, electrical, and barrier properties of the OCAs will continue to evolve very quickly, moving to evermore specialized materials. Against these trends and needs, we expect that the use of OCAs is here to stay and as newer designs are looking for unique assembly solutions, the bonding processes, their associated hardware, and the material characteristics of the adhesives will all continue to improve in the coming years.

References

1 Goetz, R. and Ouderkirk, A. (2001). Light diffusing adhesive. United States patent 6,288,172.
2 Matano, T., Nagamoto, K., Aso, Y., and Kusuma, K. (2007). Pressure-sensitive adhesive for applying optically functional film and production process for the same. United States patent application 2007/0267133A1.

3 Xia, J. and Everaerts, A. (2008). *Proceedings of the 31st Annual Meeting of the Adhesion Society.* Austin, TX.

4 Everaerts, A. and Xia, J. (2013). Cloud point-resistant adhesives and laminates. United States patent 8,361,633.

5 Inanaga, M. (2012). Acrylic group transparent adhesive film or sheet for vehicles contains ultrafine particles which have hydrophilic hydroxyl group on surface. Japanese patent 4,937,463.

6 Kishioka, H., Fumoto, H., Okamoto, M., and Niwa, M. (2012). Optical pressure-sensitive adhesive sheet. United States patent application 2012/0328873 A1.

7 Chiche, A., Dollhofer, A., and Creton, C. (2005). Eur. Phys. J. E17: 389–401.

8 Lindner, A., Lestriez, B., Mariot, S., Creton, C., Maevis, T., Luhmann, B., and Brummer, R. (2006). Adhesive and rheological properties of lightly crosslinked model acrylic networks. *J. Adh.* 82 (3): 267–310.

9 Niimi, K., Yoshikawa, H., Yamaoka, T., and Inenaga, M. Transparent double-sided self-adhesive sheet. United States patent application 2012/0156456 A1.

10 Yoon, C., Kim, J., Song, M., Park, E., and Jung, B. Adhesive composition having high flexibility. PCT patent publication WO 2013/176364 A1.

11 Nakajima, M., Sannomiya, I., and Sannomiya, Y. (2008). Interlayer for laminated glass comprises thermoplastic resin sheet having embossed pattern comprising grooves and ridges on each side. United States patent 7.378,142.

12 Sherman, A., Winkler, W., Mazurek, M., Garcia, H., and Toro, D. (2007). Method for preparing microstructured laminating adhesive articles. United States patent application 2007/0212535 A1.

13 Nguyen, T. (2013). Method for bonding a transparent substrate to a liquid crystal display and associated device. United States patent 8,468,712.

14 Shinya, Y., Kamiya, K., Toyoda, T., Endo, Y., Soudakova, N., and Kondo, H. (2009). A new optical elasticity resin for mobile LCD modules. J. Soc. Inf. Dis. 17: 331–336.

15 Smothers, W., Held, R., and Farah, H. Thermally and actinically curable adhesive composition. PCT patent publication WO 2009/086492 A1.

16 Kobayashi, S., Miwa, H., and Ishii, K. (2009). Manufacturing method of display device. United States patent application 2009/0215351 A1.

17 Busman, S., Ong, C., Sim, Y., Chan, W., Chen, X., and Toy, M. (2013). Display panel substrate assembly and an apparatus and method for forming display panel substrate assembly. United States patent application 2013/0295337 A1.

18 Shinya, Y., Hayashi, K., Ogawa, K., Kamiya, K., Toyoda, T., and Hayashi, N. (2013). 68.2: *Invited paper*: Development of novel optical bonding process and materials for flat panel display modules. SID Symp. Digest Tech. Papers 44: 946–948.

19 Pennington, B., O'Hare, J., Stensvad, K., Carlson, D., Jerry, G., Schlepp, A., and Campbell, C. (2014). Vacuumless lamination of printable LOCA. SID Symp. Digest Tech. Papers 45: 28–31.

20 Everaerts, A., Determan, M., Purgett, M., and Sura, R. (2014). *Proceedings of the 36th Annual Meeting of the Adhesion Society.* San Diego, CA.

7

Self-healing Polymer Substrates

Progyateg Chakma, Zachary A. Digby, and Dominik Konkolewicz

Department of Chemistry and Biochemistry, Miami University, Oxford, OH

7.1 Introduction

There have been numerous advances in the field of conductive technologies, photoelectronic materials, and display technologies [1–7]. However, it is also important to note that despite advances in conductive component technology, the long-term viability of the matrix that supports the flexible electronics also needs to be considered. Without a robust and durable matrix, the utility of the flexible display or electronic material will be greatly reduced due to shortened useful lifetime. Creating materials with increased toughness, or the ability to absorb energy without fracturing, will increase the longevity and useful lifetime of materials [8]. In contrast, brittle materials fracture easily and are not well suited to mechanically demanding applications [9]. However, eventually all materials fail and fracture. For many materials, such as thermoset crosslinked materials, this type of fracture can represent the end of its useful lifetime, since many cannot be easily recycled back to their original monomers [10]. An alternative approach is to engineer in the capacity for material recovery after damages or fractures are sustained. Materials with the ability to recover from otherwise detrimental damage are often termed "self-healing," "re-healable," or "mendable" materials [11–14]. Throughout this chapter the term self-healing material will be used exclusively, although other terms are used in the literature. An interesting corollary of designing materials with self-healing character is that its toughness is often increased, further increasing the material's longevity.

Significant research efforts have been devoted to designing and characterizing materials with self-healing properties [15]. Potential applications for these types of self-healing materials range from coatings to bulk materials. Some of the earliest work in the field of "self-healable" materials includes the work of Wudl et al., utilizing the well-known dynamic chemical bonds [14], and the vascular-like systems of White et al., which enable cracks to be filled upon rupture [16]. These strategies for designing self-healing materials will be discussed in greater detail in the subsequent sections of this chapter. This chapter will focus exclusively on polymeric materials with self-healing properties. This is in part because flexible electronics and displays often use a polymeric component or matrix as part of the material, even if the electro active component is often inorganic in nature [17]. In addition, the advances in conductive polymers lend themselves to efficient self-healing polymer materials [18]. The ability to self-heal or recover from otherwise catastrophic damages is of particular relevance to high-margin or high-functionality materials such as conductive materials and flexible electronics or displays. In general, if a support for a

Flexible Flat Panel Displays, Second Edition. Edited by Darran R. Cairns, Dirk J. Broer, and Gregory P. Crawford.
© 2023 John Wiley & Sons Ltd. Published 2023 by John Wiley & Sons Ltd.

high-value electronic has an extended useful lifetime, enabled by self-healing, then the overall device should also have improved long-term performance.

This chapter will highlight key properties and structure function trade-offs that exist in self-healing materials. A focus on design principles needed to create efficient materials will be presented, and a library of potential chemical scaffolds will be discussed. This enables the device designer to select from a wide range of functionalities that best meets the requirements of the targeted application. Finally, the chapter will discuss self-healing composite type materials and considerations needed to design efficient conductive-type materials, which would be essential in flexible electronics that could lead to flexible display technologies. Additional focus is placed on potential advances enabled by three-dimensional printing technology that enables complex geometric objects to be created, which could enable the next generation of flexible electronics [19]. Finally, conclusions and outlook are presented for self-healing high-end materials suitable for the next generation of high-end applications. This chapter does not provide a comprehensive review of all technologies in the field, but instead highlights developments and possibilities.

7.2 General Classes of Self-healing Polymers

Self-healing polymers are generally classified into two types: (i) materials with microcapsules or microvascular network having extrinsic healing mechanism [16, 20–24] and (ii) materials with dynamic linkages that give exchange reaction with or without stimuli via intrinsic healing mechanism [15, 20, 25, 26]. Microencapsulated materials contain healing agents, which in response to mechanical damage rapture and mix with the polymer matrix. Healing agents often get activated with help of an embedded catalyst as shown in Figure 7.1a [16]. Although these materials showed considerable recovery of mechanical properties, they have limitation of healing only a single time in the damaged area. Materials with three-dimensional microvascular network, on the other hand, mimic biological system and provide multiple healing cycles by reacting the healing agent of the capillaries with the embedded catalyst [12]. Still, reduction of catalyst or capsule content over time only provides limited healing cycles in these materials.

Dynamically crosslinked materials have received significant interest, as these materials have functional moieties in the polymeric materials and can theoretically give infinite healing cycle due to dynamic exchange of reversible bonds without losing any functional moieties unlike the former vascular or capsule technologies [20]. Due to their intrinsic healing mechanism, functional moieties in the polymer system provide the self-healing properties without using any external healing agents or catalysts. These materials have facile synthesis procedures and show crucial properties like malleability, reprocess ability, recycle ability, and biodegradability, which cannot be attained with microencapsulation or microvascular-based materials [20, 25–30]. Dynamic properties in these materials are based on either supramolecular chemistry or dynamic covalent reactions. Dynamic exchange can occur rapidly under ambient conditions or in response to a stimulus like light, heat, or pH and, using the rapidity of the dynamic exchange in different conditions, materials with numerous applications can be developed [25, 27]. An example of dynamically crosslinked materials reported by is Guan et al. shown in Figure 7.1b [31], where the self-healing mechanism is based on metal-ligand (zinc-imidazole) supramolecular or non-covalent interactions in the polymer system. A key consideration to the design and choice of groups for self-healing dynamically crosslinked materials is the trade-off between dynamic character that enables self-healing and the long-term mechanical stability, or creep resistance [32]. Creep is a measure of

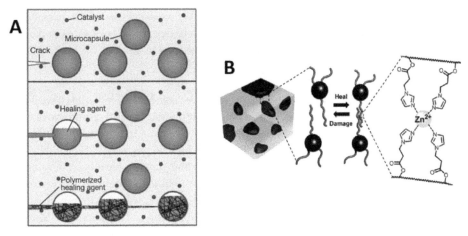

Figure 7.1 (A) Autonomic healing mechanism of a microencapsulated polymer composites reported by White et al. where upon rapture healing agent reacts with Grubb's catalyst to form crosslinked network. Reprinted by permission from Ref [16] Springer Nature: Wiley Flexible Flat Panel Displays: 2nd Edition by Gregory P. Crawford, Darran R. Cairns, Dirk J. Broer (Eds) (2001). (B) A multiphasic self-healing polymer system, where a highly dynamic metal-ligand (zinc-imidazole) interactions were used as healing source. Reprinted with permission from [31]. Copyright 2014 American Chemical Society.

a material's tendency to deform, permanently or temporarily, under stresses or external forces. Often materials with the greatest self-healing character due to rapid exchange of dynamic bonds also have creep susceptibility [33], which could limit the long-term utility of the material. Therefore, bonds must be chosen so that the material is mechanically stable under the application yet dynamic either intrinsically or when exposed to external stimulus to enable self-healing. As a general rule, non-covalently crosslinked materials tend to have weaker associations that are more creep susceptible, while dynamic-covalently crosslinked materials tend to require external stimulus such as heat, pH, or light to activate the bond and therefore are more creep susceptible. In addition, non-dynamic bonds can be introduced to increase creep resistance [33].

7.2.1 Types of Dynamic Bonds in Self-healing Polymers

Depending on the nature of the dynamic crosslinkers, self-healing materials are of two classes: (i) supramolecularly crosslinked, or (ii) dynamic-covalently crosslinked materials [34, 35]. These two classes of materials have their own advantages and disadvantages. Developing materials considering unique properties of supramolecular or dynamic covalent bonds is essential.

7.2.2 Supramolecularly Crosslinked Self-healing Polymers

Supramolecularly crosslinked self-healing polymeric materials are based on non-covalent bonds like hydrogen bond, hydrophobic interactions, ionic interactions, metal-ligand co-ordination, π–π stacking interactions, and host–guest interactions, among others [20, 25, 36–39]. These materials can be rapidly and reversibly remodeled compare to dynamic-covalently crosslinked materials especially if the material has a very low glass transition temperature (T_g) [25]. Dynamic exchange in these materials typically occurs rapidly in ambient conditions, but healing at different temperatures and in response to stimuli is also possible. Exploiting the time scale of the dynamic exchange and healing time, materials with strong or weak interaction can be developed. Selected examples of self-healing polymers with supramolecular interactions are discussed in Sections 7.2.2.1–7.2.2.3.

7.2.2.1 Hydrogen Bonding

Hydrogen-bonding interactions in supramolecularly crosslinked self-healing materials have been extensively studied and utilized due to the highly dynamic nature and stimuli responsiveness of this interaction [40, 41]. Although hydrogen bonding generally is a weak interaction, tunability of different hydrogen-bonding moieties can give significant mechanical properties [25]. Large varieties of hydrogen-bonding moieties with different dynamics and stability have been developed since the 1990s. Among them ureidopyrimidone (UPy) moiety-based supramolecularly cross-linked polymers have been extensively studied and commercialized as well due to unique properties such as high dimerization constant [20, 42]. UPy-containing copolymers have shown recovery of mechanical properties both in ambient conditions and in elevated temperatures [41, 43]. Fatty acid-based hydrogen-bonding interactions have also been utilized for thermoplastic elastomers. Poly(dimethyl siloxane) (PDMS)-based elastomers use multiple hydrogen-bonding interactions and self-healing is attained in low temperatures in a very short healing period [44]. PDMS are functionalized with urea and diethylene triamine to form soft rubbery material [45]. Other self-healing polymers that use hydrogen-bonding interactions include Urazole functionalized polymers and bis-urea-based polymers [20]. Hydrogen bonding has also been incorporated in the dynamic covalent system with dual self-healing functionality. Foster et al. reported sequential synthesis of a double dynamic interpenetrating network (IPN) system with a highly dynamic hydrogen-bonded UPy moiety and a thermoresponsive Diels–Alder adduct by free radical polymerization [46]. They found significant enhancement of mechanical properties including self-healing and malleability in IPN polymer to a single-network (SN) polymer. Synthesis of IPN and SN polymers with sequential and one pot strategies is shown in Figure 7.2a.

7.2.2.2 π–π Stacking

π–π interaction is an extensively used supramolecular chemistry for self-healing purposes in non-hydrophilic systems [20, 25]. Although π–π interactions are in typically weaker than hydrogen bonds or ionic interactions, careful design of stacked π systems with interactions between π elec-

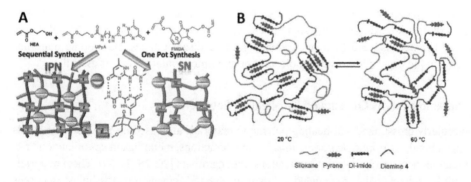

Figure 7.2 (A) Sequential and one pot synthesis of IPN and SN networks using hydrogen-bonding moiety (UPyA) and Diels–Alder adduct. Reprinted with permission from [46]. Copyright 2017 American Chemical Society. (B) Self-healing mechanism by π–π interaction of pyrene and di-imide containing polymers in response to heat. Reproduced from [47] with permission of The Royal Society of Chemistry.

tron deficient end groups with π electron-rich aromatic units resulted in self-healing polymeric materials. Hayes et al. are one of the earliest groups who developed thermoresponsive self-healing systems using only π–π interactions between π electron deficient groups containing polyimide chains and π electron rich phenyl-end groups containing siloxane polymers as shown in Figure 7.2b [47]. π–π interactions crosslinkers were disrupted in elevated temperatures and upon cooling new π–π stacking resulted in new network formation. In addition, π–π interactions have been combined with other supramolecular interactions like hydrogen bonds and metal-ligand interactions for efficient self-healing systems. Rowan et al. reported a self-healing polymer blend between pyrenemethylurea end-capped polymer and chain-folding polyimide, which utilizes both π–π interactions and inter-polymer hydrogen bonds for thermally healable elastomers [48].

7.2.2.3 Ionic Interactions

Polymeric chains incorporated with up to 20% ionic contents are known as ionomers. These materials generally use commercially available starting materials and have unique properties [49]. Ionic groups used in these materials either act as pendant groups or end groups in the polymer chains and properties of the materials can be tuned using the overall ionic contents [25, 49]. Ionic groups have significant influence in the thermal and mechanical properties of the materials by forming ionic aggregation and changing the morphology of the polymer networks [50]. According to the order to disorder transition model provided by Tadano et al., ionomers have three temperature-controlled phases – ordered state, disordered state, and molten state – and transformation between these states are important for the self-healing [51]. The most commonly used ionic group in self-healing polymers is carboxylate group, which is neutralized by metal ions or quaternary ammonium ions to give thermoplastic materials [20]. Carboxylate groups containing unsaturated monomers are often co-polymerized with n-butyl acrylate or tert-butyl acrylate. There are reports of ionomers improving mechanical properties like fracture toughness and tensile strength. Due to these unique properties, ionomers have vast commercial applications such as ballistics coatings [20, 50].

7.2.3 Dynamic-covalently Crosslinked Self-healing Polymers

Self-healing polymers based on dynamic covalent bonds have received significant interest since the 1990s and early 2000s due to the combination of robustness of covalent bond with adaptability [52]. Dynamic-covalently crosslinked materials often but not always need external stimuli such as heat, pH, or light for dynamic exchange in the polymer network. As a result, these materials tend to have the advantage of being mechanically stable in absence of stimuli, with some exceptions. These materials either have a *dissociative* or an *associative* crosslink exchange mechanism for dynamic properties [30]. The reversible bonds used in these materials can be classified by condensation type, exchange type, or addition type [15]. The most commonly used dynamic covalent chemistry are based on the Diels–Alder adducts, boronic esters, anthracene dimers, boronates, imines, transalkylation, and disulfides, thiol-Michael reaction, among others [15, 25, 27, 29, 53]. Normally dynamic covalent functional groups are incorporated in the side chains or main chains of the polymer backbone and by forming reversible polymer networks [29].

7.2.3.1 Cycloaddition Reactions

Among the cycloaddition reactions used for self-healing polymers, Diels–Alder reaction has been the most extensively used dynamic covalent reaction for self-healing polymers [54]. This is a [4 + 2] cycloaddition reaction between electron-deficient dienophiles and electron-rich dienes to form cyclohexene adducts and reversible via retro [4 + 2] cyclo-reversion. This reaction has been very important in organic chemistry due to its "click" nature and good control over stereochemistry. The most commonly used Diels–Alder adducts are based on furan and maleimide derivatives, although other adducts of anthracene-maleimide, cyclopentadiene-cyclopentadiene, and fulvene-cyanoolefine have also been used [20,26]. The earliest examples of using thermoreversible Diels–Alder reaction for dynamic polymeric materials were reported by Wudl et al. [14]. They have reported a thermally re-mendable polymeric material using furan-maleimide-based adducts.

They found about 25% of the furan-maleimide linkages in the polymeric materials disconnect by heating at 150°C for 15 minutes and reconnect upon cooling, and as a result showing healing without using any additional healing ingredients. Although most of the Diels–Alder-based self-healing materials reported in the literature need elevated temperatures, there are also reports of materials showing healing at room temperature [27]. Lehn et al. have reported a self-healing material that uses fulvene and dicyanoethylenes-based Diels–Alder adducts [55]. Fulvene and dicyanoethylenes-based Diels–Alder adducts showed reversibility in room temperature, resulting in self-healing also being attained in room temperature without applying any thermal stimulus.

[2 + 2] cycloaddition reaction is a photochemically driven reaction between two unsaturated compounds and this reaction has been used for self-healing polymers by irradiating light in different wavelengths for photodimerization and photocleavage [20, 56, 57]. Coumarin- and cinnamate-based monomers are largely used for formation of reversible cyclobutene ring in self-healing materials. In addition, anthracene dimerization has been used for light-stimulated self-healing of polymer materials [58]. [4 + 4] cycloaddition reaction is another dynamic covalent reaction, which is both photochemically and thermally driven and has been used for self-healing polymers. Unlike [2 + 2] cycloaddition, [4 + 4] cycloaddition reaction is stable to hydrolysis and isomerization [15,20].

7.2.3.2 Disulfides-based Reversible Reactions

Apart from cycloaddition reactions, disulfides-based reactions have been used significantly in self-healing polymeric materials [59, 60]. Reversibility of disulfide linkages can be attained by applying light, thermal, or different redox condition [20, 28, 56, 61]. Disulfides-based dynamic reactions are rapid and as disulfide bonds can be integrated in low glass transition temperature (T_g) polymers, healing can happen in considerably low temperatures [15, 25]. Dynamic disulfides reactions are disulfide exchange, thiol-disulfide exchange, reduction-oxidation, and radical disulfide fragmentation [20]. One of the earliest examples of the use of reversible disulfide reaction in polymer network was reported by Matyjaszewski et al., where a polymer backbone with disulfide moiety was synthesized by atom-transfer radical polymerization [62]. The disulfide bond was cleaved by reducing to thiols by dithiothreitol and then again oxidized back to disulfide linkages by reacting with FeCl$_3$. Amamoto et al. reported a polyurethane self-healing polymer that uses thiuram disulfide (TDS) moieties for radical disulfide exchanges in ambient visible light and in absence of catalysts and solvent [63], following from earlier work using UV-activated trithiocarbonate reshuffling [64]. TDS moieties form S-based radicals by visible light irradiation, which introduce dynamic self-healing in the polymer network. Although in a 24-hour healing period the materials showed significant healing, S-based radicals can give some side reactions by migration of TDS radicals to the crosslinked polymer network and by the reduction of dithiocarbamyl radical content. In their study, Odriozola et al. have reported a self-healing elastomer that undergoes reversible disulfide exchanges in room temperature [65]. They integrate aromatic disulfide linkages in poly (urea-urethane) thermoset elastomers where both disulfide metathesis and H-bonding are responsible for self-healing properties.

7.2.3.3 Acylhydrazones

The condensation reaction between acyl hydrazine and aldehyde forms an acylhydrazone bond. The formation of acylhydrazones is facilitated in acidic condition. This reaction is dynamic in response to thermal stimulus and in different pH value [29]. Schubert et al. reported different methacrylate-based self-healing polymer using reversible acyl hydrazone crosslinkers free radical polymerization [68]. Upon interaction of water with damaged surface, some fraction of acyl hydrazone crosslinkers dissociate. After heating at 100°C for 64 hours scratch was completely removed by dynamic exchange between two acylhydrazone moieties. Acylhydrazone moieties have also been utilized with other dynamic covalent chemistry like Diels–Alder reaction disulfides. Chen et al. reported an environmentally adaptive hydrogel that utilizes both acylhydrazone bonds and disulfide bonds for self-healing purposes [66]. The synthesized hydrogel shows reversible sol-gel transition by acylhydrazone metathesis in acidic condition and disulfide exchange in basic condition as shown in Figure 7.3a. Self-healing can also be achieved in neutral condition by accelerating acylhydrazone exchange by using aniline as catalyst.

7.2.3.4 Boronate Esters

Boronate ester formation is a reversible reaction between a boronic acid and diol in aqueous medium [69], or in the bulk [70]. Boronate esters have been used significantly in responsive hydrogel formation because of the facile exchange mechanism and reaction conditions. The reversibility of this reaction is dependent of the pH of aqueous medium and pKa and acidity of the boronic acid and diol respectively [20, 71]. These properties have been used in developing self-healing hydrogels that are responsive to different conditions. The Sumerlin group reported boronate ester crosslinked hydrogels that showed self-healing properties in neutral and acidic conditions [67]. They used an intramolecularly coordinating boronic acid monomer to form stable crosslinkers in both neutral and acidic pH solution. Boronic acid monomer-containing copolymer crosslinked with diol-containing copolymers to form boronate ester hydrogels. The healing mechanism of these hydrogels is shown in Figure 7.3b. Guan et al. reported dynamic

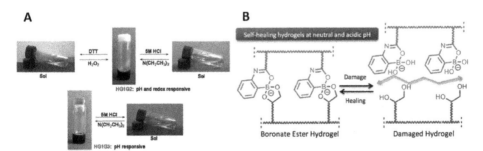

Figure 7.3 (A) sol-gel transition of disulfide and acylhydrazone containing hydrogels in response to pH and redox. Reprinted with permission from [66] Copyright 2012 American Chemical Society. (B) Healing mechanism of self-healing polymers using intramolecularly coordinating boronic acid for boronic ester exchange reaction. Reprinted with permission from [67] Copyright 2015 American Chemical Society.

polymers with tunable self-healing and malleability properties by controlling the reactivity of the dynamic boronate ester exchange in the network [72]. They synthesized two divalent boronic esters crosslinkers with different neighboring groups for slow and fast reactivity. The crosslinked materials showed dynamic properties in respect to the reactivity of the crosslinkers. Materials with fast exchanging crosslinkers showed more efficient healing and malleability compared to the materials with slow exchanging crosslinkers.

7.3 Special Considerations for Flexible Self-healing Polymers

One of the main focuses of this book is flexibility within materials. When focusing on the flexibility of self-healing polymers, there are a few special considerations that need to be taken into account such as the toughness and glass transition temperature (T_g) of the materials. Toughness is, however, a slightly ambiguous term, with multiple definitions and measurements. Likely the most widely recognized means of defining material toughness is by the area under the stress strain curve from a tensile test [8]. There also exists fracture toughness, notch toughness, etc., however, the central theme of all toughness is the amount of energy a material can absorb or disperse before it deforms to the point it fractures due to stress [8, 73, 74]. Toughness and brittleness go hand-in-hand, with brittleness being inversely related to the toughness of a material. If a material is tough, it may absorb a large amount of energy before it fractures, however, if a material is brittle it will be prone to fracture under a relatively low amount of stress.

Brittleness can be a result of being under the T_g of a material. T_g is a transition in an amorphous material from a slowly relaxing or moving "glass" state to a viscous/malleable state as temperature increases [82]. At room temperature, if a material is below its T_g it will be hard and relatively inflexible such as polyvinyl chloride (PVC) used in piping. However, if a material at room temperature is above its T_g it will be softer and more flexible like the rubber in car tires. Therefore, if designing a flexible polymer that would operate in room temperatures, the ideal candidates would have a T_g under ~ 23°C.

Table 7.1 lists a variety of common polymers used in materials, with some above and some below their T_g at room temperature. It is possible to lower the T_g of a polymer by adding a plasticizer, which increases the free rotation of the chain by spacing out the monomer units. It is

Table 7.1 Common polymers and their T_gs.

Material	T_g (°C)	T_g (°F)	Reference
Tire rubber	−70	−94	[75]
Polyvinylidene fluoride	−35	−31	[76]
Polyethyl acrylate	−25	−13	[77]
Polyvinyl fluoride	−20	−4	[76]
Polyethylene terephthalate	67 (amorphous material)	153	[78]
Polyvinyl chloride	80	176	[79]
Polystyrene	100	212	[80]
Polytetrafluoroethylene	115	239	[81]

important to note that most uncrosslinked polymers above their glass transition temperature will flow to some extent, particularly at lower molecular weights. Therefore, crosslinking either with a static crosslinker such as an amine-epoxy linkage [83] or a dynamic linkage can lead to soft flexible yet stable materials. Typically the T_g increases as the crosslink density increases, or equivalently the distance between crosslinkers decreases for both essentially static and dynamic crosslinkers [83, 84]. These crosslinked low T_g materials are elastomeric in nature. Indeed, to have optimal exchange of dynamic linkers needed for self-healing, a low T_g material is desirable to facilitate dynamic exchange of crosslinkers [32, 64, 85]. However, it is important to note that even a material below its T_g can be flexible, however, it is anticipated to be only flexible over a relatively narrow range of curvatures and bending angles. In contrast and elastomeric materials that is covalently or non-covalently crosslinked and above its T_g can be expected to be very flexible over a range of torsional, compressive, or tensile strains. Regardless, the ability of a material to self-heal is a significant advantage for flexible applications, since even if damage is sustained upon bending or flexing, this can be recovered by the self-healing nature of the material.

7.4 Incorporation of Electrically Conductive Components

Thus far, this chapter has focused on design principles for a flexible and self-healable polymer matrix. The ability for a material to be re-healed is quite an amazing feat, although polymer matrix materials tend to be electrical insulators with very poor electrical conductivity, often making the material alone a poor choice for flexible electronics that could be part of a flexible display [86, 87]. However, some electrically conductive materials exist that can re-heal, creating materials with the potential to enhance the lifetime of electronic devices. This electrical conductivity is typically introduced by adding an electrically conductive reinforcement, or nanoscale filler, or a conductive polymer [88–90]. Potential applications for self-healing conductive materials include coatings, adhesives, electronic skins, sensors, solar cells, and superconductors (Figure 7.4) [91]. However, there is not one self-healing or conductive mechanism that fulfills the demands made by every potential application. Following from a discussion on the general classes of self-healing, this section will focus primarily on the known conductive mediums in self-healing materials. As most polymers do not have a conjugated system (aside from conductive polymers, which we will discuss shortly) in which electrons may freely travel providing electrical conductivity, some form of conductive medium is usually required. Currently, the conductive mediums being explored are metals, carbon materials, conductive polymers, and ionic liquids [91]. Table 7.2 summarizes key self-healing conductive materials, specifying their composition, healing mechanism, electrical properties, healing conditions, and healing efficiency [91]. Sections 7.4.1– 7.4.4 highlight some key developments in each type of conductive component that can be incorporated into a polymer matrix, including a self-healing polymer material.

7.4.1 Metallic Conductors

The first types of conductive medium we will discuss are metallic-based conductors. One such type are Ag nanowires (AgNWs), which allow high conductivity in materials [92]. AgNWs can be incorporated into materials with different self-healing mechanisms such as Diels–Alder-based networks [93] and polyelectrolyte multilayer films [94]. Different materials provide different advantages and disadvantages. The Diels–Alder-based material was able to survive 1000 cycles of

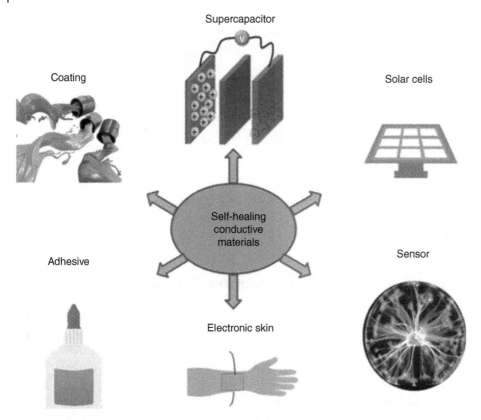

Figure 7.4 Applications of self-healing conductive materials. Reprinted by permission from [91] Springer Nature: Wiley Flexible Flat Panel Displays: 2nd Edition by Gregory P. Crawford, Darran R. Cairns, Dirk J. Broer (Eds) (2018).

bending–unbending testing with limited conductivity decrease; however, high heat is required to re-heal thereby limiting the applications of the material [91]. The polyelectrolyte system, on the other hand, can re-heal at room temperature in the presence of water, which as a hydrogel would be advantageous. Moving away from nanowires, metallic conductors can take the shape of micro particles as well. Networks can be designed to form hydrogen bonds with metallic particles such as nickel particles that then provide self-healing and conductivity mechanisms [95]. However, micro particles are not required to provide a self-healing mechanism, silver particles have been used as a conductive medium in materials employing an extrinsic microcapsule method of self-healing [96]. Microchannels can also be used as a channel full of a liquid metal, such as eutectic gallium indium alloy (EGaIn) [97]. When broken pieces are reconnected the metal merges together reforming the channel allowing electrical conductivity once again. Overall, metallic-based conductors function extremely well as a conductive medium for materials, however, they usually do not contribute to self-healing ability.

7.4.2 Conductive Polymers

While most organic polymers act as insulators, conductive polymers are unique – as their name suggests, they conduct electricity. This is due to the backbone of these materials being a conjugated

Table 7.2 A summary of typical electrical properties of conductive self-healing materials. Reprinted by permission from [91] Springer Nature: Wiley Flexible Flat Panel Displays: 2nd Edition by Gregory P. Crawford, Darran R. Cairns, Dirk J. Broer (Eds) (2018).

Self-healing material	Composition	Healing mechanism	Electrical properties[a]	Healing conditions	Electrical healing efficiency	References
Self-healing conductor	μNi particles/ supramolecular polymer	Hydrogen bonding	$40\ S\ cm^{-1}$	r.t., with pressure, 15 S	~90%	[30]
Self-healing conductor	AgNWs/D–A polymers	Diels–Alder reaction	$18.6\ \Omega\ sq^{-1}$	110 C, 1 h	97%	[37]
Self-healing conductor	AgNWs/ polyelectrolyte	Hydrogen bonding	$0.38\ \Omega\ sq^{-1}$	r.t., with droplet of water, 2 min	$0.42\ \Omega\ sq^{-1}$	[39]
Self-healing conductor	AgNWs/PCL/PVA	Low T_g	$0.25\ \Omega\ sq^{-1}$	65 C for 2 min	$0.6\ \Omega\ sq^{-1}$	[40]
Self-healing circuit wiring	EGaIn/PU	Liquid metals and self-handling PU	$10^{-3}\ \Omega\ sq^{-1}$	r.t., 1 min	Not specified	[46]
Self-healing conductive hybrid gel	PPy/G–Zn–tpy	Sol-gel transition and metal-ligand bond	$12\ S\ m^{-1}$	r.t., 1 min	$10\ S\ m^{-1}$	[50]
Self-healing conductive hydrogel	PPy/agarose-based hydrogel	Sol-gel transition	$1.95 \times 10^{-1}\ S\ cm^{-1}$	120 C, ~ 30 min	Not specified	[51]
Self-healing conductor	Graphite/bPEI	Hydrogen bonding	$1.98\ S\ cm^{-1}$	r.t. with pressure, 10 S	98%	[55]
Self-healing conductor	CNT/PBS	B–O bond	$1.21\ S\ cm^{-1}$	r.t. with pressure, 10 S	~98%	[56]
Self-healing conductor	PHEMA–CNT–(β-CD)	Host-guest interactions	$5.25 \pm 0.11\ S\ m^{-1}$	r.t., with 5 min	94.6%	[68]
Self-healing conductor	CNT/HPAMAM	Hydrogen bonding and disulfide bonding	$1.16\ K\Omega$	r.t.	~100%	[71]
Self-healing conductive hydrogel	rGO/SAP	Hydrogen bonding	$267\ \Omega\ m$	r.t., with pressure, 30 S	~100%	[75]
Self-healing ionic conductor	Supramolecular ionic polymer	Hydrogen bonding	$10^{-8}\ S\ cm^{-1}$	50 C, 2 h	Not specified	[80]
Self-healing electrode	CNT/TiO₂/ supramolecular polymer	Hydrogen bonding	$4.8\ \Omega$	r.t., with pressure	$9.8\ \Omega$ after 1st healing	[88]
Self-healing conductor	CNT/PBDMS	B–O bond	$2.1\ \Omega\ m$	r.t., with pressure	Not specified	[102]
Self-healing conductor	PDA–pGO–PAM	π–π interactions	$0.18\ S\ cm^{-1}$	r.t., 1 day	95%	[103]

(Continued)

Table 7.2 (Continued)

Self-healing material	Composition	Healing mechanism	Electrical properties[a]	Healing conditions	Electrical healing efficiency	References
Self-healing sensor	CNTs/chitosan/CNC	Hydrogen bonding	$0.59\ M\Omega$	r.t., 15 S	$1.42\ M\Omega$	[110]
Self-healing sensor	CNTs/PEI/CNC	Hydrogen bonding	$2.6 \times 10^{-3}\ S\ m^{-1}$	160 C, 10 min	Not specified	[114]

[a]Resistivity, resistance, conductivity.
r.t. room temperature.
Note[a] references refer to those of the original manuscript [91] and citations are given in footnote.
[a]References in Table 7.2.

[30] Bao et al. Nat Nanotechnol **2012** 7, pp 825–832
[37] Gong et al. (2012) Adv Mater **2012**, 25, pp 4186–4191
[39] Sun et al. Adv Mater **2012** 24, pp 4578–4582
[40] Sun et al. ACS Appl Mater Interfaces **2014**, 6, pp 16409–16415
[46] Dickey et al. Adv Mater **2013**, 25, pp 1589–1592
[50] Yu wt al. Nano Lett **2015**, 15, pp 6276–6281
[51] Park et al. ACS Nano **2014**, 8, pp10066–10076
[55] Wu and Chen J Mater Chem C **2016**, 4, pp 4150–4154
[56] Wu and Chen ACS Appl Mater Interfaces **2016**, 8, pp 24071–24078
[68] Zhang et al. Angew Chem Int Ed **2015**, 127, pp 12295–12301
[71] Hao et al. J Mater Chem A **2015**, 3, pp 12154–12158
[75] Tang et al. RSC Adv **2014**, 4, pp 35149–35155
[80] Mecerreyes et al. Macromol Rapid Commun **2012**, 33, pp 314–318
[88] Chen et al. Adv Mater **2014**, 26, pp 3638–3643
[102] Gong et al. J Mater Chem A **2015**, 3, pp 19790–19799
[103] Yuan et al. Small **2017**, 13, 1601916
[110] Tao et al. Angew Chem. Int. Ed **2017**, 56, pp 8795-8800
[114] Zhang et al. J Mater Chem A **2017**, 5, pp 9824–9832

system; there is a continuous system of p-orbitals allowing free movement of electrons in the system resulting in electrical conductivity. Some examples of conductive polymers include polyaniline (PANI), polypyrrole (PPy), and poly(3,4-ethylenedioxythiophene) [91]. However, one of the limitations of conductive polymers is that they tend to be brittle. A method to improve the mechanical qualities of these materials is to incorporate them as a composite with another material. An example of this is using PPy as a base material and introducing a supramolecular gel as a gluing agent that provides a reversible sol–gel transition [98]. This incorporation provides a mechanism for self-healing as well as decreased brittleness of the material. Other work on conductive polymers includes injectable hydrogels with self-healing based on the dynamic covalent schiff base between the amine group of chitosan-graft-aniline tetramer and benzaldehyde groups of dibenzaldehyde-terminated poly(ethleneglycol) [99]. Another focus is the addition of volatile solvents such as ethylene glycol or dimethyl sulfoxide to increase the flexibility of the material [91]. Without some method to increase flexibility of the conductive polymer, these materials are not especially suited for use in a flexible material. As the backbone carries the electric charge, any injury to the backbone such as bending a brittle material will result in a decrease in conductivity.

7.4.3 Carbon Materials

One of the more popular focuses in the development of electrically conductive self-healing materials are carbon-based materials such as carbon black, carbon nanotubes (CNTs), and graphene nanosheets. Incorporation of these materials provides increased mechanical strength and electrical conductivity [91]. CNTs can be oxidized and incorporated into hydrogel materials, which provides a mechanism for self-healing based on hydrogen bonding [100]. Similar to metals, graphene oxide can be synthesized into nanoparticles that provide electrical conductivity and a mechanism for self-healing based on hydrogen bonding with a polymer network [101]. Due to the rich range of conductive and semiconductive carbon-based materials, for instance, conductive or metallic carbon nanotubes, as well as semiconductive carbon nanotubes or graphene materials [102, 103], an almost limitless range of these materials targeted to many different applications can be envisaged.

7.4.4 Polymerized Ionic Liquids

Ionically conductive materials have been explored as the electrostatic interaction between cations and anions provides methods of self-healing and ionic conductivity [104]. These poly(ionic liquids) can also be combined with Au nanoparticles to increase the conductivity, self-healing, and mechanical properties of the material [105]. Supramolecular ionic compounds can by derived from carboxylic acids and alkyl amines as well [106]. These materials have the potential to be used as batteries while avoiding some drawbacks of liquid electrolyte batteries such as leakage and flammability [107]. However, a major disadvantage is that when they are crosslinked there is a decrease in conductivity. This issue can be remedied by incorporation of conductive nanoparticles.

7.5 Additional Possibilities Enabled by Three-dimensional Printing

As time and technology move forward, we constantly see new techniques develop in STEM fields. Arguably one of the most significant fields of advancement currently is three-dimensional (3D) printing, also known as additive manufacture. 3D printing allows macroscopic materials' complex architectures and structures to be synthesized. This is enabled through the range of novel 3D printing technologies continuously being developed [108].

The simple advantage of 3D printing is the ability to create macroscopic materials with complex structures and facilitate rapid prototyping [110, 111]. This chapter does not attempt to describe all 3D printing technologies, although we will highlight some elegant examples of 3D printing for self-healing materials [109, 112, 113], or 3D printing for flexible electronics [114]. In general there are several strategies for 3D printing: sequential deposition of a matrix material, which is also known as fused deposition modeling (FSM); stereolithography (SLA), where a photoactive material is 3D printed using masks; as well as other technologies such as powder bed inject printing where an inject head fuses powders together, and direct writing of a viscous liquid [110]. As highlighted in Figure 7.5, highly complex and intricate shapes can be generated using 3D printing. A clear advantage of 3D printing is that each of the structures in Figure 7.5 can be generated from one batch of material, with no need for a distinct mold or human operator to create the structures shown. In addition, due to the additive nature of 3D printing, in principle no material is wasted in 3D printing, unlike manufacture techniques that would sequentially remove material from a regular geometric object. 3D printing is extremely useful in extrinsic polymer systems in which self-healing is mediated through vascular systems filled with healing agents. Figure 7.6 demonstrates how complex the architecture of these materials can be. However, there are limitations to what 3D printing can achieve. Some limitations include shear thinning of material as it prints causing the material to move [113]. To prevent material from

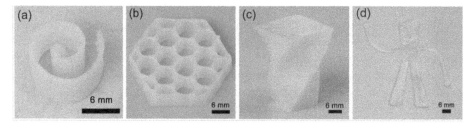

Figure 7.5 Demonstrates a variety of shapes 3D printing can create. Adapted with permission from [109]. Copyright 2018 American Chemical Society.

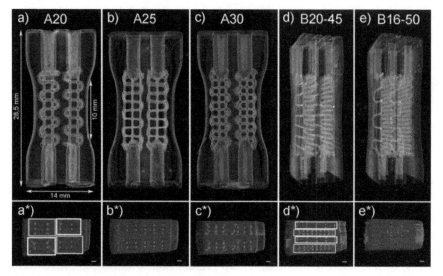

Figure 7.6 CT scans of dumbbell-shaped samples with empty microchannels. Reprinted with permission from [112]. Copyright 2017 American Chemical Society.

moving, it is possible to print into a sacrificial gelatin mold, which can be washed away. Not only can 3D printing create extrinsic healing materials it is also possible to synthesize intrinsic systems as well. Ferric ions (Fe^{3+}) have been incorporated into the material solution prior to printing providing both self-healing and conductive properties [113]. In this instance it is an example of an intrinsic system as self-healing occurs due to the polymer backbone consisting of polypyrrole-grafted chitosan with an interpenetrating network of polyacrylic acid (providing NH and carboxylic groups respectively) forming ionic interactions with Fe^{3+} [113].

In addition to the 3D printing of polymer matrix materials toward self-healing applications, there is the additional possibility of 3D printing electrically conductive components into a flexible material [115]. In particular it is possible to 3D print an electrically conductive ink to create an overall conductive composite material [116, 117]. Interestingly, work from Lewis et al. created 3D-printed flexible electronic devices by direct ink writing of both the conductive and insulating materials to be used in the device as well as the device components. The key advance of this technology was that the device components could be deposited onto a complex circuit, which goes beyond 3D printing to the insulating polymer matrix and the conductive channels [114]. With advances in 3D printing of conductive materials and 3D printing of self-healing materials it is anticipated that new powerful materials with self-healing and conductive properties will be 3D printed, including promising work from the Dickey group [118] and the electronic skin devices developed by Bao et al. [89].

7.6 Concluding Remarks

Self-healing materials have the ability to extend a product's useful lifetime and enable it to perform at a higher level. High-end applications are particularly well suited for self-healing materials, since the added cost of introducing self-healing components is best justified in a high-value product. A range of polymers can be made self-healing by incorporation of an appropriate dynamic bond that enables self-healing by repeated exchange of the dynamically associated groups. Alternatively, healing can be achieved through the catalyst promoted polymerization of liquids contained in vascular networks or capsules, which rupture upon damage to the material. Materials that contain capsules or vascular networks have good mechanical stability, although they are not necessarily best suited to high-strain applications such as those that specifically seek flexibility, as the strain could unintentionally damage the vascular network. Polymer matrix materials embedded with dynamic or exchangeable bonds as crosslinkers are well suited to this type of application. In general, the dynamic or exchangeable bond can be non-covalent (or supramolecular) or dynamic covalent in nature, and a range of commonly used chemical bonding motifs were highlighted. Care must be taken to choose an exchangeable bond that is sufficiently dynamic to enable self-healing, yet still maintain material stability. Typically, non-covalently associated groups lead to relatively rapid exchange of units, while dynamic covalent units lead to mechanical stability but often require external stimulus to activate the dynamic exchange of the covalent bonds. Polymers alone tend to have low electrical conductivity, although a variety of nanoscale reinforcements can be added to improve the electrical properties of the material. The polymer matrix can have very good mechanical properties, therefore conductive materials and polymer composite materials can have excellent electrical properties while maintaining self-healing from the polymer matrix. Additional possibilities enabled by additive manufacture or 3D printing enable a strong, robust, and adaptable framework for the next generation of advanced materials for a range of applications. With these design technologies in hand, it should be possible to create advanced polymer and polymer composite materials with self-healing properties for long-service lifetimes and excellent performance.

References

1 Gibson, R.F. (2010). A review of recent research on mechanics of multifunctional composite materials and structures. *Compos. Struct.* 92(12): 2793–2810.
2 Law, K.Y. (1993). Organic photoconductive materials: Recent trends and developments. *Chem. Rev.* 93 (1): 449–486.
3 Lau, A.K.-T. and Hui, D. (2002). The revolutionary creation of new advanced materials—carbon nanotube composites. *Compos. Part B: Eng.* 33 (4): 263–277.
4 Crawford, G.P. (2005). Flexible flat panel display technology. In *Flexible Flat Panel Displays*, 1st ed. (ed. G.P. Crawford). John Wiley & Sons.

5 Kumar, D. and Sharma, R.C. (1998). Advances in conductive polymers. *Eur. Polym. J.* 34(8): 1053–1060.

6 Inzelt, G., Pineri, M., Schultze, J.W., and Vorotyntsev, M.A. (2000). Electron and proton conducting polymers: recent developments and prospects. *Electrochim. Acta* 45(15): 2403–2421.

7 Park, S., Vosguerichian, M., and Bao, Z. (2013). A review of fabrication and applications of carbon nanotube film-based flexible electronics. *Nanoscale* 5(5): 1727–1752.

8 Brostow, W., Hagg Lobland, H.E., and Khoja, S. (2015). Brittleness and toughness of polymers and other materials. *Mater. Lett.* 159: 478–480.

9 Brostow, W. and Hagg Lobland, H.E. (2009). Brittleness of materials: implications for composites and a relation to impact strength. *J. Mater. Sci.* 4591: 242.

10 Pickering, S.J. (2006). ecycling technologies for thermoset composite materials—current status. *Compos. Part A Appl. Sci. Manuf.* 37(8): 1206–1215.

11 Hager, M.D., Greil, P., Leyens, C., Van Der Zwaag, S., and Schubert, U.S. (2010). Self-healing materials. *Adv. Mater.* 22(47): 5424–5430.

12 Toohey, K.S., Sottos, N.R., Lewis, J.A., Moore, J.S., and White, S.R. (2007). Self-healing materials with microvascular networks. *Nat. Mater.* 6(8): 581.

13 Bergman, S.D. and Wudl, F. (2008). Mendable polymers. *J. Mater. Chem.* 18(1): 41–62.

14 Chen, X., Dam, M.A., Ono, K., Mal, A., Shen, H., Nutt, S.R., Sheran, K., and Wudl, F. (2002). A thermally re-mendable cross-linked polymeric material. *Science* 295(5560): 1698–1702.

15 Dahlke, J., Zechel, S., Hager, M.D., and Schubert, U.S. (2018). How to design a self-healing polymer: General concepts of dynamic covalent bonds and their application for intrinsic healable materials. *Adv. Mater. Interf.* 0(0): 1800051.

16 White, S.R., Sottos, N.R., Geubelle, P.H., Moore, J.S., Kessler, M.R., Sriram, S.R., Brown, E.N., and Viswanathan, S. (2001). Autonomic healing of polymer composites. *Nature* 409: 794.

17 Wong, W.S. and Salleo, A. (2009). *Flexible Electronics: Materials and Applications*. Springer Science & Business Media.

18 Facchetti, A. (2010). π-Conjugated polymers for organic electronics and photovoltaic cell applications. *Chem. Mater.* 23(3): 733–758.

19 Yang, H., Leow Wan, R., and Chen, X. (2018). 3D printing of flexible electronic devices. *Small Methods* 2(1): 1700259.

20 Kuhl, N., Bode, S., Hager, M.D., and Schubert, U.S. (2016). Self-healing polymers based on reversible covalent bonds. In *Self-healing Materials* (eds. M.D. Hager, S. Van Der Zwaag, and U.S. Schubert), 1–58. Cham: Springer International Publishing.

21 Shchukin, D.G. (2013). Container-based multifunctional self-healing polymer coatings. *Polym. Chem.* 4(18): 4871–4877.

22 Toohey, K.S., Sottos, N.R., Lewis, J.A., Moore, J.S., and White, S.R. (2007). Self-healing materials with microvascular networks. *Nat. Mater.* 6: 581.

23 Hansen, C.J., Wu, W., Toohey, K.S., Sottos, N.R., White, S.R., and Lewis, J.A. (2009). Self-healing materials with interpenetrating microvascular networks. *Adv. Mater.* 21(41): 4143–4147.

24 Jin, H., Mangun, C.L., Stradley, D.S., Moore, J.S., Sottos, N.R., and White, S.R. (2012). Self-healing thermoset using encapsulated epoxy-amine healing chemistry. *Polymer* 53(2): 581–587.

25 Yang, Y. and Urban, M.W. (2013). Self-healing polymeric materials. *Chem. Soc. Rev.* 42(17): 7446–7467.

26 Imato, K. and Otsuka, H. (2017). Self-healing polymers through dynamic covalent chemistry. In *Dynamic Covalent Chemistry: Principles, Reactions, and Applications* (eds. W. Zhang and Y. Jin), chapter 9. John Wiley & Sons.

27 Roy, N., Bruchmann, B., and Lehn, J.-M. (2015). DYNAMERS: dynamic polymers as self-healing materials. *Chem. Soc. Rev.* 44(11): 3786–3807.

28 Lehn, J.-M. (2005). Dynamers: dynamic molecular and supramolecular polymers. *Prog. Polym. Sci.* 30(8): 814–831.

29 García, F. and Smulders, M.M.J. (2016). Dynamic covalent polymer. *J. Polym. Sci. A Polym. Chem.* 54(22): 3551–3577.

30 Denissen, W., Winne, J.M., and Du Prez, F.E. (2016). Vitrimers: permanent organic networks with glass-like fluidity. *Chem. Sci.* 7(1): 30–38.

31 Mozhdehi, D., Ayala, S., Cromwell, O.R., and Guan, Z. (2014). Self-healing multiphase polymers via dynamic metal–ligand interactions. *J. Am. Chem. Soc.* 136(46): 16128–16131.

32 Zhang, B., Digby, Z.A., Flum, J.A., Foster, E.M., Sparks, J.L., and Konkolewicz, D. (2015). Self-healing, malleable and creep limiting materials using both supramolecular and reversible covalent linkages. Polym. Chem. 6(42): 7368–7372.

33 Cash, J.J., Kubo, T., Dobbins, D.J., and Sumerlin, B.S. (2018). Maximizing the symbiosis of static and dynamic bonds in self-healing boronic ester networks. *Polymer Chemistry*. 15: doi:10.1039/C8PY00123E.

34 Wojtecki, R.J., Meador, M.A., and Rowan, S.J. (2010). Using the dynamic bond to access macroscopically responsive structurally dynamic polymers. *Nat. Mater.* 10: 14.

35 Enke, M., Döhler, D., Bode, S., Binder, W.H., Hager, M.D., and Schubert, U.S. (2016) Intrinsic self-healing polymers based on supramolecular interactions: State of the art and future direction. In *Self-healing Materials* (eds. M.D. Hager, S. Van Der Zwaag, and U.S. Schubert), 59–112. Cham: Springer International Publishing.

36 Herbst, F., Döhler, D., Michael, P., and Binder Wolfgang, H. (2013). Self-healing polymers via supramolecular forces. *Macromol. Rapid Commun.* 34(3): 203–220.

37 Aida, T., Meijer, E.W., and Stupp, S.I. (2012). Functional supramolecular polymers. *Science* 335(6070): 813.

38 Yan, X., Wang, F., Zheng, B., and Huang, F. (2012). Stimuli-responsive supramolecular polymeric materials. *Chem. Soc. Rev.* 41(18): 6042–6065.

39 Campanella, A., Döhler, D., and Binder, W.H. (2018). Self-healing in supramolecular polymers. *Macromol. Rapid Commun.* 0(0): 1700739.

40 Cordier, P., Tournilhac, F., Soulié-Ziakovic, C., and Leibler, L. (2008). Self-healing and thermoreversible rubber from supramolecular assembly. *Nature* 451: 977.

41 Cui, J. and Campo, A.D. (2012). Multivalent H-bonds for self-healing hydrogels. *Chem. Commun.* 48(74): 9302–9304.

42 Söntjens, S.H.M., Sijbesma, R.P., Van Genderen, M.H.P., and Meijer, E.W. (2000). Stability and lifetime of quadruply hydrogen bonded 2-ureido-4[1H]-pyrimidinone dimers. *J. Am. Chem. Soc.* 122(31): 7487–7493.

43 Oya, N., Ikezaki, T., and Yoshie, N. (2013). A crystalline supramolecular polymer with self-healing capability at room temperature. *Polym. J.* 45: 955.

44 Cho, S.H., Andersson, H.M., White, S.R., Sottos, N.R., and Braun, P.V. (2006).Polydimethylsiloxane-based self-healing materials. *Adv. Mater.* 18(8): 997–1000.

45 Montarnal, D., Tournilhac, F., Hidalgo, M., Couturier, J.-L., and Leibler, L. (2009). Versatile one-pot synthesis of supramolecular plastics and self-healing rubbers. *J. Am. Chem. Soc.* 131(23): 7966–7967.

46 Foster, E.M., Lensmeyer, E.E., Zhang, B., Chakma, P., Flum, J.A., Via, J.J., Sparks, J.L., and Konkolewicz, D. (2017). Effect of polymer network architecture, enhancing soft materials using orthogonal dynamic bonds in an interpenetrating network. *ACS Macro Lett.* 6(5): 495–499.

47 Burattini, S., Colquhoun, H.M., Greenland, B.W., and Hayes, W. (2009). A novel self-healing supramolecular polymer system. *Faraday Discuss.* 143(0): 251–264.

48 Burattini, S., Greenland, B.W., Merino, D.H., Weng, W., Seppala, J., Colquhoun, H.M., Hayes, W., Mackay, M.E., Hamley, I.W., and Rowan, S.J. (2010). A healable supramolecular polymer blend based on aromatic $\pi-\pi$ stacking and hydrogen-bonding interactions. *J. Am. Chem. Soc.* 132(34): 12051–12058.

49 Hohlbein, N., Von Tapavicza, M., Nellesen, A., and Schmidt Annette, M. (2013). Self-healing Ionomers. In *Self-Healing Polymers* (ed. W.H. Binder), chapter 13. Wiley-VCH.

50 Zwaag, S. (2008). *Self Healing Materials: An Alternative Approach to 20 Centuries of Materials Science*. Springer Science+ Business Media BV.

51 Tadano, K., Hirasawa, E., Yamamoto, H., and Yano, S. (1989). Order-disorder transition of ionic clusters in ionomers. *Macromolecules* 22(1): 226–233.

52 Kloxin, C.J. and Bowman, C.N. (2013). Covalent adaptable networks: smart, reconfigurable and responsive network systems. *Chem. Soc. Rev.* 42(17): 7161–7173.

53 Chakma, P., Rodrigues Possarle, L.H., Digby, Z.A., Zhang, B., Sparks, J.L., and Konkolewicz, D. (2017). Dual stimuli responsive self-healing and malleable materials based on dynamic thiol-Michael chemistry. *Polym. Chem.* 8(42): 6534–6543.

54 Liu, Y.-L. and Chuo, T.-W. (2013). Self-healing polymers based on thermally reversible Diels-Alder chemistry. *Polym. Chem.* 4(7): 2194–2205.

55 Reutenauer, P., Buhler, E., Boul, P.J., Candau, S.J., and Lehn, J.M. (2009). Room temperature dynamic polymers based on diels–alder chemistry. *Chem. Eur. J.* 15(8): 1893–1900.

56 Habault, D., Zhang, H., and Zhao, Y. (2013). Light-triggered self-healing and shape-memory polymers. *Chem. Soc. Rev.* 42(17): 7244–7256.

57 Chung, C.-M., Roh, Y.-S., Cho, S.-Y., and Kim, J.-G. (2004). Crack healing in polymeric materials via photochemical [2+2] cycloaddition. *Chem. Mater.* 16(21): 3982–3984.

58 Froimowicz, P., Frey, H., and Landfester, K. (2011). Towards the generation of self-healing materials by means of a reversible photo-induced approach. *Macromol. Rapid Commun.* 32(5): 468–473.

59 Canadell, J., Goossens, H., and Klumperman, B. (2011). Self-healing materials based on disulfide links. *Macromolecules* 44(8): 2536–2541.

60 Pepels, M., Filot, I., Klumperman, B., and Goossens, H. (2013). Self-healing systems based on disulfide-thiol exchange reactions. *Polym. Chem.* 4(18): 4955–4965.

61 Yoon, J.A., Kamada, J., Koynov, K., Mohin, J., Nicolaÿ, R., Zhang, Y., Balazs, A.C., Kowalewski, T., and Matyjaszewski, K. (2012). Self-healing polymer films based on thiol–disulfide exchange reactions and self-healing kinetics measured using atomic force microscopy. *Macromolecules* 45(1): 142–149.

62 Tsarevsky, N.V. and Matyjaszewski, K. (2002). Reversible redox cleavage/coupling of polystyrene with disulfide or thiol groups prepared by atom transfer radical polymerization. *Macromolecules* 35(24): 9009–9014.

63 Amamoto, Y., Otsuka, H., Takahara, A., and Matyjaszewski, K. (2012). Self-healing of covalently cross-linked polymers by reshuffling thiuram disulfide moieties in air under visible light. *Adv. Mater.* 24(29): 3975–3980.

64 Amamoto, Y., Kamada, J., Otsuka, H., Takahara, A., and Matyjaszewski, K. (2011). Repeatable photoinduced self-healing of covalently cross-linked polymers through reshuffling of trithiocarbonate units. *Angewandte Chemie International Edition* 50(7): 1660–1663.

65 Rekondo, A., Martin, R., De Luzuriaga, A.R., Cabanero, G., Grande, H.J., and Odriozola, I. (2014). Catalyst-free room-temperature self-healing elastomers based on aromatic disulfide metathesis. *Materials Horizons* 1(2): 237–240.

66 Deng, G., Li, F., Yu, H., Liu, F., Liu, C., Sun, W., Jiang, H., and Chen, Y. (2012). Dynamic hydrogels with an environmental adaptive self-healing ability and dual responsive sol–gel transitions. *ACS Macro Lett.* 1: 275.

67 Deng, C.C., Brooks, W.L.A., Abboud, K.A., and Sumerlin, B.S. (2015). Boronic acid-based hydrogels undergo self-healing at neutral and acidic pH. *ACS Macro Lett.* 4(2): 220–224.

68 Kuhl, N., Bode, S., Bose, R.K., Vitz, J., Seifert, A., Hoeppener, S., Garcia, S.J., Spange, S., Van Der Zwaag, S., Hager, M.D., and Schubert, U.S. (2015). Acylhydrazones as reversible covalent crosslinkers for self-healing polymers. *Adv. Funct. Mater.* 25(22): 3295–3301.

69 Nishiyabu, R., Kubo, Y., James, T.D., and Fossey, J.S. (2011). Boronic acid building blocks: tools for self assembly. *Chem. Commun.* 47(4): 1124–1150.

70 Cash, J.J., Kubo, T., Bapat, A.P., and Sumerlin, B.S. (2015). Room-Temperature Self-Healing Polymers Based on Dynamic-Covalent Boronic Esters. *Macromolecules* 48(7): 2098–2106.

71 Brooks, W.L.A. and Sumerlin, B.S. (2016). Synthesis and applications of boronic acid-containing polymers: From materials to medicine. *Chem. Rev.* 116(3): 1375–1397.

72 Cromwell, O.R., Chung, J., and Guan, Z. (2015). Malleable and self-healing covalent polymer networks through tunable dynamic boronic ester bonds. *J. Am. Chem. Soc.* 137(20): 6492–6495.

73 Zhu, X.-K. and Joyce, J.A. (2012). Review of fracture toughness (G, K, J, CTOD, CTOA) testing and standardization. *Eng. Fract. Mech.* 85: 1–46.

74 Zhao, X. (2014). Multi-scale multi-mechanism design of tough hydrogels: building dissipation into stretchy networks. *Soft Matter* 10(5): 672–687.

75 Galimberti, M., Caprio, M., and Fino, L. (2005). Tyre comprising a cycloolefin polymer tread band and elastomeric composition used therein. Google Patents.

76 Ibeh, C.C. (2011). *Thermoplastic Materials: Properties, Manufacturing Methods, and Applications.* CRC Press.

77 Andreozzi, L., Castelvetro, V., Faetti, M., Giordano, M., and Zulli, F. (2006). Rheological and thermal properties of narrow distribution poly(ethyl acrylate)s. *Macromolecules* 39(5): 1880–1889.

78 Demirel, B., Yaraş, A., and Elçiçek, H. (2016). Crystallization behavior of PET materials. *Balikesir Üniversitesi Fen Bilimleri Enstitüsü Dergisi* 13(1): 26–35.

79 Wilkes, C.E., Summers, J., Daniels, C., and Berard, M. (2005). *PVC Handbook 2005.* Hanser, München, 379–384.

80 Fox, T.G. and Flory, P.J. (2003). The glass temperature and related properties of polystyrene. Influence of molecular weight. *J. Polym. Sci.* 14(75): 315–319.

81 Nicholson, J. (2017). *The Chemistry of Polymers.* Royal Society of Chemistry.

82 Gibbs, J.H. and DiMarzio, E.A. (1958). Nature of the glass transition and the glassy state. *J. Chem. Phys.* 28(3): 373–383.

83 Lesser, A.J. and Crawford, E. (1998). The role of network architecture on the glass transition temperature of epoxy resins. *J. Appl. Polym. Sci.* 66(2): 387–395.

84 Zhang, B., Digby, Z.A., Flum, J.A., Chakma, P., Saul, J.M., Sparks, J.L., and Konkolewicz, D. (2016). Dynamic Thiol–Michael Chemistry for Thermoresponsive Rehealable and Malleable Networks. *Macromolecules* 49(18): 6871–6878.

85 Garcia, S.J. (2014). Effect of polymer architecture on the intrinsic self-healing character of polymers. *Eur. Polym. J.* 53: 118–125.

86 Spitalsky, Z., Tasis, D., Papagelis, K., and Galiotis, C. (2010). Carbon nanotube–polymer composites: Chemistry, *processing, mechanical and electrical properties. Prog. Polym. Sci.* 35(3): 357–401.

87 Winey, K.I., Kashiwagi, T., and Mu, M. (2007). Improving electrical conductivity and thermal properties of polymers by the addition of carbon nanotubes as fillers. *Mrs Bulletin* 32(4): 348–353.

88 Williams, K.A., Boydston, A.J., and Bielawski, C.W. (2007). Towards electrically conductive, *self-healing materials. J. Royal Soc. Interf.* 4(13): 359–362.

89 Benight, S.J., Wang, C., Tok, J.B., and Bao, Z. (2013). Stretchable and self-healing polymers and devices for electronic skin. *Prog. Polym. Sci.* 38(12): 1961–1977.

90 Siddique Jamal, A., Ahmad, A., and Mohd, A. (2018). Self Healing Materials and Conductivity. In *Electrically Conductive Polymer and Polymer Composites* (eds. A. Khan, M. Jawaid, A.A.P. Khan, and A.M. Asiri), chapter 18. Wiley-VCH.

91 Zhang, Q., Liu, L., Pan, C., and Li, D. (2018). Review of recent achievements in self-healing conductive materials and their applications. *J. Mater. Sci.* 53(1): 27–46.

92 Xu, F. and Zhu, Y. (2012). Highly conductive and stretchable silver nanowire conductors. *Adv. Mater.* 24(37): 5117–5122.

93 Gong, C., Liang, J., Hu, W., Niu, X., Ma, S., Hahn, H.T., and Pei, Q. (2013). A healable, *semitransparent silver nanowire-polymer composite conductor. Adv. Mater.* 25(30): 4186–4191.

94 Li, Y., Chen, S., Wu, M., and Sun, J. (2012). Polyelectrolyte multilayers impart healability to highly electrically conductive films. *Adv. Mater.* 24(33): 4578–4582.

95 Tee, B.C.K., Wang, C., Allen, R., and Bao, Z. (2012). An electrically and mechanically self-healing composite with pressure- and flexion-sensitive properties for electronic skin applications. *Nat. Nanotechnol.* 7: 825.

96 Odom, S.A., Chayanupatkul, S., B.B.J., Zhao, O., Jackson, A.C., Braun, P.V., Sottos, N.R., White, S.R., and Moore, J.S. (2012). A self-healing conductive ink. *Adv. Mater.* 24(19): 2578–2581.

97 Chiech, I.R.C., Weiss, E.A., Dickey, M.D., and Whitesides, G.M. (2007). Eutectic gallium–indium (egain): a moldable liquid metal for electrical characterization of self-assembled monolayers. *Angewandte Chemie International Edition* 47(1): 142–144.

98 Shi, Y., Wang, M., Ma, C., Wang, Y., Li, X., and Yu, G. (2015). A conductive self-healing hybrid gel enabled by metal–ligand supramolecule and nanostructured conductive polymer. *Nano Lett.* 15(9): 6276–6281.

99 Dong, R., Zhao, X., Guo, B., and Ma, P.X. (2016). Self-healing conductive injectable hydrogels with antibacterial activity as cell delivery carrier for cardiac cell therapy. *ACS Appl. Mater. Interf.* 8: 17138.

100 Du, R., Wu, J., Chen, L., Huang, H., Zhang, X., and Zhang, J. (2013). Hierarchical hydrogen bonds directed multi-functional carbon nanotube-based supramolecular hydrogels. *Small* 10(7): 1387–1393.

101 Annabi, N., Shin, S.R., Tamayol, A., Miscuglio, M., Bakooshli, M.A., Assmann, A., Mostafalu, P., Sun, J.Y., Mithieux, S., Cheung, L., Tang, X., Weiss, A.S., and Khademhosseini, A. (2016). Highly elastic and conductive human-based protein hybrid hydrogels. *Adv. Mater.* 28(1): 40–49.

102 Ebbesen, T., Lezec, H., Hiura, H., Bennett, J., Ghaemi, H., and Thio, T. (1996). Electrical conductivity of individual carbon nanotubes. *Nature* 382(6586): 54.

103 Stankovich, S., Dikin, D.A., Dommett, G.H., Kohlhaas, K.M., Zimney, E.J., Stach, E.A., Piner, R.D., Nguyen, S.T., and Ruoff, R.S. (2006). Graphene-based composite materials. *Nature* 442(7100): 282.

104 Aboudzadeh, M.A., Zhu, H., Pozo-Gonzalo, C., Shaplov, A.S., Mecerreyes, D., and Forsyth, M. (2015). Ionic conductivity and molecular dynamic behavior in supramolecular ionic networks; the effect of lithium salt addition. *Electrochim. Acta* 175: 74–79.

105 He, X., Zhang, C., Wang, M., Zhang, Y., Liu, L., and Yang, W. (2017). An Electrically and Mechanically Autonomic Self-healing Hybrid Hydrogel with Tough and Thermoplastic Properties. *ACS Appl. Mater. Interf.* 9(12): 11134–11143.

106 Aboudzadeh, M.A., Muñoz, M.E., Santamaría, A., Marcilla, R., and Mecerreyes, D. (2012). Facile synthesis of supramolecular ionic polymers that combine unique rheological, *ionic conductivity, and self-healing properties. Macromol. Rapid Commun.* 33(4): 314–318.

107 Yuan, J. and Antonietti, M. (2011). Poly(ionic liquid)s: Polymers expanding classical property profiles. *Polymer* 52(7): 1469–1482.

108 Khoo, Z.X., Teoh, J.E.M., Liu, Y., Chua, C.K., Yang, S., An, J., Leong, K.F., and Yeong, W.Y. (2015). 3D printing of smart materials: A review on recent progresses in 4D printing. *Virtual Phys. Prototyp.* 10(3): 103–122.

109 Kuang, X., Chen, K., Dunn, C.K., Wu, J., Li, V.C.F., and Qi, H.J. (2018). 3D printing of highly stretchable, shape-memory, and self-healing elastomer toward novel 4d printing. *ACS Appl. Mater. Interf.* 10(8): 7381–7388.

110 Wang, X., Jiang, M., Zhou, Z., Gou, J., and Hui, D. (2017). 3D printing of polymer matrix composites: A review and prospective. *Compos. Part B: Eng.* 110: 442–458.

111 Schubert, C., Van Langeveld, M.C., and Donoso, L.A. (2014). Innovations in 3D printing: a 3D overview from optics to organs. *Br. J. Ophthalmol.* 98(2): 159–161.

112 Postiglione, G., Alberini, M., Leigh, S., Levi, M., and Turri, S. (2017). Effect of 3D-printed microvascular network design on the self-healing behavior of cross-linked polymers. *ACS Appl. Mater. Interf.* 9(16): 14371–14378.

113 Darabi, M.A., Khosrozadeh, A., Mbeleck, R., Liu, Y., Chang, Q., Jiang, J., Cai, J., Wang, Q., Luo, G., and Xing, M. (2017). Skin-inspired multifunctional autonomic-intrinsic conductive self-healing hydrogels with pressure sensitivity, stretchability, and 3d printability. *Adv. Mater.* 29(13): 1700533.

114 Valentine, A.D., Busbee, T.A., Boley, J.W., Raney, J.R., Chortos, A., Kotikian, A., Berrigan, J.D., Durstock, M.F., and Lewis, J.A. (2017). Hybrid 3D printing of soft electronics. *Adv. Mater.* 29(40): 1703817.

115 MacDonald, E. and Wicker, R. (2016). Multiprocess 3D printing for increasing component functionality. *Science.* 353(6307). doi:10.1126/science.aaf2093.

116 Adams, J.J., Duoss, E.B., Malkowski, T.F., Motala, M.J., Ahn, B.Y., Nuzzo, R.G., Bernhard, J.T., and Lewis, J.A. (2011). Conformal printing of electrically small antennas on three-dimensional surfaces. *Adv. Mater.* 23(11): 1335–1340.

117 Ahn, B.Y., Duoss, E.B., Motala, M.J., Guo, X., Park, S.-I., Xiong, Y., Yoon, J., Nuzzo, R.G., Rogers, J.A., and Lewis, J.A. (2009). Omnidirectional printing of flexible, stretchable, and spanning silver microelectrodes. *Science.* 323(5921): 1590–1593.

118 Palleau, E., Reece, S., Desai, S.C., Smith, M.E., and Dickey, M.D. (2013). Self-healing stretchable wires for reconfigurable circuit wiring and 3d microfluidics. *Adv. Mater.* 25(11): 1589–1592.

8

Flexible Glass Substrates

Armin Plichta[1], Andreas Habeck[1], Silke Knoche[1], Anke Kruse[1], Andreas Weber[1], and Norbert Hildebrand[2]

[1] *Schott AG*
[2] *Schott North America Inc.*

8.1 Introduction

Since the development of display technology, glass has played an important role. From the early days of cathode-ray tube (CRT) bulbs to today's special thin-film transistor (TFT) liquid crystal display (LCD) and plasma display panel (PDP) glass types, glass development has gone hand in hand with display technology advances.

Today, in order to be competitive in the traditional CRT market, LCD technology has improved by processing glass substrates with a size of roughly 4 m^2 (Gen 7). A new trend for the smaller displays is under discussion: How can displays be made thinner to accommodate a new product characteristic such as flexibility? Despite the fact that flexibility is not defined, four different grades can be distinguished according to application: (a) ultra-thin and flat displays (mobiles, PDAs, laptops), (b) ultra-thin and curved displays (mobiles, automotive), (c) ultra-thin and bendable displays (smartcards), and last but not least (d) ultra-thin and "highly" flexible displays (e.g. rollable or wearable displays). Since glass substrates are the standard substrate material at the moment, flexibility can only be achieved by reducing the substrate thickness significantly. This chapter will discuss in detail the pros and cons of the next generation of glass substrates in the thickness range of 0.2 mm–30 μm.

8.2 Display Glass Properties

8.2.1 Overview of Display Glass Types

With the multitude of material and product properties required by the different display technologies there exists not one but several display glass types.

Glass describes a material in a "frozen liquid" state [1]. From a practical viewpoint, soda-lime glass for hollow glass and for window glass applications is produced in the highest volume today. Soda-lime glass has a high content of SiO_2 (70–75%) (network builder) and lower amounts of Na_2O (12–16%) and CaO (10–15%) (network modifier). Mixed in the right proportions during melting, the mixture will not crystallize during cooling and will form a new material – glass. Adding B_2O_3 or Al_2O_3 to the SiO_2 network creates new glass types, the borosilicates or aluminosilicates, or if both oxides are present the aluminoborosilicates.

Flexible Flat Panel Displays, Second Edition. Edited by Darran R. Cairns, Dirk J. Broer, and Gregory P. Crawford.
© 2023 John Wiley & Sons Ltd. Published 2023 by John Wiley & Sons Ltd.

Within these glass types the composition of the network builders and the modifiers may still be varied, resulting in a large variety of glass types with different material properties. Most glass types used for flat panel displays are in the borosilicate and aluminosilicate glass families.

Plastic substrates are used today in various applications, but glass is the favorite choice for high-tech applications. Glass has many advantages over plastic:

- Optical properties:
  - high transmission rate in the visual range
  - high homogeneity of the refractive index
  - no retardation
  - high ultra-violet (UV) resistance.
- Thermal properties:
  - high temperature stability
  - high dimensional stability
  - low thermal expansion coefficient
  - no outgassing.
- Chemical properties:
  - o high resistance to chemical attack
  - o excellent barrier properties against water and oxygen.
- Mechanical properties:
  - o high scratch resistance
  - o high scrub resistance.

8.2.2 Glass Properties

From our day-to-day experiences we are familiar with some glass properties that make this material so important for display technologies. We can easily look through it, and it is strong and passive enough to encapsulate the more sensitive components of a display. When we look in more detail, we see that glass exhibits many different properties that make it the material of choice for all types of display technology today.

8.2.2.1 Optical Properties

Glass has been used in optical components for centuries. Clear glass types show a very high internal transmittance, typically over 98% for a 1-mm thick glass sheet in the visible part of the electromagnetic spectrum. With a refractive index of ~1.5, a luminous transmittance of > 90% ($t = 1$ mm) is typical for all display glass types.

Figure 8.1 shows the optical transmission for D263. For some special applications, glass with optical filter properties is used. A typical example would be the use of neutral optical density or color enhancement filters for improving daylight readability.

Another important optical glass property is the very low stress birefringence for glass. Even very thin glass sheets (down to 30 μm thickness) show these very low values, making them the ideal substrate material for LCD technologies.

8.2.2.2 Chemical Properties

During the display manufacturing process, many different chemicals are used to form specific display components. Resistance to any form of chemical attack is important for any display sub-

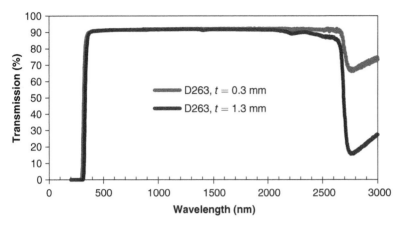

Figure 8.1 Transmission of D263 in the visible and near-IR spectrum.

strate material. Silicates, especially borosilicates and aluminoborosilicates, have a very high resistance to chemicals.

The chemical durability of glass as it concerns various chemicals is measured in material loss per exposed area, typically expressed in milligrams per square centimeter (mg/cm^2). The chemical attack is a function of temperature, time, and concentration of the respective chemical. For glass, three types of chemical attack are generally considered. Acid, base, and water attack describe the reaction of a glass type to aqueous solutions with different pH values.

As an example, Figure 8.2 shows the chemical durability of a standard borosilicate glass (Borofloat) against various chemicals. Although glass is not completely inert to chemical attack, typically it does not cause any significant degradation of the substrate material. Based on an SiO_2 network, glass typically shows the lowest chemical durability against chemicals containing fluorine followed by strong base chemicals; different glass types will react differently.

8.2.2.3 Thermal Properties

During the manufacturing process, glass is brought into the liquid state (melting) and subsequently cooled to room temperature. Since the glass does not solidify like other materials, it does not show a long-range molecular order. Scientifically, glass is typically not in its thermodynamically lowest energy state at room temperature, consequently, some material properties are not only a function of the glass composition but also a function of the cooldown speed. The slower the glass is annealed, the denser or more compact it becomes. Further processing steps in the manufacture of displays may require elevated temperatures. During such a temperature cycle, glass may show a further dimensional change – thermal shrinkage. The amount of thermal shrinkage depends on temperature, time, and the glass type.

The control of thermal shrinkage is very important for thin display technologies. If the thermal shrinkage is too large, processing of multiple mask steps in photolithographic processes may become difficult (misalignment of registration marks), leading to lower yield. The same is true if the front and the back substrate show any differences in thermal shrinkage. For example, AF37 (Schott) shows shrinkage below 10 ppm.

Another important and more commonly known thermal property is the coefficient of thermal expansion (CTE). As a solid, glass exhibits a linear thermal expansion up to a temperature around

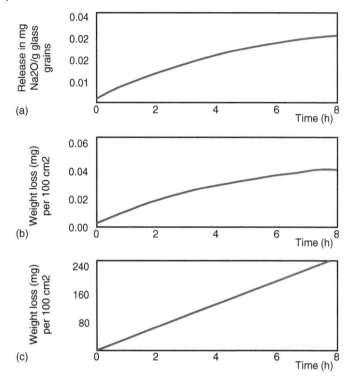

Figure 8.2 Chemical durability of Borofloat against (a) water attack, (b) acid attack, and (c) alkali attack.

the transformation temperature T_g. Looking at the length change ($\Delta\,l/l$) of a piece of glass, one can see the change in behavior around T_g in Figure 8.3. For most glass types used today in the flat panel display industry, a linear behavior can be assumed from $-50°C$ to $> 450°C$.

Some display technologies require high-temperature steps. This is pushing the development of high-strain-point glass types with a $T_g > 600°C$. The CTE value depends on the glass composition. Most borosilicate and aluminoborosilicate glass types used in the flat panel industry have a CTE of approximately $4 \times 10^{-6}\,K^{-1}$ (AF45 Schott, CTE $= 4.5 \times 10^{-6}\,K^{-1}$), which allows a better CTE match to Si layers used in active-matrix display technologies. "TFT glasses" such as AF37(Schott), 1737 (Corning), and AN100 (Asahi) show similar CTE values of approximately $3.7 \times 10^{-6}\,K^{-1}$. Display technologies not requiring active-matrix backplanes use glass types with higher CTE values $> 7 \times 10^{-6}\,K^{-1}$.

8.2.2.4 Surface Properties

Based on the composition, structure, and manufacturing process, glass has a very smooth surface, which is created by surface tension during solidification. The silicate structure gives glass a hard surface that can be polished to a very low surface roughness if necessary (e.g. glasses produced by float technology must be polished). Typical roughness values for display substrates are $R(RMS) \leq 1$ nm.

Depending on the production process, the glass substrate may exhibit other longer-range surface deformations such as warp, waviness, or thickness deviation. These features differ with the various manufacturing processes; for the downdraw (DD) process some features exhibit unidirectional structures, whereas for the float process the structures are bidirectional.

Figure 8.4 shows the contribution of the single quality features to the real appearance of a display glass substrate.

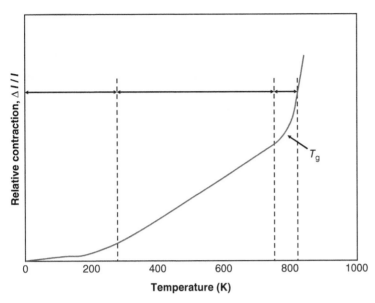

Figure 8.3 Length change of glass as a function of temperature.

Based on a silicate structure, glass surfaces have a high tendency to absorb water molecules. As this process starts directly after manufacturing, the glass surface changes over time and this surface change may influence thin-film deposition processes at a later stage.

8.2.2.5 Permeability

Water and oxygen easily attack all kinds of organic materials, leading to chemical changes within a display thus limiting its lifetime. However, glass has a very low permeability to gases and moisture – an important factor for all display technology. Very low permeation is inherent in the silicate structure, which is not only very dense but also forms very strong bonds to oxygen and water. Permeation values for oxygen and water through glass sheets are therefore so low that they are below the detection limit and, typically, it is assumed that glass sheets do not show any oxygen or water permeation.

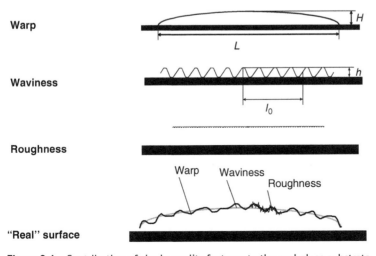

Figure 8.4 Contribution of single quality features to the real glass substrate.

8.3 Manufacturing of Thin "Flexible" Glass

A clear definition of "thin" does not exist, however, due to the progress of TFT LCD technology, which uses 0.7 mm-"thick" glass, some people are using this thickness as the upper limit for thin glass. The lower limit is defined by the so-called microsheet with a thickness of 50 μm [2].

In principle, glass becomes what is commonly understood as "flexible" below a thickness of approximately 200 μm [3]. The production of thin glass for the flat panel display modules has to keep pace with the increase in size and quality levels, therefore several process developments and upscaling procedures are ongoing. However, only two production principles have survived and are in the frame: float technology and DD technology.

Float technology can produce glass substrates with a thickness down to 0.4 mm but DD technology offers the potential to produce glass down to 50 μm, or even 30 μm, which is the technical limit today. Both production technologies will be described in section 8.3.1.

8.3.1 Float and Downdraw Technology for Special Glass

Figure 8.5 shows the principal flow of a glass production facility. The common processes of all glass productions are: mixing of raw materials (I), melting (II), refining (III), and homogenization (IV) of the glass melt; and annealing (VI) and cutting (VII) of the glass sheet. For float technology (Va), the hot forming is performed horizontally on a liquid tin bath, whereas for DD technology, the hot forming takes place in a vertical direction (Vb). In general, float technology offers throughputs of over 500 tons/day, whereas current draw technologies are limited to approximately 8 tons/day.

After hot forming, annealing (VI) is necessary to cool down the continuous glass ribbon and to relax the mechanical stress induced during hot forming. The refining step (III) reduces the bubble content of the glass melt. Depending on the glass type, various refining agents such as antimony oxide, tin oxide, or alkali halogenide can be used. Homogenization is performed with a stirrer. All tubes and hot-forming equipment for molten glass after the refining are made of precious metal alloys.

The DD processes for special glasses are developed mainly at Corning and Schott although they are also used by other glass manufactures around the globe. Corning uses the so-called over-

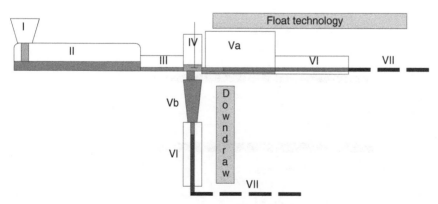

Figure 8.5 Basic process steps of modern glass production: I = mixing raw materials, II = melting, III = refining, IV = homogenization, Va = float technology, Vb = downdraw (DD) technology, VI = annealing, VII = cutting.

flow fusion process [4]. Schott uses a special nozzle slot design for the production of thin and ultra-thin glasses with high-quality surface features, as well as a float technology for the production of large-size standard TFT glasses.

Production of ultra-thin glass of thickness 100 μm and below is now a reality. Every year several hundred tons are produced by the DD technology for use as microscopy cover slides, glass touch panels, and for wafer-level chip size packaging.

8.3.2 Limits

Display glass substrates must fulfill a multitude of requirements. For LCDs the surface quality is decisive to ensure exact control of LCD geometry. The quality features of the end product can be controlled from the glass production processes. Table 8.1 summarizes the relation of the glass production processes to important quality parameters required for display technology.

8.3.2.1 Thickness Limits for Production

The float technology is limited for thin-glass production in terms of thickness and surface quality. The thinnest glass currently available is 0.4 mm thick, whereas the DD technology limit for production is currently at 30–50 μm. This shows that only the DD technology is in a position to follow the trend to thinner and even flexible displays. Figure 8.6 gives an impression of the state of the art, in glass production using DD technology. Figure 8.7 shows the thickness data for a thin glass sheet, 50-μm thick, produced by DD technology over a net width of > 600 mm.

Table 8.1 Relation of production process and corresponding quality features.

Process	Quality features	General requirements
Raw materials	Glass composition including optical, thermal, and electrical properties	Alkaline content < 0.1% for TFT displays Optical transmission visible range > 90% CTE < 4 ppm/K for TFT
Melting	Inclusions (stones, quartz)	< 30 μm
Refining	Bubbles	<30 μm
Homogenization	Uniform glass properties	
Hot forming	Surface properties: waviness, wave, surface roughness, and thickness	Waviness < 100 nm Wave < 20 nm Surface roughness < 1 nm Thickness = 0.7 mm Thickness deviation within a sheet < 10 μm, from sheet to sheet < 40 μm
Annealing	Flatness (warp) Shrinkage	Warp < 150 μm Shrinkage < 10 ppm
Cutting into oversized sheets	Size tolerance	±0.5 mm

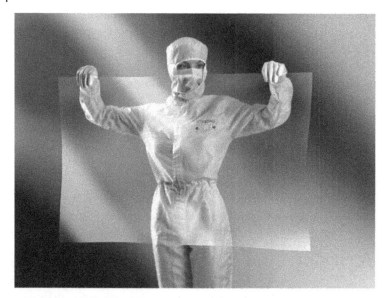

Figure 8.6 Large thin glass sheet, 50-μm thick, produced by the DD technology.

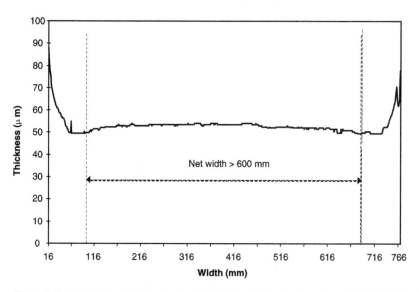

Figure 8.7 Thickness data for a large 50-μm thin glass sheet produced by the DD technology.

8.3.2.2 Surface Quality Limits for Production

Glass sheets produced by the float technology need to be polished to achieve the required surface quality features, whereas current DD technologies are providing excellent surface quality without any postprocessing.

The allowed surface deviations for LCD technology are in the submicron range. Figure 8.8 illustrates in detail the surface quality requirements. An evaluation window is defined (in this case 20 mm) within the scan length of the measurement equipment. Within the evaluation window, the maximum wave and slope of the sample are determined over shorter spans. Here are some typical values of the specification for LCD glass substrates:

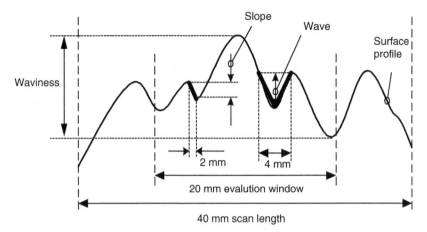

Figure 8.8 Detailed diagram of surface quality requirements for LCD technology.

- waviness 20 pv (peak-to-valley): < 0.100 μm
- wave $20^*/4^\dagger$: < 0.020 μm
- slope $20^*/2^\ddagger$: < 0.025 μm.

Moreover, the available advanced DD technologies offer excellent roughness quality features. The drawn surface shows roughness data in the range of 1 nm RMS roughness without any special treatment. In summary, DD technology is the sole production technology achieving economic production of high-quality flexible glass sheets.

8.4 Mechanical Properties

8.4.1 Thin Glass and Glass/Plastic Substrates

System manufacturers tend to regard the glass strength as a subordinate property. "Hard" product features such as optical transmission, dimensional stability within the system design and during processing, chemical barrier qualities, or electric material characteristics such as functional sizes take precedence over mechanical properties. Nevertheless, the glass strength is a highly important property because it determines the mechanical strength of glass parts. For this reason, the mechanical peculiarities of glass in manufacturing or during usage are discussed in many articles [5, 6]. However, the manufacturing of curved, bendable, or rollable displays will require glass sheets with a thickness of <100 μm.

8.4.2 Mechanical Test Methods for Flexible Glasses

The surface of thin sheet glass usually has a very good mechanical quality. Microdefects may be induced in the glass surface by mishandling it, in the sense of mechanically damaging it or causing abrasion. This applies to handling during the manufacturing process, as well as to the typical

---

* Scan length in mm
† Wave length in mm
‡ Slope length in mm

wear and tear by customer usage. In short, the strength of a glass part depends on its entire history, which involves mechanical or chemical traces placed on the glass surface. The stronger these traces, the more reduced the strength and the lifetime, and the higher the loss probability of a glass part. Of course, it is assumed as a basic condition that in the final product the mechanical requirements for the use of brittle materials have been considered.

In most mechanical laboratories, the surface strength of brittle materials is evaluated by a standard procedure, the so-called ring-on-ring procedure (DIN EN 1288). Care must be taken that reproducible tension profiles in the test objects are obtained by choosing test setups with suitable dimensions in particular for the testing of ultra-thin sheet glass applications (Section 8.5).

Much more critical than weakening of the surface strength by surface damages is the breaking edge produced when the glass sheets are cut into smaller pieces. In most cases, this breaking edge is responsible for the failure of a glass component. We will primarily deal with this type of failure.

Low-cost manufacturing is usually carried out by the time-parallel processing of many single structures on a large-area glass substrate. The necessary separation of these single structures, e.g. displays, induces serious, strength-relevant injuries in the glass surface, or glass body when carried out with standard separation procedures such as diamond scribing, wheel cutting, dicing, and laser cutting.

Figure 8.16 shows the broken edge characteristic of an ultra-thin glass sheet cut into smaller pieces by diamond scribing (substrate thickness 100 μm). Looking from above, we see the typical fine trace of the initial cracking with strong injuries in the form of microcracks or chips, followed by mist and twist hackle marks. In the lower part, opposite the initial cracking, the breaking edge is almost smooth. With unfavorable breaking, this part may also be damaged by chips or cracks. From each of the surface structures displayed here, a crack may originate under local tension, hence causing the failure of the glass component.

The heterogeneity of the fracture patterns with respect to the correlated sheet surfaces required different mechanical tests to be carried out for a complete characterization of the edge quality. As a standard test procedure, static (three-) four-point bending tests (Figure 8.9) are carried out in a first step. Depending on the position of the test bodies, the already optically visible directional dependence of the mechanical edge quality can be taken into account.

For glass, strength is not single valued (in contrast to other materials, e.g. metals), but it varies statistically. It is therefore represented as a probability of fracture under a given load and envi-

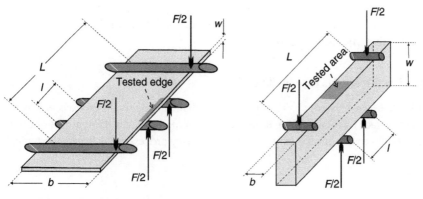

Figure 8.9 Four-point bending test.

ronmental condition. After the destructive tests have been carried out, the acquired data are analyzed with various statistical standard tools (DIN 55303) [7]. It is very important – and must be emphasized to the manufacturers of the final devices – that the necessary statistical test procedures (e.g. χ^2 or t-test) that verify the statistical hypotheses can only be executed when a sufficient number of valuable breaking events are available. This is the only way to acquire solid results.

In addition to the described static tests, it is useful to investigate in the laboratory the impact of short-term loads, such as when a PDA is dropped, or during processing on the production line (e.g. use of stoppers to position thin sheets). The main focus is again on the critical breaking edge of the glass substrate. A test setup is shown in Figure 8.10.

When the mallet is equipped with a cladding and the thin sheet glass is placed on a cantilever support, it is possible to simulate the conditions for transport on a production line and for display assembly. It is also possible to generate "wear and tear damage" and its relevance for the durability of the glass item may then be investigated with the static test methods described earlier.

Ultra-thin sheet glasses naturally tend to break under smaller loads than "thick" sheets, because mechanical stresses, assuming the same load, are approximately inversely proportional to the square of the glass thickness.

In fact, ultra-thin sheet glasses are at least "as strong" as thick sheets, but their dimensions make them more sensitive to mishandling in the production line or at the customer.

Consequently, the demands on the mechanical stability are higher for flexible substrates because the substrates, and therefore the breaking edges, are much more stressed. The two-point bending procedure is suitable for testing such stresses (Figure 8.11). This test procedure has been established for telecommunication fibers [8] and adapted for ultra-thin sheet glasses. In this test, ultra-thin sheet glasses are placed between two traveling jaws, which are then tightened until the test piece fails. As can be seen in Figure 8.11, today's cutting technologies and suitable polymer coatings, applied especially to the breaking edge region, enable durable glass bending radii of 30 mm with a predicted loss probability of < 1%. Corresponding results, for example, the process influence and the enhancement of the already high breaking edge quality by effective edge coating, are explained in Section 8.5.1.

Figure 8.10 Edge impact test.

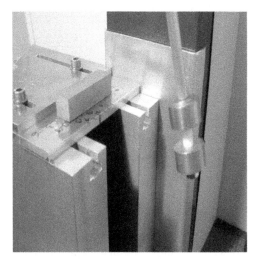

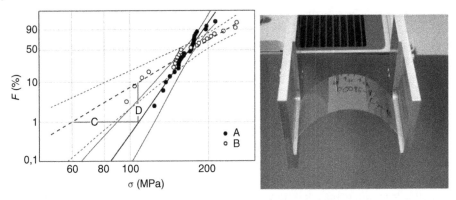

Figure 8.11 Set-up for two-point bending procedure and the influence of edge positioning.

As an example for the demonstrated heterogeneity of cutting edges (Figure 8.16) and the influence on the fracture strength, Figure 8.11 shows the strength distribution of the upper edge (A, initial scribing) and the lower edge (B). As can be seen, the edges of the ultra-thin sheet glasses have high strength (parameter of the Weibull distribution $\sigma_0 > 160$ MPa). However, depending on the positioning, strength σ differs by almost 100% for an identical failure probability $F = 1\%$ (C) or the failure probability at the edge bottom is expected to increase by a factor of 10 under a load of 110 MPa (D). These results indicate that attention should be paid to the positioning of the glass sheets when bend ultra-thin sheet glasses are being processed.

8.5 Improvement in Mechanical Properties of Glass

The properties of glass in Section 8.2.1 are important for different display technologies. The most important advantage of plastic is the low breakage probability. Even the flexibility, which plastic is said to have, can be achieved by glass. Thus, bending radii of 31 mm for glass of thickness 100 µm and 12 mm for 50 µm-thin glass (at a failure rate of 1%) were measured by Schott [9].

There are several methods to achieve the lowest possible breakage rate for glass:

- careful handling with suitable equipment
- thermal or chemical hardening
- use of multilayer and compound glasses
- coating.

Of these general possibilities for glass treatment, only coating can be considered for thin glasses. Hardening can only be used for thick glasses, and compound glasses reduce the flexibility. Therefore, coating is the only chance to reduce the breakage rate and to keep the flexibility of the thin glass.

8.5.1 Reinforcement of Glass Substrates

The reinforcement coating has two positive effects on the glass strength. First of all, the possible damage to the glass can be blocked by a surface coating, which means the number of microcracks is reduced. Secondly, the coating results in a compressive stress on the glass surface. Therefore, the growth of already existing cracks can be reduced or even stopped.

8.5.1.1 Principal Methods of Reinforcement

A reinforcement of glass for display applications can be achieved by:

- lamination with plastic foil
- thin layer coating such as a plasma despoiled hardcoat
- thick layer coating, e.g. of polymers.

When using these coating processes, the advantageous properties of both materials can be combined. That means the excellent barrier properties and the dimensional stability of the glass can be combined with the resistance of damage to plastic.

Strength tests have shown that the breakage rate of glass depends not only on the damage of the surface, but also or even more heavily on the damage at the edges of the substrate. The protection of the substrate surface is important, but the protection of the edges is even more critical as approximately 90% of the breakage is caused by damaged edges.

Figure 8.12 shows the influence of the edge and surface coating on the strength. When the surface is coated, the edges are also coated. It shows the positive influence of edge coating against the uncoated substrate. However, the highest strength values were achieved by a combined coating of the edge, as well as of the surface.

8.5.1.2 Materials for Reinforcement Coatings

The influence of the coating material on the achievable edge strength σ is shown in Figure 8.13. In the histogram, uncoated glass substrates (Ref.) are compared with polymercoated glass substrates (P1, P2, and P3). P1, P2, and P3 represent different polymer materials. Notice that the strength of uncoated substrates is decreased due to handling (Proc.). In this case the processed substrates went through the coating process without applying a polymer. Substrates coated with a polymer show a higher strength after the same handling steps when compared to the uncoated substrates.

Due to a suitable polymer coating, significant protection of the glass can be achieved, thus reducing the damaging influences of processing and handling steps. Such a coating can also reduce the influence of already existing defects. Figure 8.14 shows the "healing" effect of a polymer coating. In this test, the substrate surfaces were sandblasted and the surface strength σ was measured before and after coating. The amount of glass breakage in the manufacturing process is reduced if the glass surfaces are coated, and the process becomes more stable to minor handling mistakes in production. Both effects are crucial for increased "robustness" of ultra-thin glass applications [3].

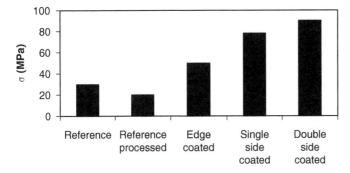

Figure 8.12 Influence of edge and surface (including edges) coating on the edge strength (P2 coating) at a failure probability of 1%.

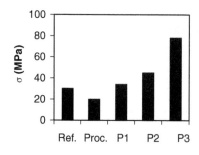

Figure 8.13 Edge strength σ of processed (coated) substrates at a failure probability of 1%: Ref. = reference glass (not processed); Proc. = processed glass without coating; P1, P2, P3 = processed glass with different coatings.

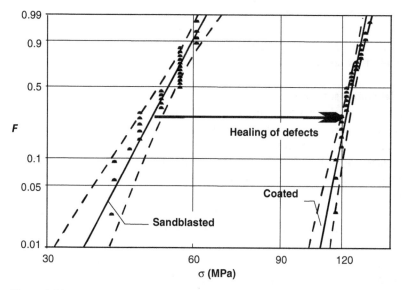

Figure 8.14 Improved surface strength of damaged (sandblasted) glass surface by coating.

8.6 Processing of Flexible Glass

When manufacturing displays, there are some important aspects for both the substrate supplier and the display manufacturer, such as transportation issues, cleaning methods and separation methods. Moreover, a mass production manufacturing process is not yet available. This holds for any kind of flexible substrate, glasses as well as plastics. The current situation for mass production of displays involves the use of standard manufacturing equipment, which allows only minor changes in material properties to keep the product cost competitive. An important material property is the so-called sagging, which depends mainly on the Young modulus E of the substrates. Sagging values describe how much a substrate is allowed to sag under special conditions of the handling equipment.

Sagging specifications and substrate flexibility run counter to each other. This means if flexible substrates are to be processed with standard manufacturing equipment there are only two choices: modify all automated handling steps or modify the substrate so that its stiffness becomes comparable to that of thicker standard substrates. The first choice involves building a new production line. This is the reason why new display technologies have a greater advantage in manufacturing flexible displays, as standard equipment is not yet available (e.g. organic light-emitting diode [OLED]).

On the other hand, advanced LCD technology must also be able to deal with "flexible" substrates if we consider the new Gen 7 glass substrates. The traditional sagging value for these substrates is nearly 1 m instead of several millimeters of sagging tolerated for smaller substrate sizes. Nevertheless, the LCD industry has developed solutions to overcome this problem with new handling concepts.

From the viewpoint of plastic substrates, the idea of roll-to-roll processing is favored as it is more cost-effective. Following this concept, a lot of other issues need to be resolved and can't be discussed here. In any case, glass-based flexible displays can only use batch processes, because even if rolling glass ribbons is possible, the practical use is limited due to low yield related to glass breakage. In summary, these are the reasons why current efforts are made to manage the glass substrates only with standard equipment benefiting from possible cost savings and therefore making the product cost competitive. With this background information the focus of activity is on cleaning, cutting and transportation of flexible glass substrates.

8.6.1 Cleaning

Every substrate must be cleaned before processing. How intensive the cleaning has to be will depend on the contamination of the substrates. Such contamination can be particles caused by the environment such as dust, glass particulates due to the cutting process, or organic compounds such as fingerprints or paper prints due to handling, storage, and transportation. In order to remove such contamination, different cleaning techniques need to be combined.

The basic cleaning methods are:

- brush cleaning
- megasonic (400 kHz–3 MHz) and ultrasonic (25–100 kHz) cleaning
- cleaning by hydrodynamic forces (laminar and turbulent flow, drag and lift forces, spray)
- dry ice cleaning
- UV/ozone cleaning
- laser cleaning.

When cleaning standard glass sizes, good cleaning results are achieved by using commercially available equipment. However, for cleaning ultra-thin glasses or very large sizes like Gen 7 and above, new concepts, e.g. a substrate holder, are necessary. The challenge here is the flexibility of those substrates.

Based on commercially available ultrasonic washing equipment, very good cleaning results have been achieved for coated substrates at a thickness of 100 μm. For 300 mm × 300 mm substrates super-twisted nematic (STN) LCD quality has been achieved.

8.6.2 Separation

There are a number of methods of separation of glass and displays. Depending on the accuracy and shape, several methods are used:

- sawing
- guillotining

- cutting:
  - water jet cutting (rarely used)
  - laser cutting (full-body cut, laser scribing)
  - cutting by wheel (hard-metal wheel or diamond wheel)
  - cutting by diamond.

At present the most commonly used separation methods are conventional cutting by wheel or diamond and by laser.

When speaking of conventional cutting, this implies a mechanical scribe process followed by a breaking step. This process causes fragmentation, as can be seen in Figure 8.15, which may generate glass particles. In order to avoid such glass particles, the full-body laser cut was developed. The full-body cut uses a laser beam to apply heat to the glass. The glass is then immediately cooled, which generates a stress-induced cut that results in complete separation.

For thin glass sheets < 200 μm, the conventional cutting process is being utilized in a newly developed process. When properly adjusted, this conventional cutting equipment produces edge quality and strength results as good as those of a laser-cut substrate. After the substrate is diamond scribed, the following breaking step results in a high edge quality as shown in Figure 8.16.

8.7 Current Thin Glass Substrate Applications and Trends

In today's world there are numerous buzzwords such as "miniaturization" and "multifunctional" forcing the industry to look for alternative material compositions. One decisive point for all of the

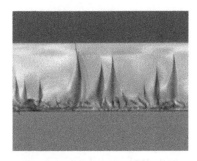

Figure 8.15 Edge quality of standard cutting equipment (using a diamond wheel).

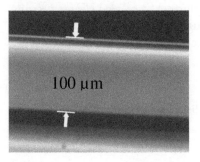

100 μm

Figure 8.16 Edge quality of improved cutting equipment (using a diamond).

handheld and energy-saving applications is the thickness, and thus the weight, of the materials used. Talking about display applications, and microelectro mechanical systems (MEMS), the substrate material glass has played and still plays an important role. Thanks to different production technologies (drawing, floating, casting, etc.) and different chemical compositions (soda-lime, borosilicate, aluminosilicate, etc.) the substrate material glass can serve a huge variety of applications and sometimes it might not even seem that glass is present inside the product.

8.7.1 Displays

The leading technology now and in the future is TFT (active matrix), and the second is the family of TN/STN (passive matrix). Besides these two major technologies there are smaller markets for OLED and PolyLED, liquid crystal on silicon (LCOS), electroluminescent display (ELD), PDP, and many more. One demanding area is the handheld communication market (e.g. mobile phones, PDAs, and laptops), which requires higher functionality, easier handling, and reduced weight. The mobile phone market began some years ago with passive matrix black and white displays and substrate thicknesses of 1.10 mm and 0.70 mm. Today no new mobile phone is equipped with such thick substrate materials. The main market is driven with 0.50 mm- and 0.40 mm-thick substrates for the main displays and 0.30 mm for the sub-displays. The leading mobile phone makers are now focusing on a thickness < 0.30 mm for both displays. This request can easily be fulfilled for the passive matrix displays using STN/CSTN technology by using thinner glass substrates; the rising TFT makers use alternatives like etching or polishing down the TFT panels (which are mainly being produced in 0.70 mm glass thickness). Both ways are acceptable to the industry, as both technologies can comply with the requirement for lightweight and good picture quality.

The mainstream TN technology passive matrix is still using a 0.70 mm thickness with the exception of watch displays, which for quite some time have already used a 0.30 mm-thick substrate. And, who knows how much thinner these displays will become in the future. The same trend can be seen for laptops, for example, using TFT substrates. First prototypes have been presented to the public already with glass substrates as thin as 0.30 mm and even thinner (for curved displays).

8.7.2 Touch Panels

Touching is one of the most instinctive human actions. Therefore touch panels (Figure 8.17) can be found in many industrial environments, hospitals, tourist centers, schools, retail, and automotive industries. Touch screens make access to using computers easier, as clearly defined menus eliminate the need for keyboards. Thin glass touch panels with low reflectance are used in car navigation systems, as well as for other applications. Thin glass substrates with thickness of about 200 µm offer the advantage of furnishing the device with the required flexibility to recognize the "touch" of a wide variety of objects.

Flexible glass sheets provide touch panels with better scratch resistance than plastic, which is beneficial in rough environments.

8.7.3 Sensors

The market for sensors is very large and growing, as sensors can be found in industry and in our daily lives as well. Small to ultra-small sensors are offered from all manufacturers. Again, a suit-

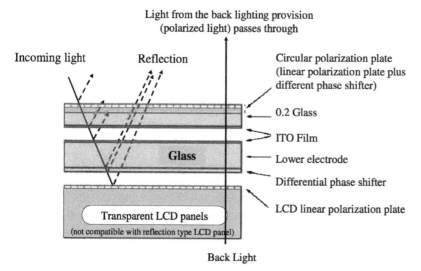

Light from the back lighting provision
(polarized light) passes through

Incoming light   Reflection   Circular polarization plate
(linear polarization plate plus
different phase shifter)

0.2 Glass

ITO Film

Glass   Lower electrode

Differential phase shifter

Transparent LCD panels   LCD linear polarization plate
(not compatible with reflection type LCD panel)

Back Light

Figure 8.17 Principal structure of a touch panel.

able material for this application is glass. Sensors used, for example, in the automotive industry for airbags temperature control, gasoline control, etc., are already very small parts, which is important. They need to become smaller and lighter in weight to help reduce gasoline or energy consumption and keep the environment cleaner.

Promising growth can be seen in charge-coupled device (CCD) and complementary metal-oxide-semiconductor (CMOS) image sensors, especially with applications for new cell phones with or without camera systems. Digital cameras also require small parts with low weight and high transparency as well as a number of high-technology features combined in a small part with special glass covers or windows using glass thicknesses of approximately 0.30 mm. Many manufacturers are involved in this market, but the real volume is done by only a few.

8.7.4 Wafer-level Chip Size Packaging

An alternative technology using a glass–silicon–glass sandwich developed by Shellcase/Israel uses glass substrates as thin as 100 μm. This enables the manufacturer to produce miniaturized packages well suited for use in handheld communication devices such as portable computers, mobile phones, and wireless communication as well as in the medical field. A total package thickness of 330–730 μm definitely requires highly precise thin glass substrates.

References

1 Scholze, H. (1988). *Glas*. Berlin: Springer Verlag.
2 Kessler, T., Wegener, H., Togawa, T., Hayashi, M., and Kakizaki, T. (1997) Large microsheet glass for 40-in class PALC displays. *Proceedings of the 4th International Display Workshops*, 347–349.
3 Plichta, A., Weber, A., and Habeck, A. (2003). Ultra thin flexible glass substrates. *MRS Proceedings*. 269: H9.1.

4 Lapp, J.C., Bocko, P.L., and Nelson, J.W. (1994). Advanced glass substrates for flat panel displays. *SPIE Proceedings* 2174: 129.

5 Rawson, H. (1988). Why do we make the glass so weak? A review of research on damage mechanisms. *Glastechnische Berichte* 61 (9): 231–246.

6 ICG (2003) Advanced course on the strength of glass: basics and test procedures. Research Association of the German Glass Industry (HVG).

7 Sachs, L. (2002). *Angewandte Statistik*. Berlin: Springer Verlag.

8 Matthewson, M.J., Kurkjian, C.R., and Gulati, S.T. (1986). Strength measurement of optical fiber bending. *J. Am. Ceram. Soc.* 69 (11): 815–821.

9 Plichta, A., Deutschbein, S., Weber, A., and Habeck, A. (2002). Thin glass-polymer systems as substrates for flexible displays. *SID Proceedings* 53–55.

9

Toward a Foldable Organic Light-emitting Diode Display

Meng-Ting Lee, Chi-Shun Chan, Yi-Hong Chen, Chun-Yu Lin, Annie Tzuyu Huang, Jonathan H. T. Tao, and Chih-Hung Wu

AUO Display Plus Corp., Taiwan

9.1 Panel Stack-up Comparison: Glass-based and Plastic-based Organic Light-emitting Diode

As consumers become more dependent on electronics in the digital age, the demand for these electronics to provide an increasing amount of information also grows stronger. The development in new display technologies has enabled the size of bulky cathode-ray tube displays to be reduced to thinner and smaller flat liquid crystal displays (LCDs) over the past decades; this allows information to be presented in a more compact manner. Now, nearly everyone has a thin and small handheld device that is capable of displaying a limitless amount of information from the internet. However, consumer demand and the desire for the ultimate user experience continue to push the display industry toward products with higher display resolution, wider color gamut, higher contrast ratio, and a larger active display size; some are now multifunctional with integrated touch capabilities, low power consumption, and a long battery life. In addition to these device specifications, the devices even take on a curved form. To make them more fashionable, the device makers are now aiming for flexible displays that could be bendable, foldable, or even rollable like a piece of paper. What was once a science fiction fantasy is now becoming a reality.

The key differences between the traditional rigid active-matrix light-emitting diode (AMOLED) display and the flexible AMOLED display lie between the kind of substrate and the encapsulation technology. As shown in Figure 9.1, after replacing glass substrate and glass encapsulation by plastics and thin-film encapsulation (TFE) for the display (shown in Figure 9.1(a)), the thickness of the AMOLED display is greatly reduced and the device is no longer rigid [1]. Figure 9.1(b) shows the schematics of the stacking structure of the flexible AMOLED device. Thin-film transistor (TFT)/OLED/TFE layers were fabricated on a plastic substrate; a circular polarizer and a touch-sensing film are then applied to the panel. Because OLED materials degrade rapidly in the environment of oxygen and moisture, an effective encapsulation layer is required to produce a reliable device.

In general, the encapsulation needs to achieve a water vapor transmission rate (WVTR) of $\sim 10^{-6}$ g/m^2/day for an AMOLED device to pass its reliability test. One of the most well-known approaches so far is to form multiple layers of organic and inorganic thin films, or the so called Vitex structure [2–4]. It was reported that flexible encapsulation layers made of alternating Al$_2$O$_3$ and polyacrylate layers can achieve 10^{-6} g/m^2/day. The inorganic layers act as the primary moisture barrier while the organic layers decouple the pinholes of the inorganic layers. As a result, the diffusion length for water to permeate through the stack is extended. In addition, the organic layer

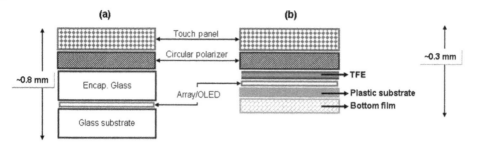

Figure 9.1 (a) Structure of conventional glass-based AMOLED display. (b) Structure of current flexible AMOLED display.

works as a planarization layer that provides a smoother surface for the deposition of the subsequent inorganic layer and for an easy lamination process after the TFE structure is complete. With the structure similar to Figure 9.2, compared to a single inorganic layer, this kind of alternative stacking structure reduced WVTR dramatically from 10^{-3} to 10^{-4} or 10^{-6} g/m²/day.

The function of a circular polarizer (CPL) is to filter out ambient light. Because the top of the OLED region consists of an electrode with high reflection, e.g. Ag, as shown in Figure 9.3, the region of the OLED electrode without the CPL reflects incident light and the off state does not appear dark. As a result, the contrast ratio suffers. As shown in Figure 9.4, a CPL consists of a linear polarizer and a 1/4λ retarder film. A linear polarizer first allows only linearly polarized light to pass through. The 1/4λ retarder film then turns linear wave to circular wave. When circular rays strike a reflecting surface (e.g. the electrode of a top emission OLED) and are reflected, the phase relationships between the two rays are reversed. Therefore, the light subsequently passing through the linear polarizer does not include reflected light. Utilization of the circular polarizer thus reduces glare, internal reflections, and most importantly increases contrast ratio of the display. However, the drawback is that the emitted light also reduces to less than 40%, requiring higher power consumption than initially predicted. In addition, the usage of thicker CPLs (~140 µm) limits the flexible application of OLED.

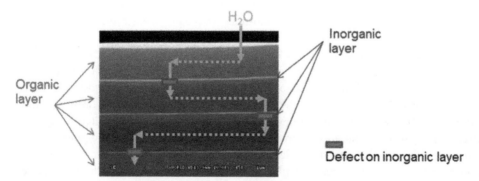

Figure 9.2 Example of inorganic/organic stacking layer used in TFE.

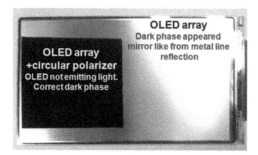

Figure 9.3 Appearance of OLED panel with and without the usage of a CPL.

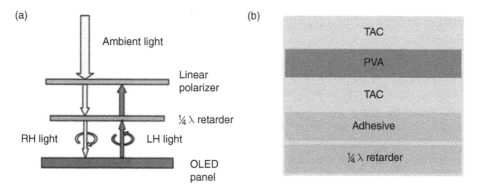

Figure 9.4 (a) Basic component of a circular polarizer. With the combination of a linear polarizer and a 1/4λ retarder, all ambient light was absorbed. (b) Typical layer structure of a circular polarizer.

9.1.1 Technology for Improving Contrast Ratio of OLED Display

Since many flexible displays are targeted for outdoor use, having a high-contrast ratio under strong ambient light is one of the most significant features for displays. However, the strong reflection from the metal electrodes of conventional OLEDs would cause considerable reduction of the contrast ratio. Therefore, reducing ambient reflectance for OLED panels is an important issue. The ambient reflected light from the AMOLED panels can be expressed mathematically and divided into three parts:

$$R = R_S + R_{AA} + R_{non\text{-}AA}$$

where R_S is the reflectance of the first surface, usually caused by the difference of refraction index between air and the top film. The part of incident light that passes through the top film into the cell of the device would be reflected or absorbed by materials of OLED panels. Besides the reflected light from the first surface, all the reflected light from the device can be separate into R_{AA} and $R_{non\text{-}AA}$, which are shown in Figure 9.5. R_{AA} is the reflected light from the active areas of OLED. The characteristic of its reflectance spectrum is affected by different micro-cavity length design. $R_{non\text{-}AA}$ is the reflected light from non-emissive areas. Several approaches to reduce the various types of reflected light from OLED displays will be discussed.

Various types of anti-reflection (AR) coatings have been developed including single-layer AR coatings, multilayer AR coatings, gradient refraction index AR coatings, and inhomogeneous

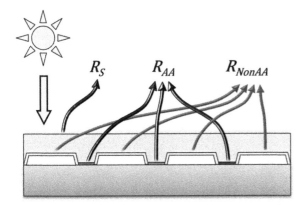

Figure 9.5 Schematic of individual reflected light from OLED panels.

layer AR coatings [5–7]. Based on multi-wave interference, AR coatings are used to create the phenomenon of destructive interference within interfaces on the reflected side. For a simple example, when light incident into a single-layer AR coating, the relative phase-shift between the two reflected waves at the upper and lower boundary of the thin film must be π for a given wavelength to achieve, destructive interference between the two reflected waves. A single-layer AR coating is generally limited to a specific wavelength and so it can be useful in the case of a monochromatic reflection. To eliminate reflection of a wide range of ambient light, double or multilayer AR coatings must be used. The AR coating method can easily reduce the reflected light from the first surface (R_S), but the major difficulty, the reflection from the metal electrode in OLEDs, cannot be solved.

To reduce the majority of the reflected ambient light from the metal of OLED devices, R_{AA} and R_{non-AA}, the use of CPL is one of the most common technologies [8, 9]. The advantage of using CPL in displays is that the reflected light is nearly completely blocked; that is, it can eliminate almost all the reflected light from the metallic region in displays (R_{AA} and R_{non-AA}). Meanwhile, the light emitted by OLED can be transmitted. Unfortunately, in this process nearly 60% of OLED light is also blocked. Furthermore, the CPL is generally too thick to bend for use in flexible applications.

The use of black electrodes had also been published, and several categories, such as light-absorbing cathodes or black cathodes with destructive interference structure [10, 11] were demonstrated. In the case of light-absorbing cathodes, some materials with strong absorption were inserted between the metallic cathode and electron injection layers, which prevent reflection of both ambient and OLED light. In the case of black cathodes with a destructive interference structure, the cathode is composed of a thin semitransparent metal layer, a phase-changing layer, and a thick reflective metal layer. The light wave reflected from the thin semitransparent metal layer and the one reflected from the thick reflective metal layer interfere destructively, as the latter has a π-phase difference with respect to the former due to the phase-changing layer. The component of the OLED-emitted light directed toward the black cathode is also not reflected as in the case of normal reflecting cathodes. Hence, both the absorbing cathode and the black cathode with destructive interference structure cause large OLED-brightness reduction: the luminescent efficiency decreased to 50% of the conventional OLED.

A combination of color filter (CF) and black matrix (BM) may be one of the best approaches for reducing reflection of the flexible OLEDs because of its inherent advantage [12, 13]. That is, the CF will only allow the incident light with a specific wave band to pass through the CF and into the OLED device. Similarly, the CF will absorb the reflected ambient light again as the ambient light is reflected from the device electrode. Meanwhile, the CF does not absorb the wavelength emitting by the OLEDs because the CF is well aligned with the corresponding primary colors of the OLED pixel. Furthermore, the incident light in the transparent band of the CF will be trapped in the OLEDs, resulting low reflectance because of the Fabry–Pérot effect in micro-cavity OLEDs. Additionally, the BM with high optical density was positioned above the non-emissive areas to block any ambient light being reflected by the metal. Unlike CPL, the combination of CF/BM can be readily applied to flexible displays.

In summary, the use of AR coatings is the best solution for reducing surface ambient reflection (R_S), but it cannot reduce the majority of the ambient light reflected from the metal of the OLED device, R_{AA} and $R_{non\text{-}AA}$. To reduce R_{AA} and $R_{non\text{-}AA}$, the use of CPLs, black electrodes, and a combination of CF/BMs are excellent solutions. However, the CPL causes inevitable power loss of 50–60% due to its finite transmittance. It may also be unsuitable when it comes to highly flexible devices due to its relatively large thickness. Similarly, the black electrode causes large OLED brightness reduction due to its strong absorption. In contrast, using CF/BM technology will substantially reduce the majority of the ambient light reflected from the metal of OLED devices while keeping the luminescent efficiency of the OLED devices. Hence, to realize high-contrast ratio flexible OLEDs, the CF/BM technology has been regarded as one of the best solutions.

9.2 CF–OLED for Achieving Foldable OLED Display

A foldable display that utilizes CPL has three major drawbacks as shown in Figure 9.6. Firstly, a display with a typical total thickness of 300 μm can result a large stiffness and low flexibility. Secondly, the deformation of CPL can easily occur when the panel is tested under the environmental test condition of high temperature and high humidity (60°C/90%). Thirdly, the asym-

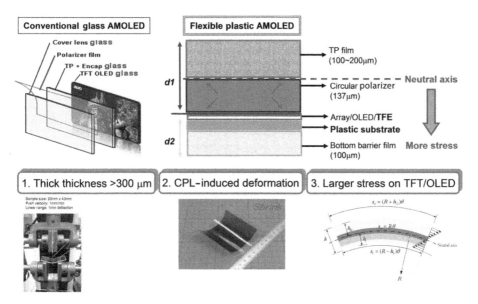

Figure 9.6 Major drawbacks of plastic OLED with CPL for foldable application.

metric panel structure (APS) allows the neutral axis (the location with zero stress) to be far from the TFT, OLED, and TFE layers in the panel. All three issues may be attributed to the CPL having a thickness of approximately 140 µm. In order to realize a foldable display, AU Optronics proposed a new structure, the so-called symmetric panel stacking (SPS) structure, by using a 10-µm thick CF/BM plate to replace the CPL while maintaining the same function of CPL [14–19]. The total panel thickness is less than 100 µm and the SPS structure can minimize stress on the weak layers like TFT, OLED, and TFE when the panel is folded.

9.2.1 Mechanism of the AR coating in CF–OLED

The foldable display based on SPS CF–OLED panel stacking structure consisted of two main parts: the CF or the upper part, and AMOLED or the lower part. For the CF part, good optical properties are crucial. The CF, BM, and AR layers were applied consecutively on a colorless transparent plastic-coated glass substrate. For the OLED part, the low-temperature polysilicon thin-film transistors (LTPS-TFT) need to be fabricated in order to operate the OLED display; therefore, a high-temperature-resistant (HTR) plastic was chosen to be coated on glass. The OLED layers were then evaporated onto the LTPS-TFTs with a fine metal mask to define red–green–blue (RGB) sub-pixels. Next, the two portions were assembled with 5 µm thick adhesive using a high-precision lamination system with the accuracy of assembly smaller than 3 µm. Glass was then de-bonded by a laser-scanning technique. Functional plastic films were then attached to the upper and lower surfaces of the assembled device. Accordingly, the chromatic aberration and color mixing at large viewing angles can be avoided. The flow chart of the assembling process is shown in Figure 9.7.

Due to the intrinsic optical properties, cavity OLED possesses low reflectivity around the emission peak. The reflectivity of RGB cavity OLED is shown in Figure 9.8. Therefore, external illumination would experience low reflection inside the cavity OLED. At the same time, color filters can only allow transmission of a small range of light around the emission peak while

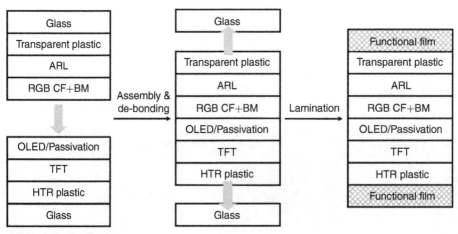

* ARL: Anti-reflection layer
* HTR: High-temperature-resistant

Figure 9.7 Process flow of plastic-based AMOLED with SPS panel structure.

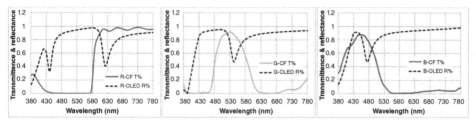

Figure 9.8 Reflectance of cavity RGB OLED and transmittance of RGB color filter.

absorbing all other ranges of light. Therefore, if an incident light has the wavelength that is around the cavity emission peak, it would pass the color filter but experience low reflection inside the cavity, resulting in low reflection; if the incident light is not around the cavity emission peak, then it would be absorbed by the color filter and show low reflection for the observer as well. The BM on CF also improves the contrast due to reduced reflection at the non-emissive area of the OLEDs. Thus, an overall low reflection without the assistance of CPL over the visible light wavelength range is achieved by combining the cavity structure and CF with BM.

In addition to low ambient light reflection, a good display also needs a reflective hue of neutral black. In order to reduce the limitation of the fine metal mask, current OLED products use a pixel rendering design, which means the number of RGB sub-pixels is not equal in a panel. Normally the number of green sub-pixels is twice that of the red and the blue sub-pixels. For a display that adopts the SPS CF–OLED panel stacking structure, the panel would look "greenish" or "yellowish" rather than appearing as the "neutral black" to the observer when the device is turned off. In order to realize a reflective hue of neutral black, the transmittance spectrum of each color of the color filters, the overlap among the transmittance spectrum, the total size and number of the sub-pixels of each color, and the micro-cavity height need to be carefully designed. One of the approaches is to reduce the transmittance of the green CF to suppress the reflection in the green region. Then a reflective hue of delta-a*b* of 2.17 can be achieved. Detailed optical performances are summarized in Table 9.1.

9.2.2 Optical Performance of CF–OLED

The SPS CF–OLED incorporating cavity RGB OLED and CF has many advantages over the traditional structure incorporating CPL. The most noticeable one should be the improvement on power consumption. Since the RGB sub-pixels were monochromatic and were already saturated

Table 9.1 Reflective luminance and hue of SPS CF–OLED with different transmittance of green CF (measured by a spectrophotometer, model CM2600d by Konica Minolta).

	L*	a*	b*	Δa*b*
CF–OLED with STD G–CF	47.32	−11.85	5.91	13.24
CF–OLED with low T% G–CF	40.35	−2.17	−0.01	2.17
CF–OLED with low T% G–CF and AR film on the surface	31.44	−1.93	2.74	3.35

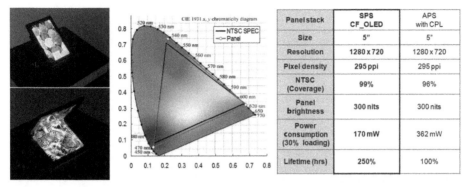

Panel stack	SPS CF_OLED	APS with CPL
Size	5"	5"
Resolution	1280 x 720	1280 x 720
Pixel density	295 ppi	295 ppi
NTSC (Coverage)	99%	96%
Panel brightness	300 nits	300 nits
Power consumption (30% loading)	170 mW	362 mW
Lifetime (hrs)	250%	100%

Figure 9.9 Panel specification of a 5" foldable OLED display with the SPS CF–OLED structure.

by the micro-cavity, we adopted high transparency CF materials with 70–90% total transmittance to increase efficiency. Without the use of CPL (with a transmittance of ~ 42%), the power consumption of a 5-inch panel with high-definition resolution is only about 170 mW under 300 nits panel brightness with 30% image loading. The power consumption was only 47% of that consumed by the bare panel with CPL. Meanwhile, the lifetime of the panel with the SPS CF–OLED panel structure can be projected to be 2.5 times longer than that of a display using CPL. This result also can be attributed to the higher transmission of CF, which means the panel needs lower driving current to achieve target brightness. Detailed panel specifications are summarized in Figure 9.9.

In addition, the CF can serve as a spectrum purifier to improve the color space performance and color variation over large viewing angles. The panel possesses a wide coverage of 99% NTSC (National Television System Committee) color space which would be sufficient for high quality screens. The color variation issue of cavity OLED was also resolved by the CF. The emission spectra of cavity OLED at large viewing angles are usually strongly affected due to the change of optical path length. However, the CF would "filter out" the undesired portion and reduce the color variation. That is to say, the distortion of the electroluminescence (EL) spectra at large viewing angles would be suppressed. Therefore, the display can possess a wider viewing angle with consistent quality. By further incorporating new OLED emitters and optimizing OLED thickness by using a capping layer on top of the cathode, a panel with BT.2020 color space can be obtained, as summarized in Table 9.2.

Table 9.2 Optimal device thickness and emitter set with CF structure to achieve BT.2020 color space.

Emitter set	Device	CPL	BT. 2020 coverage ratio	Power	RGB JNCD (60 degree)
NTSC	Standard	Y	78%	100%	12.1 / 7.2 / 6.5
NTSC	Strong cavity	Y	88%	120%	31.1 / 7.6 / 3.3
BT. 2020	Standard	Y	88%	98%	12.9 / 2.0 / 5.3
BT. 2020	Modified capping layer	Y	90%	103%	15.5 / 2.5 / 7.7
BT. 2020	Modified capping layer + CF	N	95%	56%	8.9 / 1.8 / 7.5

9.3   Mechanical Performance of CF–OLED

In a multilayer system, when the structure is under bending, as shown in Figure 9.10, the neutral axis can be calculated using the following equation:

$$G_x = \frac{\sum_i E_i h_i y_i}{\sum_i E_i h_i} \tag{1}$$

where G_x is the distance between the neutral axis and the top surface. E_i is Young's modulus of layer i, h_i is the thickness of layer i, and y_i is the distance of layer i from the top surface. Additionally, strain at any distance away from the neutral axis can be calculated by

$$\varepsilon = \frac{y}{R} \tag{2}$$

where ε is the strain at distance y away from the neutral axis and R is the bending radius.

For the SPS CF–OLED, the thickness and composition of each layer of panel have been thoroughly considered and defined. As mentioned earlier, the complete panel structure primarily consist of the CF part and the OLED/TFT part. Figure 9.11(b) shows the strain distribution within the panel from the top to the bottom layers of an OLED device with the SPS structure. As shown in the figure, the neutral axis inside the symmetric panel is very close to the OLED/TFT layer. Therefore, even when the rolling radius of the panel is small, the OLED/TFT layer does not have to withstand large stress, thus maintaining good device performance after numerous foldings or even rolling. On the contrary, in the conventional panel stacking of the APS structure with CPL, the neutral axis is far from the weak layers like the TFT and TFE layers. The outermost film also suffers from a very large strain, which is already over the yield point of 1% for coventional plastics. In a real folding experiment, a panel with the APS structure after 100 000 folding cycles at the folding radius of 4 mm could only be lit up partially, meaning that the TFT layer is damaged, as shown in Figure 9.12. During the folding test, the panel was observed periodically to record any film damages. Before the completion of 100 000 folding cycles, we already found a crack on the surface of the CPL and buckling occurred between the

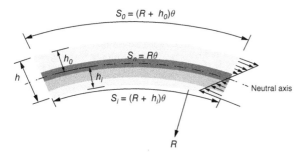

Figure 9.10   Schematic of a multilayer device under bending.

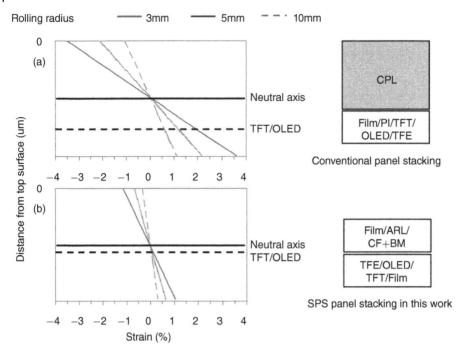

Figure 9.11 Strain distribution of plastic-based AMOLED panels with (a) conventional panel stacking with CPL and (b) SPS CF–OLED panel stacking.

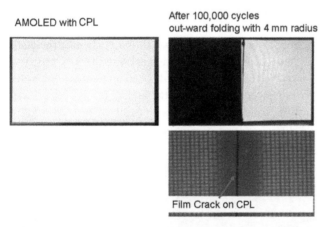

Figure 9.12 AMOLED panel with conventional CPL before and after 100 000 folding cycles. Only a partial area of the display region can be lit up after folding 100 000 cycles. The OM image shows that the crack was observed on the surface of the CPL.

adhesive and the film on the backside of the panel. Based on the visual observation, we could determine that the maximum folding cycles that a panel with CPL could practically endure are less than 100 000 folding cycles.

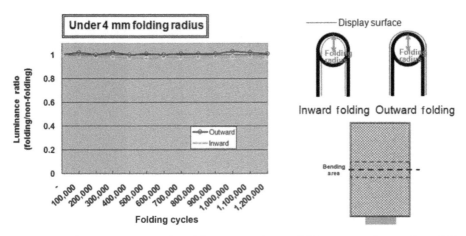

Figure 9.13 Luminance ratio of the folding area to the non-folding area as the panel was folded with display inward and outward condition.

9.3.1 Bi-directional Folding Performance and Minimum Folding Radius of SPS CF–OLED

The folding ability of the SPS CF–OLED panel was judged by measuring the luminance ratio between the folding and non-folding areas, corresponding to the areas that are folded and the areas that remain flat during the test, respectively. The ratio remained identical after millions of folding–unfolding cycles at a folding radius of 4 mm, indicating that the emission properties of the panel were not affected, and the folding process did not accelerate the degradation of the display whether the repeat folding was performed in an inward or outward direction, as shown in Figure 9.13. We have further placed the panel that completed millions of folding cycles for additional environmental test at $60°C/90\%$ relative humidity (RH). Figure 9.14 shows that the luminance ratio still remains close to 1 after 500 hours of environmental test, and no additional defects or pixel shrinkage were found. These phenomena implied that all the critical functional layers including TFT, OLED, and TFE were not damaged during folding. The panel with a 10-μm thick CF outperformed the conventional thick CPL design by essentially relieving the overall

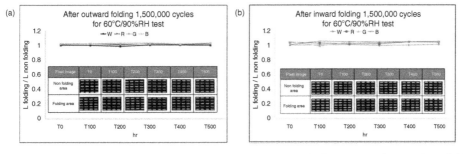

Figure 9.14 Luminance (L) ratio of the folding area to the non-folding area as the panel was folded. The inserted photographs showed the pixel images with test time: (a) outward folding direction and (b) inward folding direction.

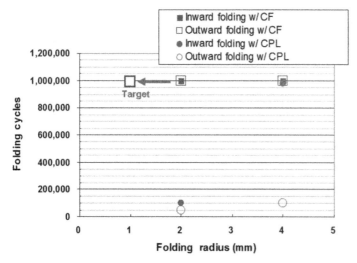

Figure 9.15 Comparison of the folding cycles achieved in AMOLED panels with conventional CPL versus the CF structure under different folding radius.

strain and preventing the most fragile part of the flexible display from being damaged. Figure 9.15 shows the summary of the folding results for panels with the conventional CPL versus the SPS CF–OLED. The display with the CF structure was clearly shown to have better flexibility than the display with the CPL structure at either folding radius of 4 mm or 2 mm. With continuous effort and persistence, we would expect to obtain a display with the SPS CF–OLED structure to be folded at the 1 mm folding radius in the near future.

9.4 Touch Panel Technology of CF–OLED

For most smartphones and electronic consumer products, a satisfying user experience not only requires a high-quality display but also a smooth touch function. However, the distance between the touch sensor and the OLED cathode in very thin flexible displays is very narrow, with the sensor and the cathode having strong interference with each other. In order to integrate touch function into flexible displays, the touch sensor is fabricated using a metal mesh instead of a typical indium tin oxide (ITO) layer in ITO-based touch sensors. To have the same touch signal, the metal mesh requires a smaller pattern area compared to the ITO layer in an ITO-based sensor. Due to the small overlap of the touch sensor and the OLED cathode, the metal mesh sensor has the advantage of a much smaller parasitic capacitive loading. Additionally, metal-based touch sensors has demonstrated better flexibility compared to ITO-based sensors. Therefore, the usage of a metal mesh touch sensor could result in a very good signal-to-noise ratio.

The drawback of constructing metal mesh touch sensors is that metal lines are visible to the eyes because the metal reflects the ambient light easily, and the Moiré effect could be easily observed when the metal mesh overlaps with the pixel arrangement of a similar pattern. However, both issues can be solved by fabricating a BM layer above the metal mesh. Figure 9.16 shows the struc-

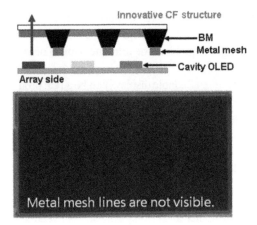

Figure 9.16 Schematics of the touch sensor being fabricated using metal mesh, as integrated with the SPS CF–OLED panel structure. Because the metal mesh is under the BM layer, the ambient light would not be reflected.

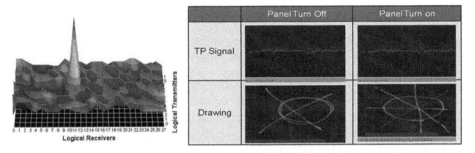

Figure 9.17 Touch signal and drawing pattern on the SPS CF–OLED panel.

Table 9.3 Criteria and specification for flexible cover film.

	Rigid type cover	Flexible type cover
Material	Glass	Plastic
Tactile experience	Excellent	Average
Transmittance	>99%	>90%
Impact resistance	Pass ball drop test	Pass ball drop test
Pencil hardness	9 hours	>6 hours
Anti-scratch	No visual damage after steel wool test	No visual damage after steel wool test
Anti-reflection	V	V
Anti-finger print	V	V
Flexibility	-	Folding Test

ture of the touch sensor being fabricated between the color filter and the display, and the picture shows metal lines are not visible. The touch sensor was successfully integrated into the flexible display, as evidenced by Figure 9.17 which shows the touch signal and the drawing pattern on the displays.

Window cover films play an important role in determining if foldable displays can be a real product in the market. The traditional function of rigid-type cover films must not only protect the panel from damages such as impacts, scratches, or fingerprints but also possess high transmittance for good visual quality. The difficulty in selecting a good foldable cover is that the cover has to keep the same characteristics of a traditional rigid-type cover, but must also have excellent flexibility. However, it is hard to find a material that can be easily folded while maintaining good impact resistance. Table 9.3 shows the criteria for both rigid-type and flexible-type cover film. The adhesive between the panel and the cover film also affects the characteristics of the cover film. The adhesive in a traditional rigid-type display should have high transmittance and strong adhesion between the cover and the panel. In a foldable display, if the adhesive is too soft, it will affect the pencil hardness of the cover film. On the other hand, if the adhesive is too hard, buckling is easily observed when the panel folds. The key criteria of selecting a foldable adhesive is that the storage modulus has to be larger than 10^4 Pa, which ensures the panel has good flexibility and hardness.

9.5 Foldable Application

9.5.1 Foldable Technology Summary

In the previous sections, the basic panel stack-up has been described; the reasons for using technologies such as SPS CF–OLED to reduce unwanted ambient light reflections and the reasons for stacking the various panel layers in a specific order were also discussed. In this section, other research efforts that enable foldable display panels will be summarized.

9.5.1.1 Polymer Substrates and Related Debonding Technology

While the laser lift-off process was previously mentioned as the technology to separate the polyimide (PI) substrate from the glass backing [20], small dust particles and scratches could easily reduce the yield of the separation process. J. Koezuka et al. proposed using a thin buffer layer, approximately 2–3 μm in thickness, between the PI substrate and the glass backing to reduce the effect of foreign substances on the separation process [21]. Using a linear laser system also developed by the group, Koezuka was able to show that even with foreign substances (in the form of a pen marking) that could block the laser light, the PI substrate could still be completely separated from the glass backing without visible damages. In addition, the electrical properties of the field-effect transistor (FET) show no significant change. Another debonding method that has been commercialized is the FlexUP technology [22, 23], in which a PI or polyacrylic acid substrate is coated onto a debonding layer (DBL), and the entire structure is attached to a glass carrier. Once the device fabrication process is complete, the substrate can be mechanically removed from the glass carrier by cutting around the edge of the DBL, which has weak adhesion with the PI substrate.

9.5.1.2 Alternative TFT Types to LTPS

Display makers prefer the LTPS manufacturing process due to the performance in terms of mobility and well-established production line. Yet display prototypes using oxide semiconductors and TFTs have also been produced to showcase the comparable mobility and good threshold voltage uniformity [24]. Oxide backplanes have also been used to produce flexible OLED proto-types [25]. Since its discovery, C-axis aligned crystalline indium-gallium-zinc oxide (CAAC-IGZO) & cloud-aligned composite indium-gallium-zinc oxide (CAC-IGZO) FETs have been shown to be more resistant to plasma than nano-crystalline IGZO while having lower resistivity and shorter channel length, thereby reducing its size and allowing a narrower bezel in the display. In the same work, the FET structure was then applied in a functional 8K × 4K flexible display [26].

9.5.1.3 Encapsulation Systems to Protect Devices against Moisture

Traditionally, glass-based display panels were not designed to be foldable or flexible; metal layers or glass-sealing technologies were used together with desiccants to protect them against environmental moisture. However, as panels become flexible, deformation-induced surface defects become a concern. To prevent defects in the foldable devices during the deformation process, their thicknesses must be reduced to minimize stress and strain on the surface and within the structure. Because the devices also need to be protected from oxygen and moisture in the environment, the TFE process has been a popular solution to keep the device thin while maintaining adequate protection. A multilayered structure can extend the travel path of moisture and oxygen, while a densely deposited thin film can also delay their damage to the device. Therefore, inorganic/organic multilayered structures produced by a combination of chemical vapor deposition (for organic or inorganic layers) and other methods such as ink jet printing or spray coating (for organic layers) have been proposed [2]. Atomic or molecular layer deposition was also proposed as a single-layer solution for encapsulation [27]. For the edges of the displays, the typical solution is to extend the border of the encapsulation to increase the traveling distance of moisture/oxygen; as a method to maximize the viewable area, a side-wall barrier type structure (Figure 9.18) was proposed to achieve the same protection effect while keeping the border small [28].

9.5.2 Novel and Next-generation Display Technologies

As rigid displays give way to fixed curve displays and foldable displays, manufacturers are continuing to push the boundary of the physical form of a display. Flexible displays become rollable and more compact as their radius of curvature continues to decrease; companies like LG even demonstrated a rollable 65-inch OLED TV at the 2018 Consumer Electronics Show. Electrophoretic displays that consume very low power also have many novel applications. When used with flexible backplane technologies, they could be used as electronic shelf labels, wearable

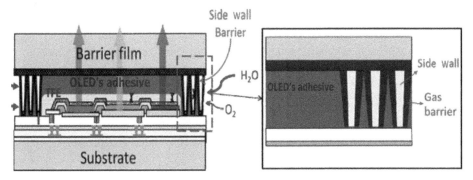

Figure 9.18 The side-wall barrier structure proposal. *Source:* ITRI [28].

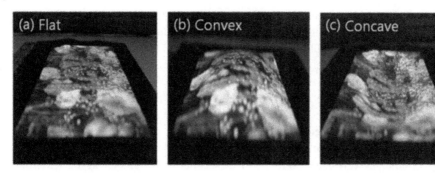

Figure 9.19 The stretchable AMOLED display made using thermoforming technique. *Source*: Samsung [35].

consumer electronics with a small bending radius, small size e-readers, and large format signage [29]. Creative ways involving Ag nanoparticles have also been developed to manufacture electrical circuits onto pre-strained substrates allowing for wearable sensors and ultimately stretchable displays [30]. Initially carbon nanotube (CNT) thin films were studied to produce transparent and stretchable TFTs, and they were paired with high-speed printing technologies to provide low-cost solutions and high-performance devices [31]. More recently, free-form AMOLED displays were demonstrated with a fixed concave and convex shape using thermo-forming processes and an innovative backplane design. The displays continued to use LTPS TFTs for their high mobility, but microcavities were built into the panel to separate sub-pixels into tiny islands and to allow for local flexibility. Furthermore, this confined each TFT and reduced the effect of V_{th} shift and leakage current on the performance of the panel. The mechanical behavior of the backplane was also similar to soft rubber, making the panel suitable for dynamically stretchable displays [32]. To make devices more intrinsically stretchable, wrinkled electrodes and elastomeric dielectric materials have been explored. However, the performance of the devices suffers because the electronic components are often encased in elastic matrices, which could decrease the overall performance.

As display technologies evolve, many of the ideas that seemed like science fiction are coming true in real life; rolling up your television when not in use is no longer a fantasy, and folding your daily digital newspaper while a new headline awaits you the next day can also be a reality. Competing technologies may exist to realize curved or flexible displays, and each has its merits over other technologies and its own obstacles to overcome. Even though the enabling technologies outlined were focused on their applications for OLED displays, they could also be useful for LCDs or micro-LED displays. Perhaps the next interpretation would be a hybrid display using a combination of technologies.

References

1 Huang, A.T., Chang, C.-S., Chan, C.-S., Wang, C.-L., Chang, -C.-C., Lai, Y.-H., Tu, C.-H., and Lee, M.-T. (2016). Flexible AMOLED displays make progress. *Inf. Disp.* 32 (2): 18–23.
2 Chwang, A.B. (2003). Thin film encapsulated flexible organic electroluminescent displays. *Appl. Phys. Lett.* 83: 413–415.

3 Weaver, M.S., Michalski, L.A., Rajan, K., Rothman, M.A., Silvernail, J.A., Brown, J.J., Burrows, P.E., Graff, G.L., Gross, M.E., Martin, P.M., Hall, M., Mast, E., Bonham, C., Bennett, W., and Zumhoff, M. (2002). A single-layer permeation barrier for organic light-emitting displays. *Appl. Phys. Lett.* 81: 2929.

4 Suen, C.S. and Chu, X. (2009). Multi thin film barrier protection of flex-electronics. *Solid State Technol.* 51: 36–39.

5 Yang, C.-J., Lin, C.-L., Wu, C.-C., Yeh, Y.-H., Cheng, C.-C., Kuo, Y.-H., and Chen, T.-H. (2005). High-contrast top-emitting organic light-emitting devices for active-matrix displays. *Appl. Phys. Lett.* 87: 143507.

6 Chen, S., Xie, J., Yang, Y., Chen, C., and Huang, W. (2010). High-contrast top-emitting organic light-emitting diodes with a Ni/ZnS/CuPc/Ni contrast-enhancing stack and a ZnS anti-reflection layer. *J. Phys. D: Appl. Phys.* 43 (36): 365101.

7 Macleod, H.A. (2010). *Thin-film Optical Filters*. Boca Raton, FL: CRC press.

8 Vaenkatesan, V., Wegh, R.T., Teunissen, J.-.P., Lub, J., Bastiaansen, C.W.M., and Broer, D.J. (2005). Improving the brightness and daylight contrast of organic light-emitting diodes. *Adv. Funct. Mater.* 15 (1): 138–142.

9 Kim, B.-C., Lim, Y.-J., Song, J.-H., Lee, J.-H., Jeong, K.-U., Lee, J.-H., Lee, G.-D., and Lee, S.-H. (2014). Wideband antireflective circular polarizer exhibiting a perfect dark state in organic light-emitting-diode display. *Opt. Express* 22 (107): A1725–A1730.

10 Cho, H. and Yoo, S. (2012). Polarizer-free, high-contrast inverted top-emitting organic light emitting diodes: Effect of the electrode structure. *Opt. Express* 20 (2): 1816–1824.

11 Man, J.-X., He, S.-J., Zhang, T., Wang, D.-K., Jiang, N., and Lu, Z.-H. (2017). Black phase-changing cathodes for high-contrast organic light-emitting diodes. *ACS Photon.* 4 (6): 1316–1321.

12 Ishibashi, T., Yamada, J., Hirano, T., Iwase, Y., Sato, Y., Nakagawa, R., Sekiya, M., Sasaoka, T., and Urabe, T. (2006). Active matrix organic light emitting diode display based on "Super Top Emission" technology. *Jpn. J. Appl. Phys.* 45 (5S): 4392.

13 Kim, S., Kwon, H., Lee, S., Shim, H., Chun, Y., Choi, W., Kwack, J., Han, D., Song, M., Kim, S., Mohammadi, S., Kee, I., and Lee, S.Y. (2011). Low-power flexible organic light-emitting diode display device. *Adv. Mater.* 23: 3511-3516.

14 Lee, M.T., Wang, C.L., Chan, C.S., Fu, C.C., Chen, C.C., Lin, K.H., Huang, W.C., Tsai, C.H., Weng, Z.X., Chan, C.C., Lin, Y.L., Huang, T.Y., Lin, P.Y., Lu, H.H., and Lin, Y.H. (2015). A symmetric panel stacking design for achieving a 3mm rolling radius in plastic-based AMOLED displays. *SID Digest* 46 (1): 958–961.

15 Lee, M.T., Wang, C.L., Chan, C.S., Fu, C.C., Chen, C.C., Lin, K.H., Huang, W.C., Tsai, C.H., Weng, Z.X., Chan, C.C., Lin, Y.L., Huang, T.Y., Lu, H.H., and Lin, Y.H. (2017). Toward a foldable display: The rise of AMOLED. *IDMC Digest* S14–2.

16 Lee, M.T., Wang, C.L., Chan, C.S., Fu, C.C., Chen, C.C., Lin, K.H., Huang, W.C., Chen, Y.H., Su, W.J., Chang, C.H., Tu, C.H., Lu, P.H., Tsai, C.H., Weng, Z.X., Tao, J.H., Lu, H.H., and Lin, Y.H. (2016). Ultra Durable Foldable AMOLED Display Capable of Withstanding One Million Folding Cycles. *SID Digest* 47 (1): 305–307.

17 Lee, M.T., Wang, C.L., Chan, C.S., Fu, C.C., Su, W.J., Weng, Z.X., and Lin, Y.H. (2016). Bi-directional foldable AMOLED display with millions repeat folding cycles. *IDW Digest*. 645–647.

18 Lee, M.T., Wang, C.L., Chan, C.S., Fu, C.C., Shih, C.Y., Chen, C.C., Lin, K.H., Chen, Y.H., Su, W.J., Liu, C.H., Ko, C.H., Weng, Z.X., Lin, J.H., Chin, Y.C., Chen, C.Y., Chang, Y.C., Huang, A.T.Y., Lu, H.H., and Lin, Y.H. (2017). Achieving a foldable & durable OLED display with BT.2020 color space using innovative color filter structure. *SID Digest* 25 (4): 229–239.

19 Lee, M.T., Wang, C.L., Chan, C.S., Fu, C.C., Shih, C.Y., Chen, C.C., Lin, K.H., Chen, Y.H., Su, W.J., Liu, C.H., Ko, C.M., Weng, Z.X., Lin, J.H., Chin, Y.C., Chen, C.Y., Chang, Y.C., Huang, A.T.Y., Lu, H.H., and Lin, Y.H. (2017). Achieving a foldable & durable OLED display with BT.2020 color space using innovative color filter structure. *J. Soc. Inf. Disp.* 25: 229–239.

20 Haskal, E.I., French, I.D., Lifka, H., Sanders, R., Sanders, P., Kretz, T., Chuiton, E., Gomez, G., Mazel, F., Prat, C., Templier, F., Campo, M.D., and Shahr, F. (2007). Flexible OLED displays made with the EPLaR process. *Eurodisplay Digest.* 36–39.

21 Koezuka, J., Idojiri, S., Shima, Y., Nakada, M., Okazaki, K., Aoyama, T., and Yamazaki, S. (2017). Flexible OLED display using C-Axis-Aligned-Crystal/Cloud-Aligned composite oxide semiconductor technology and laser separation technology. *SID Digest* 48 (1): 329–332.

22 Lee, C.C., Chang, Y.Y., Cheng, H.C., Ho, J.C., and Chen, J. (2010). A novel approach to make flexible active matrix displays. *SID Digest* 41 (1): 810–813.

23 Lee, C.C., Ho, J.C., Chen, G., Yeh, M.H., and Chen, J. (2015). Flexibility improvement of foldable AMOLED with touch panel. *SID Digest* 46 (1): 238–241.

24 Park, J.S. (2013). Oxide TFTs for AMOLED TVs. *Inf. Disp.* 29 (2): 16–19.

25 Wang, L., Ruan, C., Xu, H., Li, M., Zou, J., Tao, H., Zhou, L., Pang, J., Lan, L., Nig, H., Wu, W., Yao, R., Xu, M., and Peng, J. (2017). Flexible AMOLED based on Oxide TFT with high mobility. *SID Digest* 48 (1): 342–344.

26 Yamazaki, S., Hirohashi, T., Takahashi, M., Adachi, S., Tsubuku, M., Koezuka, J., Okazaki, K., Kanzaki, Y., Matsukizono, H., Kaneko, S., Mori, S., and Matsuo, T. (2014). Back-channel-etched thin-film transistor using c-axis-aligned crystal In-Ga-Zn oxide. *J Soc Inf Disp* 22: 55–67.

27 Ghosh, A.P., Gerenser, L.J., Jarman, C.M., and Fornalik, J.E. (2005). Thin-film encapsulation of organic light-emitting devices. *Appl. Phys. Lett.* 86: 223503.

28 Chen, J.L., Ho, J.C., Chen, G., Yeh, M.H., Lee, Y.Z., and Lee, C.C. (2016). Foldable AMOLED integrated with on-cell touch and edge sealing technologies. *SID Digest* 47 (1): 1041–1044.

29 Tsai, C.C. (2016). Advances in flexible electrophoretic displays. *IDW Digest* 47 (1): 1133–1135.

30 Lee, B., Byun, J., Oh, E., Kim, H., Kim, S., and Hong, Y. (2016). All-ink-jet-printed wearable information display directly fabricated onto an elastomeric substrate. *SID Digest* 47 (1): 672–675.

31 Ohon, Y. (2016). Flexible/stretchable electronics based on carbon nantube thin films. *IDW Digest* 1384–1386.

32 Hong, J.H., Shin, J.M., Kim, G.M., Joo, H., Park, G.S., Hwang, I.B., Kim, M.W., Park, W.S., Chu, H.Y., and Kim, S. (2017). 9.1-inch stretchable AMOLED display based on LTPS technology. *J. Soc. Inf. Disp.* 25: 194–199.

10

Flexible Reflective Display Based on Cholesteric Liquid Crystals

Deng-Ke Yang[1], J. W. Shiu[2], M. H. Yang[2], and Janglin Che[2]

[1] *Advanced Materials and Liquid Crystal Institute, Chemical Physics Interdisciplinary Program and Department of Physics, Kent State University, Ohio*
[2] *Industrial Technology Research Institute, Hsinchu, Taiwan*

10.1 Introduction to Cholesteric Liquid Crystal

Cholesteric liquid crystals (CLCs) are self-assembled systems consisting of chiral rod-like (or disk-like) organic molecules [1–4]. They possess orientational order such that locally the average direction of the long molecular axes of the constituent molecules are aligned along a common direction known as the liquid crystal (LC) director, which is denoted by a unit vector $\vec{n}$. Because of the anisotropic shape and orientational order of the molecules, CLCs exhibit anisotropic electric and optical properties. Their response to an externally electric field, namely the dielectric constant, depends on the direction of the electric field with respect to the director $\vec{n}$. When the electric field is perpendicular to the director, the dielectric constant is $\varepsilon_{\perp}$; when the electric field is parallel to the director, the dielectric constant is $\varepsilon_{//}$. Because the dielectric anisotropy $(\varepsilon_{//} \neq \varepsilon_{\perp})$, the orientation of the CLC can be changed by externally applied electric fields. CLCs also exhibit birefringence. When a light propagates through a CLC, if the polarization is perpendicular to the director, the experienced refractive index is n_o, known as the ordinary refractive index; if the polarization is parallel to the director, the experienced refractive index is n_e, known as the extraordinary refractive index.

Because the constituent molecules of CLCs are chiral, CLCs possess a spatial helical structure where the liquid crystal (LC) director twists around an orthogonal axis, known as the helical axis, as shown in Figure 10.1. On a plane perpendicular to the helical axis, the LC director is unidirectional and does not vary with position. Along the helical axis, the LC director twists. The distance for the director to rotate 360^0 is known as the pitch and denoted by P. Because CLCs are not polar, namely their physical properties are the same in the directions of $\vec{n}$ and $-\vec{n}$, the periodicity is $P/2$.

The first ever observed material to exhibit CLC phase was cholesteryl benzoate [5]. For this historical reason, the liquid crystal phase is called cholesteric liquid crystal. Nowadays, CLCs used in applications are mixtures of nematic liquid crystals and chiral dopants. Sometimes the liquid crystal phase is also called chiral nematic liquid crystal. The chemical structures of common chiral dopants (Merck) are shown in Figure.10.2. The capability of a chiral dopant to induce helical structure is characterized by its helical twisting power (HTP). The pitch of a mixture consisting of $x\%$ chiral dopant and $(1-x)\%$ nematic host is given by [1]

Flexible Flat Panel Displays, Second Edition. Edited by Darran R. Cairns, Dirk J. Broer, and Gregory P. Crawford.
© 2023 John Wiley & Sons Ltd. Published 2023 by John Wiley & Sons Ltd.

Helical axis
n
P

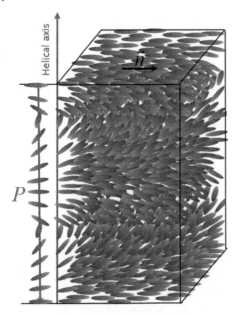

Figure 10. 1 Schematic diagram of the helical structure of CLC.

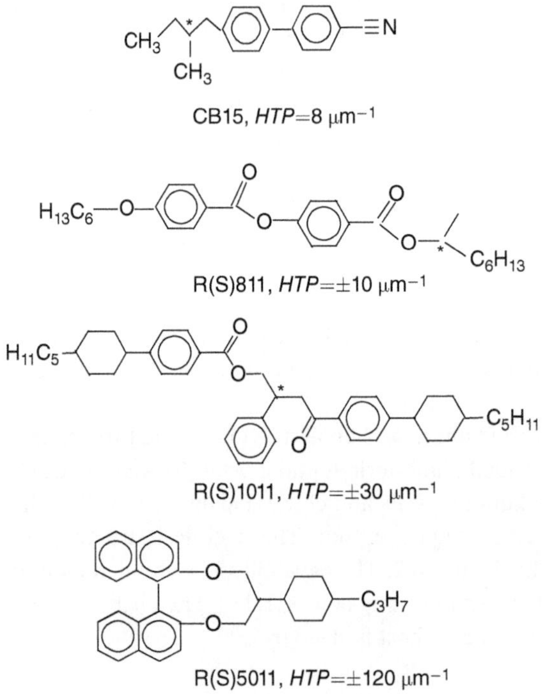

CB15, HTP=8 μm−1
R(S)811, HTP=±10 μm−1
R(S)1011, HTP=±30 μm−1
R(S)5011, HTP=±120 μm−1

Figure 10.2 Chemical structures of chiral dopants.

$$P = \frac{1}{(HTP) \cdot \chi} \qquad (10.1)$$

The unit of HTP is μm^{-1}. A positive value of HTP means the induced helical structure is right-handed, while a negative value of HTP means the induced helical structure is left-handed. For example, chiral dopant R811 has the HTP of 10 μm^{-1} and chiral dopant S811 has the HTP of $-10 \, \mu m^{-1}$. For example, when 30% R811 is mixed with 70% nematic host, the pitch is 0.33 μm. The HTPs of the chiral dopants are listed in Table 10.1. In selecting a chiral dopant, the most important property to be considered is the HTP. The HTP of a chiral dopant mainly depends on its molecular structure, and slightly depends on the molecular structure of the nematic host. Other important properties are its solubility and melting temperature. When a chiral dopant with a concentration over its solubility limit is doped to a nematic host, it phase separates from the host and causes an instability problem. The melting temperature of a chiral dopant affects the cholesteric phase temperature region. The solubility and the melting temperature are related: a chiral dopant with a high melting temperature usually has a low solubility. In constructing a CLC for devices, multiple chiral dopants may be used to obtain desired pitch, Ch phase temperature region, and stability.

10.2 Reflection of CLC

In CLCs, the LC director twists periodically along the helical axis, resulting in a periodical change of their refractive index. Therefore, they Bragg reflect light [1, 4, 6–9]. The reflection of a CLC depends on the polarization state of the incident light. It only reflect circularly polarized incident light with the same handedness as the helical structure of the CLC. If the incident light is circular polarized with the opposite handedness as the helical structure of the CLC, the light reflected by the liquid crystal at different positions on the helical axis are not coherent and cancel one another. Therefore, there is no reflection. If the incident light is circular polarized with the same handedness as the helical structure of the CLC, the light reflected by the liquid crystal at different positions on the helical axis may be coherent and enhance one another. Therefore, there is reflection when the wavelength of the light is matched to the pitch of the CLC. In this case, if the electric field of the incident light is perpendicular to the LC director and the wavelength equals $n_o P$, the electric field remains perpendicular to the LC director everywhere inside the CLC. The encountered refractive index is always n_o and the light is reflected. If the electric field of the incident light makes an angle α to the LC director and the wavelength of the light equals $[n_o n_e / (n_e^2 \sin^2 \alpha + n_o^2 \cos^2 \alpha)^{1/2}]P$, the electric field remains at the same angle to the LC director everywhere inside the

Table 10.1 Common commercially available chiral dopants.

Chiral dopant	HTP (μm^{-1})	Melting point (°C)
CB16	8	-50
R811	10	48
S811	-10	48
R1011	30	133
S1011	-30	133
R5011	120	>100
S5011	-120	>100

CLC. The encountered refractive index is always $n_{effective} = n_o n_e/(n_e^2 \sin^2 \alpha + n_o^2 \cos^2 \alpha)^{1/2}$ and the light is reflected. If the electric field of the incident light is parallel to the LC director and the wavelength equals $n_e P$, the electric field remains parallel to the LC director everywhere inside the CLC. The encountered refractive index is always n_e and the light is reflected. Therefore, reflected light has a bandwidth $\Delta\lambda = n_e P - n_o P = \Delta n P$, where $\Delta n = n_e - n_o$ is the birefringence, and the central wavelength of the reflection band is at $\lambda_c = 2[(n_o + n_e)/2](P/2) = \bar{n}P$, where $\bar{n} = (n_e + n_o)/2$ is the average refractive index. The birefringence of most CLCs is about 0.2. The reflection bandwidth is 60 nm and the color of the reflected light is pure. A typical reflection spectrum of a CLC is shown in Figure 10.3. If the incident light is unpolarized, it can be decomposed into a left-handed circular polarized light and a right-handed circular polarized light. Only the light with the same handedness as the helical structure of the CLC is reflected, and thus the maximum reflectance is 50%. If 100% reflectance of unpolarized incident light is needed, it can be achieved by two schemes. The first scheme uses a stack of two CLC layers with helical structures with opposite handednesses. One CLC layer reflects left-handed circular polarized component of the incident light and the other CLC layer reflects right-handed circular polarized component. The second scheme uses a stack of two CLC layers with the same handed (for example, right-handed) helical structure and a half waveplate in the middle. The top CLC layer reflects the right-handed circular polarized component of the incident light. When the unreflected left-handed circular polarized component propagates through the half waveplate, its polarization is converted to right handed circular polarization. It will be reflected by the bottom CLC layer. When it propagates through the half waveplate on its way back, its polarization is converted back to left-handed circular polarization. It will then pass the top CLC layer.

The peak reflectance of the reflection band of a CLC depends on the layer thickness of the CLC. The peak reflection (for circular polarized incident light with the same handedness as the helical structure of the CLC) as a function of the CLC layer thickness h is given by [4]

$$R = \left| \frac{\left(e^{2\pi h \Delta n/\bar{n}P} - 1\right)}{\left(e^{2\pi h \Delta n/\bar{n}P} + 1\right)} \right|^2 . \tag{10.2}$$

The calculated peak reflectance versus the CLC layer thickness is plotted in Figure 10.4, where the CLC has the following parameters: $P = 0.35\,\mu m$, $n_o = 1.55$ and $n_e = 1.75$. When $h = 10P = 3.5\,\mu m$, the peak reflectance is nearly 100%.

The reflection of the CLC is also dependent on the incident angle of the incident light θ. The central wavelength of the reflection band is given by

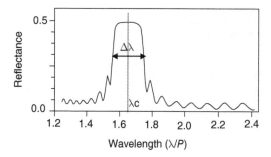

Figure 10. 3 Reflection spectrum of a CLC with $n_o = 1.55$ and $n_e = 1.75$. The incident light is unpolarized.

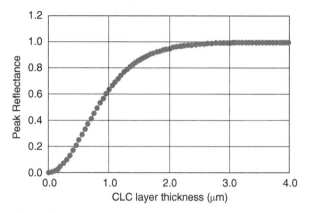

Figure 10.4 Peak reflectance as a function of layer thickness of a CLC with P = 0.35μm n_o = 1.55 and n_e = 1.75. The incident light is circular polarized with the same handedness as the helical structure of the CLC.

$$\lambda_c = \bar{n}P\cos\theta. \tag{10.3}$$

When the incident angle increases, the reflection band is shifted to shorter wavelength according to Equation (10.3), and furthermore peak reflectance increases and the polarization state of the reflected light deviates from circular polarization and becomes elliptical polarization. In a CLC display, the liquid crystal is sandwiched between two parallel substrates. The maximum incident angle outside the display is 90°. Inside the CLC, the incident angle is about arcsin[1.0 x sin(90°) /1.65] = 37°. At this incident angle, the reflection band is shifted to $\cos(37°)\bar{n}P = 0.8\bar{n}P$.

10.3 Bistable CLC Reflective Display

When a CLC is used to make a reflective display, it is sandwiched between two parallel substrates with transparent electrode, such as indium-tin-oxide (ITO). The optical properties of the CLC depends on its state, which is characterized by the orientation of the helical axis. In the absence of applied voltages, there are two possible states [4, 5, 10–16]. One is the planar state (also known as planar texture) where the helical axis is perpendicular to the display cell substrate, as shown in Figure 10.5(a). The CLC is reflecting in this state. A polarizing microphotograph of the planar state is shown in Figure 10.5(c). The lines are the oily streaks whose structure will be discussed later. The other state is the focal conic state (also known as focal conic texture) where the orientation of the helical axis is more less random throughout the display cell, as shown in Figure 10.5(b). The focal conic state has a poly-domain structure and is weakly scattering. A polarizing microphotograph of the focal conic state is shown in Figure 10.5(d).

Both the planar state and focal conic state could be stable in the absence of applied voltage, because the helical structure is preserved in both states. This bistability is very useful in many applications, which makes it possible to display information without applied voltages and thus without energy consumption. The bistability depends on the free energies of the planar and focal conic states. The involved free energies are the bulk elastic energy and surface energy at the cell substrate surface. Although the cholesteric layers are bent in the focal conic state, it costs very little energy and thus does not affect the bistability. The surface energy plays a major role in the determination of the stability [13, 15]. If there is a strong homogeneous alignment layer, such as a rubbed polyimide layer, on

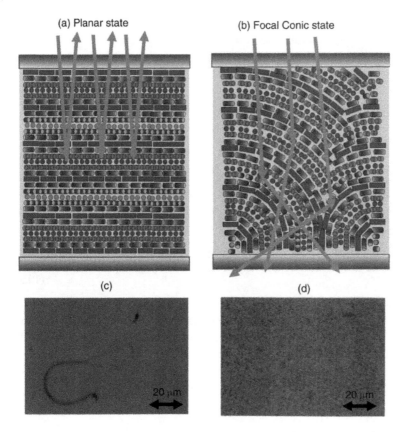

Figure 10.5 Side view of the LC director configuration of the two stable states of bistable CLC display. (a) Planar state, (b) Focal conic state, (c) microphotograph (in reflection mode) of planar state, and (d) microphotograph (in reflection mode) of focal conic state.

the two inner surfaces of the cell, in the planar state, the liquid crystal on the surface is aligned along the easy direction of the alignment layer; the surface energy is minimized. Therefore, the planar state is stable at 0 V. In the focal conic state, the liquid crystal in some regions is not parallel to the easy direction of the alignment layer, and the surface energy is higher. Therefore, the focal conic state is not stable at 0 V. If the CLC is switched to the focal conic state by an external stimulus, it will gradually relax to the planar state after the removal of the external stimulus. If the surface anchoring is weak, such as a rubbed ITO surface, the surface energy difference between the planar state and focal conic state is small, and the focal conic state becomes stable at 0 V. If a homeotropic alignment layer is used, even though the focal conic state is more compatible with the surface anchoring, the surface energy difference between the two states is also small, and both states are stable at 0 V.

In the case of strong homogeneous anchoring, the focal conic state can be stabilized by dispersing a small amount of polymer in the CLC [13, 17, 18]. In this scheme, although the surface energy of the focal conic state is still high, the dispersed polymer creates a high energy barrier between the planar state and focal conic state. Once the CLC is switched to the focal conic state by an external stimulus, the energy barrier prevents the liquid crystal domains from rotating, and therefore the focal conic state becomes stable. This material is known as polymer stabilized bistable CLC. In the preparation, a small amount of photo-polymerizable monomer is mixed with a CLC. The mixture is filled into a display cell. A voltage is applied across the cell to switch the material to the focal conic state, and then the cell is irradiated by UV light to polymerize the monomer. When the monomer is polymerized, it phase separates from the CLC to form a polymer network that stabilizes the focal conic state.

For reflective displays, specular reflection, such as the reflection from a flat mirror, is not desirable, because the displayed images can only be seen from one direction and thus the viewing angle of the displays is narrow. Diffused reflection, such as the reflection from white paper, is pleasing, because the displayed images can be seen from different angles. If a strong homogeneous alignment layer is used for CLC reflective displays and there is no dispersed polymer, the perfect planar state will be obtained, which specularly reflects light. Furthermore, the color of the reflected light changes with the viewing angle according to Equation (10.3). Therefore, the perfect planar state is not preferred. Imperfect planar state with a poly-domain structure is preferred, which can be achieved by using a weak alignment layer, or a homeotropic alignment layer or dispersing a small amount of polymer. With an imperfect planar state, the viewing angle of the display is much better. Furthermore, gray level reflectances become possible (discussed in detail in Section 10.6).

The focal conic state does not reflect light. It does, however, scatter light. In order to obtain high contrast ratio between the planar state and focal conic state, the backward light scattering must be minimized. The backward scattering depends on the size of the domains of the focal conic state. The domain size is usually larger than the wavelength of visible light. The larger the domain size, the weaker the backward scattering is. The factors that affect the domain size are the physical properties such as the pitch, elastic constants, and viscosity coefficient of the CLC, the alignment layer, the dispersed polymer, and the voltage used to obtain the focal conic state.

10.4 Color Design of Reflective Bistable CLC Display

10.4.1 Mono-color Display

In reflective CLC displays, in order to achieve high contrast ratio between the reflecting planar state and non-reflecting focal conic state, a black absorption layer is usually placed on the back plate of the display cell, which prevents ambient light from coming into the display from the bottom side and prevents incident light from being reflected from the bottom substrate–air interface [16, 17]. In this design, the planar state reflects mono-colored light and the focal conic state has no reflection. For example, the incident light is white and the CLC reflects green light. When the CLC is in the planar state, the display has a green appearance, as shown in Figure 10.6(a). When the CLC is in the focal conic state, the display has a black appearance, as shown in Figure 10.6(b).

In some applications, a white background is desired. This can be achieved by using a CLC reflecting yellow light and an absorption layer reflecting blue light, as shown in Figure 10.6 [19]. When the CLC is in the planar state, the liquid crystal reflects yellow light and the absorption layer reflects blue light; the display has a white appearance, as shown in Figure 10.6(c). When the CLC is in the focal conic state, the liquid crystal reflects no light and the absorption layer still reflects blue light; the display has a blue appearance, as shown in Figure 10.6(d). In this design, the display shows blue words on a white background.

10.4.2 Full-color Display

Full-color-reflective CLC displays can be achieved by using three CLCs reflecting red, green, and blue light. There are two schemes to build the displays. In the first scheme, three CLC layers are

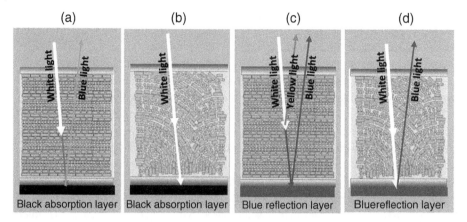

Figure 10.6 Schematic diagram of color design of reflective bistable CLC display. (a) Green state of green/black display, (b) black state of green/black display, (c) white state of blue/white display, and (d) blue state of blue/white display.

stacked, as shown in Figure 10.7(a) [20–22]. In the construction of the three CLCs, three different chiral dopant concentrations are used, such that they reflect red, green, and blue light respectively. There is a partition layer between two neighboring CLC layers, which prevents the interlayer diffusion of the chiral dopant. Each CLC layer has its own electrodes such that its reflection can be controlled independently. When the CLC layers are in the focal conic state, they scatter light. Their scatterings depend on their pitch. In order to achieve high reflection and high contrast ratio, the stacking order is that the CLC layer reflecting blue light is on the top, the CLC reflecting green light is in the middle, and the CLC reflecting red light is on the bottom.

The second scheme used to build full-color-reflective CLC displays is to use a single layer that is divided into three regions as shown in Figure 10.7(b) [23–26]. CLCs reflecting red, green, and

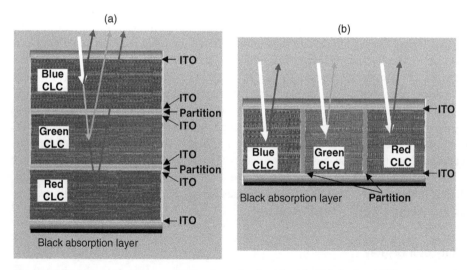

Figure 10.7 Schematic diagram of full-color-reflective bistable CLC display. (a) Three-layer scheme, (b) single-layer scheme.

blue colored light are filled into the three regions. There is a partition between two neighboring regions to stop inter-region diffusion of the liquid crystal. There are two methods to build the single-layer CLC display. In the first method, an empty cell with partitions is fabricated first. CLCs reflecting the three colored lights are then filled into the cell one by one. In the second method, an empty cell with partitions is fabricated first. A CLC with a photo-tunable chiral dopant is constructed and is then filled into the cell. At the end, the cell is irradiated by UV light to tune the color of the three regions. Under UV irradiation, the chiral dopant undergoes a molecular confirmation change or a chemical reaction and its HTP changes. In the irradiation photo-masks are used to control the irradiation time of the three regions. Different irradiation times result in different amounts of confirmation change of the chiral dopant.

As will be described later in this chapter, encapsulated CLCs are used to build flexible reflective CLC displays. The encapsulation will prevent the inter-regional diffusion and thus also plays the role of the partition.

10.5 Transitions between Cholesteric States

Bistable reflective CLC displays can be switched from one state to another by external stimuli. In this section we discuss the transitions between cholesteric states induced by externally applied voltages [4, 6, 10, 11]. We consider a CLC with a positive dielectric anisotropy ($\Delta\varepsilon > 0$). Initially the CLC is in the planar state as shown in Figure 10.8(a). When an intermediate voltage is applied across the display cell, the planar state, where the liquid crystal director is perpendicular to the generated electric field (in the cell normal direction), is unstable. The CLC is switched to the focal conic state whose electric energy is lower because in some regions the LC director is parallel to the electric field as shown in Figure 10.8(b). If the applied voltage is turned off from this state, the CLC will remain in the focal conic state. In order to switch the CLC back to the planar state, a high voltage must be applied to switch the CLC to the homeotropic state (also known as homeotropic texture), as shown in Figure 10.8(c), where the helical structure is unwound and the electric energy is a minimum, because the LC director is parallel to the electric field. The homeotropic state is unstable when the applied voltage is removed. The final state of the CLC depends on how the applied voltage is removed. If the voltage is removed slowly, the CLC will transform to the focal conic state. If the voltage is removed quickly, the CLC will transform to a transient planar state, as shown in Figure 10.8(d), whose structure is similar to that of the planar state, but its pitch is longer than that of the planar state. The transient planar state is unstable because the elastic energy is not yet minimized. The CLC will relax from the transient planar state to the planar state.

10.5.1 Transition from Planar State to Focal Conic State

When an intermediate voltage is applied across the display cell, the transition from the planar state to the focal conic state is realized through either the growing and buckling of a structure known as oily streak, whose structure is shown in Figure 10.9(a) [27], or Helfrich undulation [28–31], whose structure is shown in Figure 10.9(b). In the oily streak, the helical structure is preserved and the pitch is unchanged, but the helical axis is tilted. In some regions, the LC director is rotated toward the direction of the electric field, and thus the electric energy is reduced, which is the driving force for the transition. Because the helical structure is preserved, the twist

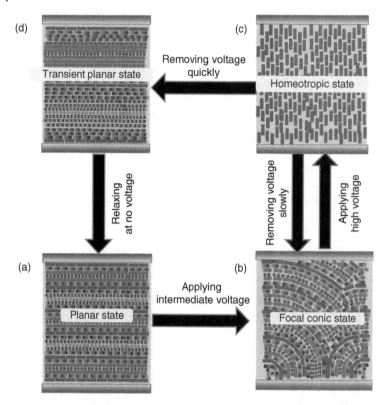

Figure 10.8 Schematic diagram of the transitions in CLC.

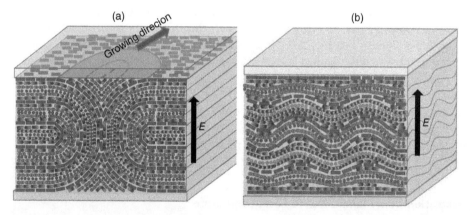

Figure 10.9 LC director configurations of oily streak and Helfrich undulation in the transition from the planar state to the focal conic state. (a) Oily streak, (b) Helfrich undulation.

elastic energy is 0. Due to the variation of the helical axis in space there is a bend elastic energy, which is against the transition. Furthermore, there is a wall energy on the interface between the domains, and there is a surface energy at the cell surface, both of which are also against the transition. For a CLC reflecting visible light, the threshold electric field E_{PF} for the transition is about 2 V/μm. The transition is a nucleation process: it starts from some irregularities, such as impurity and alignment layer defects, and is slow, on the order of 100 ms.

In the Helfrich undulation, the helical structure is preserved, but the pitch is changed and the helical axis is tilted. In some regions, the LC director is rotated toward to the direction of the electric field, and thus the electric energy is reduced, which is the driving force for the transition. Because the pitch is changed, there is a twist elastic energy. Because of the variation of the helical axis in space, there is also a bend elastic energy. Both elastic energies are against the transition. The threshold electric field for the Helfrich undulation is usually slightly higher than that of the oily streak. The transition is a homogeneous process. It does not need nucleation seeds and is fast, on the order of 10 ms. Above the threshold field, the amplitude of the Helfrich undulation increases with the applied electric field. When the applied electric field is sufficiently high, the structure collapses and the CLC is switched to the focal conic state.

10.5.2 Transition from Focal Conic State to Homeotropic State

When the applied voltage is increased, the helical axis becomes more and more parallel to the display cell substrate and the CLC is gradually switched to a state known as fingerprint state, as shown in Figure 10.10. There is no sharp boundary between the focal conic state and the fingerprint state. When the applied voltage is increased more, the LC director becomes more parallel to the applied electric field and the pitch becomes longer. The electric energy decreases, which drives the transition, while the twist elastic energy increases, which is against the transition. The pitch become infinitely long, namely, the helical structure is unwound and the CLC is switched to the homeotropic state, when the applied electric field is higher than a threshold field given by [1, 4]

$$F_{FH} = E_C = \frac{\pi^2}{P}\sqrt{\frac{K_{22}}{\varepsilon_0 \Delta\varepsilon}}, \tag{10.4}$$

where P, K_{22}, and $\Delta\varepsilon$ are the pitch, twist elastic constant, and dielectric anisotropy of the CLC.

10.5.3 Transition from Homotropic State to Focal Conic State

The homeotropic state is only stable when an electric field higher than E_c is applied. When the applied field is decreased, the CLC can transform either to the focal conic state or the transient planar state, depending on how much and how fast the electric field is decreased. We first consider the transition from the homeotropic state to the focal conic state when the applied electric field is decreased slowly. In the middle of the CLC, when the LC director starts to form a helical structure (with the helical axis parallel to the cell substrate) from the homeotropic state, there must be both left-handed and right-handed twist. Say the intrinsic helical structure of the CLC is right-handed. In the region the twist is right-handed, the twist elastic energy is decreased, while in the regions where the twist is left-handed, the twist elastic energy is increased more. The total twist elastic energy would increase, instead of decreasing. Therefore, there is an energy barrier against the formation of the helical structure. The transition is only possible if there are nucleation seeds that help reduce the energy barrier. The helical structure is restored by the growth of the twist finger, whose structure is shown in Figure 10.11 [4, 6]. It starts from a nucleation seed at which the LC director is not aligned homeotropically. Then it grows into the bulk. There is a

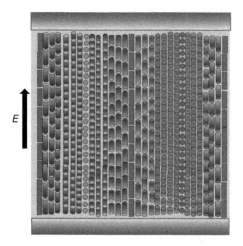

Figure 10.10 Side view of the LC director configuration of fingerprint state in the transition from the focal conic state to the homeotropic state.

Figure 10.11 Top view of the LC director configuration of the twist finger in the transition from homeotropic state to focal conic state.

hysteresis for the transition: the applied electric field must be decreased below a threshold E_{HF} in order for the twist finger to grow. The value of the threshold depends on the structure of the nucleation seeds. Usually $E_{HF} \sim 0.9E_c$. Because it is a nucleation process, the transition is slow, on the order of 100 ms.

10.5.4 Transition from Homeotropic State to Transient Planar State

Now we consider the transition from the homeotropic state to the transient planar state when the applied electric field is decreased. The CLC can restore the helical structure through an intermediate state known as conic helical structure or oblique helical structure, as shown in Figure 10.12 [4, 6, 32–35]. There are both twist and bend deformation in the conic helical structure. The pitch of the conic helical structure is P_c, which is different from the intrinsic pitch P of the planar state. The conic angle (the angle between the LC director and the helical axis) is θ. The free energy density is given by

$$f = \frac{1}{2}K_{22}\left(\frac{2\pi}{P} - \frac{2\pi}{P_c}\sin^2\theta\right)^2 + \frac{1}{2}K_{33}\left(\frac{2\pi}{P_c}\right)^2\sin^2\theta\cos^2\theta + \frac{1}{2}\Delta\varepsilon\varepsilon_o E^2\sin^2\theta \tag{10.5}$$

When the free energy is minimized with respect to P_c, we have

$$\frac{\partial f}{\partial P_c} = K_{22}\left(\frac{2\pi}{P} - \frac{2\pi}{P_c}\sin^2\theta\right)\left(\frac{2\pi\sin^2\theta}{P_c^2}\right) - K_{33}\left(\frac{2\pi}{P_c}\right)\left(\frac{2\pi}{P_c^2}\right)\sin^2\theta\cos^2\theta = 0. \tag{10.6}$$

From Equation (10.6) we will get

$$P_c = \left(\frac{K_{33}}{K_{22}}\cos^2\theta + \sin^2\theta\right)P. \tag{10.7}$$

In the beginning of the transition (close to the homeotropic state), θ is small, $\sin\theta = \theta$, $\cos\theta = 1$, $P_c = (K_{33}/K_{22})P$. Equation (10.5) becomes

$$f = \frac{1}{2}K_{22}\left(\frac{2\pi}{P}\right)^2\left(1 - \frac{K_{22}}{K_{33}}\theta^2\right)^2 + \frac{1}{2}K_{33}\left(\frac{2\pi}{P}\right)^2\left(\frac{K_{22}}{K_{33}}\right)^2\theta^2 + \frac{1}{2}\Delta\varepsilon\varepsilon_o E^2\theta^2. \tag{10.8}$$

The derivative of f with respect to θ^2 at $\theta = 0$ is

$$\frac{\partial f}{\partial\theta^2}\Big|_{\theta=0} = \left[\Delta\varepsilon\varepsilon_o E^2 - K_{22}\frac{K_{22}}{K_{33}}\left(\frac{2\pi^2}{P}\right)\right]. \tag{10.9}$$

In order for the transition to occur, the derivative must be smaller than 0, namely, the free energy decreases with θ. Therefore, the applied electric field must be smaller than the threshold field given by

$$E_{HP} = \sqrt{\frac{K_{22}}{\Delta\varepsilon\varepsilon_o}\frac{K_{22}}{K_{33}}\left(\frac{2\pi}{P}\right)^2} = \frac{2}{\pi}\sqrt{\frac{K_{22}}{K_{33}}}\frac{\pi^2}{P}\sqrt{\frac{K_{22}}{\Delta\varepsilon\varepsilon_o}} = \frac{2}{\pi}\sqrt{\frac{K_{22}}{K_{33}}}E_c \tag{10.10}$$

Usually, $K_{33}/K_{22} \sim 2$, and then $E_{HP} = 0.45E_c$. When the applied field is decreased below E_{HP}, the transition takes place, the conic angle changes quickly, but the pitch cannot. The CLC ends in a planar state with the pitch $P_c = (K_{33}/K_{22})P$. This planar state is known as the transient planar state, because its pitch is different from the intrinsic pitch and it is not a stable state. The transition from the homeotropic state to the transient planar state is a homogeneous process. Its transition time, given by $\tau_{HP} = \gamma P^2/k_{22}$, is fast, on the order of 1 ms.

10.5.5 Transition from Transient Planar State to Planar State

The transient planar state is not stable, the CLC will relax to the (stable) planar state with the intrinsic pitch. The process is also a nucleation process [36]. One example is shown in Figure 10.13. From the nucelation seed, more twisting is introduced such that the pitch decreases toward the intrinsic pitch. The number of pitches across the cell is increased by half or one. Because it is a nucleation process, it is slow, on the order of 100 ms.

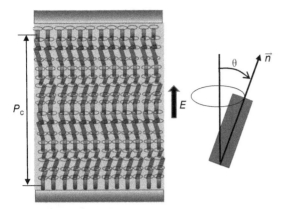

Figure 10.12 Side view of the LC director configuration of the conic helical structure in the transition from homeotropic state to transient planar state.

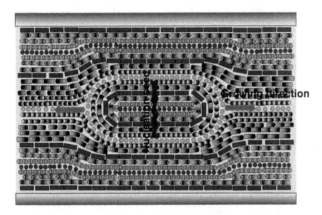

Figure 10.13 Side view of the LC director configuration in the transition from transient planar state to planar state.

As an example, a CLC reflects light at the wavelength 485 nm. Its elastic constants are $K_{22} = 4.9$ pN and $K_{33} = 8.4$ pN . The ratio between them is $K_{33}: K_{22} = 1.7$. When a high voltage is applied, it is switched to the homeotropic state with the reflectance of 0. The applied voltage is turned off at time 0 ms, the CLC transforms to the transient planar state and then to the (stable) planar state. The reflection spectra at various times are shown in Figure 10.14 [33]. At 0 ms, the CLC is in the homeotropic state and its reflection is 0. At 0.2 ms, a reflection peak appears at the wavelength $\lambda_c = 836$ nm. This reflection peak is produced by the transient planar state. As time goes on, the reflection increases. A maximum reflection is reached at about 1 ms. This indicates that the transition time from the homeotropic state to the transient planar state is about 1 ms. After that, the reflection peak of the transient planar state begins to decrease and becomes 0 at about 5 ms. At this time, a second reflection peak appears at the wavelength $\lambda_c = 485$ nm. This reflection peak is produced by the (stable) planar state. As time goes on, the reflection increases and reached to the maximum at about 7000 ms. This indicates the transition from the transient

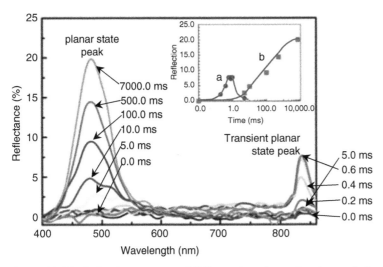

Figure 10.14 Reflection spectrum of CLC at various times after the applied voltage is turned off from the homeotropic state. Inset: Reflection as a function of time curves of the peak reflection of the planar state and transient planar state. Curve a: transient planar state, curve b: planar state.

planar state to the (stable) planar state is about 7000 ms. The ratio between the wavelengths of the two reflection peaks is λ_c: $\lambda = 836$: $485 = 1.7$, which equals to K_{33}: K_{22}.

10.6 Driving Schemes

Bistable reflective CLC displays have two stable states in the absence of applied voltage. One is the reflecting planar state and the other one is the non-reflecting focal conic state. This bistability makes it possible to make multiplexed displays by using passive matrix, which consists of two substrates with stripe electrodes. The stripe electrodes on the inner surface of the two substrates are orthogonal. The region between two orthogonal stripe electrodes is a pixel. An electric field can be applied to the LC in the pixel by applying a voltage across the electrodes. Once the pixel is switched to the planar state or focal conic state by a voltage pulse, it will remain in that state after the removal of the pulse.

10.6.1 Response to Voltage Pulse

In order to design a driving scheme for the bistable CLC display, its response to voltage pulses must be studied. A typical response is shown in Figure 10.15 where insets are used to schematically illustrate the state of the CLC [4, 6, 14]. The drawing in the left side of the insets shows the state of the CLC during a voltage pulse (when voltage is applied) and the drawing in the right side shows the state of CLC after the pulse (when voltage is removed). The horizontal axis is the voltage of the applied voltage pulse, and the vertical axis is the reflection of the CLC measured at 0 V after the pulse. When the pulse is turned off, in the beginning the reflection changes with time. In order to measure the stabilized reflection, the time interval between the end of the pulse and the time at which the reflection is measured should be long enough, typically 1 second.

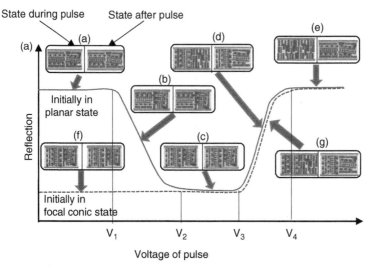

Figure 10.15 Reflection of bistable CLC display after voltage pulse versus voltage of pulse. Solid line: initially in planar state, dashed line: initially in focal conic state.

The solid line is the response of the CLC initially in the planar state. When a pulse with voltage less than V_1 is applied, it is too low to switch the LC, and thus the LC is in the planar state as shown by the left side drawing in Inset (a). After the pulse, the LC remains in the planar state as shown by the right-side drawing in Inset (a). The reflection is high (the maximum). When a pulse with voltage between V_1 and V_2 is applied, during the pulse some LC domains are still in the planar state and some LC domains are switched to the focal conic state, as shown by the left side drawing in Inset (b). After the pulse, the LC domain in the planar state remains in the planar state and the LC domain switched to focal conic state remains in the focal conic state, as shown by the right-side drawing in Inset (b). The reflection is decreased to a gray level. When a pulse with voltage between V_2 and V_3 is applied, during the pulse, all the LC domains are switched to the focal conic state, as shown by the left side drawing in Inset (c). After the pulse, all the LC domains remains in the focal conic state, as shown by the right-side drawing in Inset (b). The reflection is decreased to the minimum. When a pulse with voltage between V_3 and V_4 is applied, during the pulse some LC domains are switched to the focal conic state and some LC domains are switched to the homeotropic state, as shown by the left side drawing in Inset (d). After the pulse, the LC domain switched to the focal conic state will remain in the focal conic state and the LC domain switched to the homeotropic state will transform to the planar state, as shown by the right-side drawing in Inset (b). The reflection is increased to a gray level. When a pulse with voltage higher than V_4 is applied, during the pulse all the LC domains are switched to the homeotropic state, as shown by the left side drawing in Inset (e). After the pulse, all the LC domains will transform to the planar state, as shown by the right-side drawing in Inset (e). The reflection is increased to the maximum.

The dashed line is the response of the CLC initially in the focal conic state. When a pulse with voltage less than V_3 is applied, during the pulse the LC is in the focal conic state as shown by the left side drawing in Inset (f). After the pulse, the LC remains in the focal conic state as shown by the right-side drawing in Inset (f). The reflection is low (the minimum). When a pulse with voltage between V_3 and V_4 is applied, during the pulse some LC domain remains in the focal conic state and some LC domain is switched to the homeotropic state, as shown by the left side drawing in Inset (g). After the pulse, the LC domain in the focal conic state will remain in the focal conic state and the LC domain switched to the homeotropic state will transform to the planar state, as shown by the right-side drawing in Inset (g). The reflection is increased to a gray level. When a pulse with voltage higher than V_4 is applied, during the pulse all the LC domains are switched to the homeotropic state, as shown by the left side drawing in Inset (e). After the pulse, all the LC domains will transform to the planar state, as shown by the right-side drawing in Inset (e). The reflection is increased to the maximum.

In the bistable CLC display, the LC has a poly-domain structure because of the weak homogeneous alignment layer or homeotropic alignment layer or dispersed polymer. The size of the domains is on the order of tens of microns and cannot be seen by the naked eye. Each domain is bistable, either in the reflecting planar state or non-reflecting focal conic state. When they are in the planar state, the helical axes of the domains are not exactly perpendicular to the cell substrate, and thus can generate a diffused reflection. The responses of the domains to voltage pulses are different because of the different orientations of their helical axes and the different boundary conditions imposed on them. Voltage pulses can be used to produce gray level reflections.

The response of the CLC to voltage pulse depends on the width of the voltage pulse. The pulse width is on the order of tens of ms. When the pulse width is decreased, the required voltages to switch the LC to the focal conic state and homeotropic state increases. The pulse width cannot be arbitrarily short. If the pulse width is too short, it would be impossible to switch the LC completely to the focal conic state.

10.6.2 Conventional Driving Scheme

The conventional driving scheme for the bistable CLC display is schematically shown in Figure 10.16 [4, 6, 14]. The display is addressed one row at a time. In addressing row 1, as shown in Figure 10.16(a), the voltage applied to row 1 is V_s which is chosen to be $(V_3 + V_4)/2$. The data voltage to switch a pixel to the homeotropic state is chosen to be $-\Delta V/2 = -(V_4 - V_3)/2$. The voltage applied across the pixel is $V_{on} = V_s - (-\Delta V/2) = V_4$. Thus, the pixel is switched to the homeotropic state during the addressing and transforms to the planar state after the addressing. The data voltage to switch a pixel to the focal conic state is chosen to be $+\Delta V/2 = (V_4 - V_3)/2$. The voltage applied across the pixel is $V_{off} = V_s - (+\Delta V/2) = V_3$. Thus, the pixel is switched to the focal conic state during the addressing and remains in the focal conic state after the addressing. The voltage applied to other rows, which are not being addressed, is 0. The amplitude of the voltage applied to the pixels in the rows not being addressed is $\Delta V/2$, which must be lower than the voltage V_1 in order to prevent cross-talking. This requirement is usually satisfied. The advantage of the conventional driving scheme is that the waveform is simple and the cost of the driving circuitry is low. The drawback is that the time interval Δt to address a line is long, usually on the order of a few tens ms. The frame time to address a display equals the product of Δt and the number m of lines of the display. For a high information content display with more than 1000 lines, the frame time is longer than 10 s.

10.6.3 Dynamic Driving Scheme

The dynamic driving scheme is based on the fast transition time of the homeotropic state to transient planar state transition and the hysteresis in the transition between the focal conic state

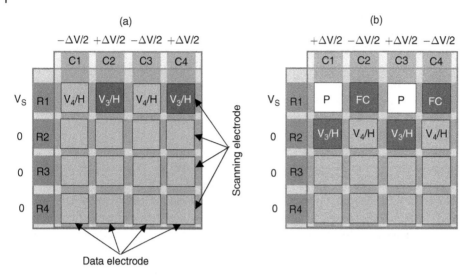

Figure 10.16 Schematic diagram of conventional driving scheme for bistable CLC display. (a) Addressing row 1, (b) addressing row 2.

and homeotropic state [4, 6, 37, 38]. It consists of three phases: preparation phase, selection phase and evolution phase as shown in Figure 10.17. The function of the preparation phase is to switch the CLC to the homeotropic state. Its voltage must be higher than hE_c, where h is the cell thickness and E_C (given by Equation 10.4) is the critical electric field for unwinding the helical structure, and the time interval Δt_P is a few tens ms. The function of the selection phase is to decide the final state of the CLC after the addressing. The time interval Δt_s of the selection phase is short, about 1 ms. If the applied voltage in this phase is higher than hE_{HP}, where E_{HP} (given by Equation 10.10) is the critical electric field for the transition from the homeotropic state to the transient planar state, the CLC remains in the homeotropic state, as shown in Figure 10.17(a). If the applied voltage is less than hE_{HP}, the CLC transforms to the transient planar state, as shown in Figure 10.17(b). The time interval should be equal to or longer than the transition time of the homeotropic state to transient planar state transition. The function of the evolution phase is to help decide the final state. The voltage of the evolution phase is between hE_{HF} and hE_c. As discussed in Section 10.6.2, $E_{HF} \sim 0.9E_c$, lower than E_c but sufficiently high to switch the CLC from the transient planar state to the focal conic state. If the CLC is held in the homeotropic state in the selection phase, it will remain in the homeotropic state in the evolution phase and transform to the transient planar state and then to planar state afterward, as shown in Figure 10.17(a). If the CLC is switched to the transient planar state in the selection phase, it will be switched to the focal conic state in the evolution phase and remains in the focal conic state afterward, as shown in Figure 10.17(b). The time interval Δt_E of the evolution phase is a few tens ms, which should be long enough to switch the CLC from the transient planar state to the homeotropic state.

Although the time intervals of the preparation phase and evolution phase are long, time sharing can be used to reduce the frame time of the dynamic driving scheme. Multiple line can simultaneously be put into the preparation phase and evolution phase. The number of lines put in the preparation phase should be an integer larger than $\Delta t_P/\Delta t_s$, and the number of lines put in the evolution phase should be an integer larger than $\Delta t_E/\Delta t_s$. The frame time to address a display with m lines is

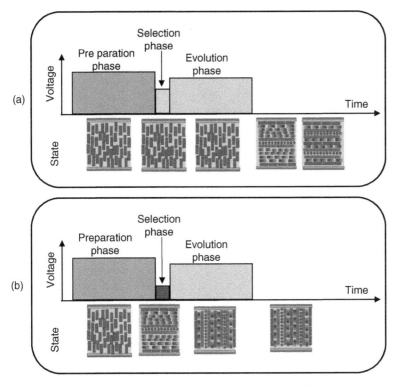

Figure 10.17 Schematic diagram of dynamic driving scheme. (a) Addressing a pixel to planar state, (b) addressing a pixel to focal conic state.

$$\Delta t = \Delta t_P + \Delta t_E + m \Delta t_s. \tag{10.11}$$

For example, $\Delta t_P = \Delta t_E = 40$ ms, $\Delta t_E = 1$ ms and $m = 1000$, the frame time is 1.08 s, which is short enough for some applications such as e-books.

10.6.4 Thermal Driving Scheme

Another cost-effective and time-effective driving scheme is thermal driving scheme, which makes use of the thermo-electro-optical effect of CLCs [39–41]. The threshold voltage V_{PF} to switch a CLC from the planar state to the focal conic state and the threshold voltage V_{FH} to switch the CLC from the focal conic state to the homeotropic state depends on $\sqrt{K/\Delta\varepsilon}$, where K is the effective elastic constant and $\Delta\varepsilon$ is the dielectric anisotropy of the LC. They are related to the orientational order parameter $K \propto S$ and $\Delta\varepsilon \propto S$. Therefore, the threshold voltages are proportional to $\sqrt{S}$. When temperature is increased, the order parameter decreases, and thus the threshold voltages decrease.

A typical reflection versus continuously applied voltage of the bistable CLC displays is shown in Figure 10.18 [39–41]. The CLC is initially in the reflecting planar state with high reflection. When the applied voltage is low, the CLC remains in the planar state and its reflection is high. When the applied voltage is increased above the threshold voltage for the planar state to focal

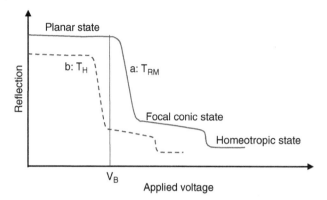

Figure 10.18 Reflection versus continuously applied voltage curve of bistable CLC display. Curve a: at room temperature, curve b: at high temperature.

conic state transition, the CLC is switched to the focal conic state and its reflection decreases. When the applied voltage is increased above the threshold voltage for the focal conic state to the homeotropic state transition, the CLC is switched to the homeotropic state, and its reflection decreases a little bit more because there is some backward scattering in the focal conic state. When temperature is increased, the reflection versus voltage curve will be shifted to low voltage. If initially the CLC is in the planar state at room temperature T_{RM}, when a voltage V_B is applied, the CLC will remain in the planar state, as shown in Figure 10.18. If the temperature is increased to T_H, the CLC will be switched to the focal conic state. It remains in the focal conic state when the temperature is decreased to room temperature.

The thermal driving scheme, also known as thermal addressing, makes use of the temperature dependence of the reflection versus voltage curve. It is suitable for bistable CLC displays made with thin plastic substrates. In the addressing, a bias voltage V_B is applied to the entire bistable CLC display, and the display goes through a thermal printer. In the regions touched by the thermal print head (TPH), the local temperature is increased, and thus the CLC is switched to the focal conic state. In the regions untouched by the thermal print head, the CLC remains in the planar state.

10.6.5 Flow Driving Scheme

The transition from the focal conic state to the planar state can also be induced by flow. This phenomenon is used to make CLC writing tablets [42, 43]. The CLC consists of a layer of CLC and two flexible plastic substrates. When the CLC cell is mechanically pressured as shown in Figure 10.19(a), a flow parallel to the cell is produced. If initially the CLC is in the focal conic state, the flow will switch it to the planar state. Qualitatively this phenomenon can be explained by the anisotropic translational viscosity property of the LC. There are three viscosity modes, depending on the relative directions of the LC director, velocity, and velocity gradient. Viscosity mode 1 is shown in Figure 10.19(b), where velocity is perpendicular to the LC director and the velocity gradient is parallel to the LC director. This mode has the largest viscosity coefficient. Mode 2 is shown in Figure 10.19(c), where velocity is perpendicular to the LC director and the velocity gradient is also perpendicular to the LC director. This mode has the medium viscosity coefficient. Mode 3 is shown in Figure 10.19(d), where velocity is parallel to the LC director and the velocity gradient is perpendicular to the LC director. This mode has the smallest viscosity coefficient. In the mechanically induced flow of the CLC, the velocity is the largest in the middle of the cell and 0 at the cell surface. The velocity gradient is perpendicular to the cell surface. When the CLC is in the focal conic state, the involved viscosity modes are 1 and 2, and thus the viscosity is larger. When the CLC is in the planar state, the involved viscosity modes are 2 and 3, and thus the viscosity is smaller. When the viscosity is larger, it produces larger torque to rotate the LC. Therefore, the CLC is switched from the focal conic state to the planar state. Words are written by pressing the tablet. An intermediate voltage is applied to erase words.

10.7 Flexible Bistable CLC Reflective Display

In order to produce flexible displays, thin plastic substrates must be used. Plastic substrates have a few issues that make them incompatible with polarizer-based displays, but not with bistable CLC reflective displays. The first issue is non-uniform birefringence of plastic substrates. In the production process of plastic substrates, non-uniform birefringence is generated due to strains. If

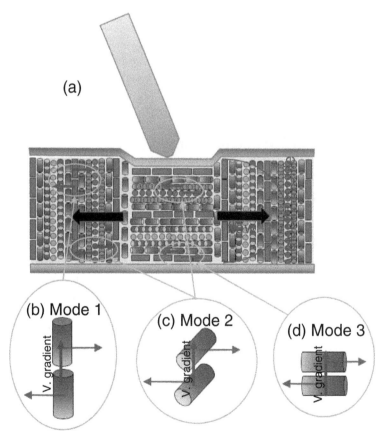

Figure 10.19 (a) Schematic diagram of mechanically induced flow writing of bistable CLC, (b) viscosity mode 1, (c) viscosity mode 2, and (d) viscosity mode 3.

they were used to make polarizer-based displays, the non-uniform birefringence would degrade the image quality. In CLC reflective displays, the reflection is produced by the CLC. They do not need polarizers, which makes it possible to use flexible thin plastic substrates. This is the first reason why bistable CLC reflective displays are compatible with plastic substrates.

The second issue is active matrix. For a LC does not exhibit bistability, it can only be used to make multiplexed displays on active matrix. In the production process of active matrix, the inorganic semiconductor thin-film-transistor (TFT) is usually fabricated at a temperature higher than 300°C. The TFT processing temperature is higher than the glass transition temperature of most plastic films. At such a high temperature, the substrates will deform. CLCs exhibit bistability. They can be used to make multiplexed displays on passive matrix. This is the second reason why bistable CLC reflective displays are compatible with plastic substrates.

The third issue is alignment layer. Some LC displays need strong homogeneous alignment layers. The alignment layers are produced by spin-coating, high temperature (~200°C) baking, and mechanical rubbing. Most plastic substrates cannot tolerate the high temperature baking. The bistable CLC displays do not need strong homogeneous alignment layers. In some cases, they do not need alignment layers at all. This is the third reason why bistable CLC reflective displays are compatible with plastic substrates.

The fourth issue is viscosity and self-adhesivity. A cost-effective way to produce large area displays is to use roll-to-roll process. In this process, first the LC is coated on one plastic substrate and then another plastic substrate is laminated on top of it. The viscosity of pure LCs is too low, making it difficult to coat them. Furthermore, it is much easier to achieve uniform cell thickness if the material sandwiched between the two plastic substrates is self-adhesive. The viscosity and self-adhesivity of pure CLCs are low. It is difficult to use them in roll-to-roll production process. This problem can be solved if CLCs are encapsulated.

10.8 Bistable Encapsulated CLC Reflective Display

In order to make CLCs suitable for production by roll-to-roll process, they should be encapsulated as shown in Figure 10.20. In order to preserve their bistability and reflectivity, the droplet size should be larger than 5 microns [44, 45]. When the encapsulated CLC droplets are in the planar state, as shown in Figure 10.20(a), they reflect light. A polarizing optical microphotograph of the droplets in the planar state is shown in Figure 10.20(c). When an intermediate voltage is applied across the cell, the CLC droplets are switched to the focal conic state, as shown in Figure 10.20(b), and they weakly scatter light. A polarizing optical microphotograph of the droplets in the focal conic state is shown in Figure 10.20(d). When a high voltage is applied, the droplets are switched to the homeotropic state. When the high voltage is removed suddenly, the droplets transform to the planar state with their helical axes perpendicular to the cell substrate. The polymer acts to increase viscosity for coating and to increase adhesivity to keep the cell substrates together. The concentration of the polymer should be less than that of the CLC to remain high reflection. The droplet size may be larger than the cell thickness and thus the droplets usually have oblate shape.

Encapsulated CLCs can be produced by two methods. The first method is polymerization induced phase separation. In this method, a CLC is mixed with a photo-polymerizable monomer (with functionality higher than 1 and a small amount of photo-initiator. The mixture is

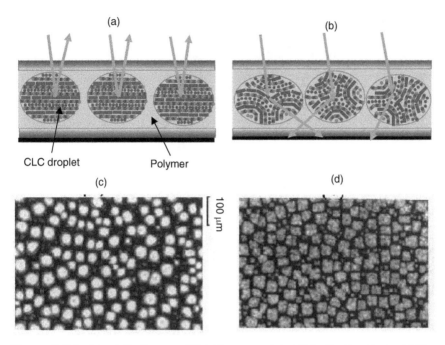

Figure. 10.20 Schematic diagram of bistable encapsulated CLC reflective display. (a) Planar state, (b) focal conic state, (c) microphotograph of planar state, (d) microphotograph of focal conic state.

sandwiched between two plastic substrates with ITO coating on their inner surfaces. The film is then irradiated by UV light to polymerize the monomer. When the monomer is polymerized, the CLC phase separates from the polymer to form droplets. In order to form large droplets, the monomer concentration should be low, less than 20%, and the UV intensity should be low, a few mW/inch^2.

The second method is emulsion. In this method, a CLC, water and a water-soluble polymer are mixed. The mixture is put into a container and stirred rapidly. The CLC and polymer form droplets with the LC inside and the polymer as the shell. In order to stabilize the droplet, a surfactant is added. The material is coated on a plastic substrate. The water is allowed to evaporate.

Then the second substrate is laminated on top. The droplet size is determined by the concentrations of the components and the stirring speed.

10.9 Production of Flexible CLC Reflective Displays

Bistable CLC reflective displays have the merits of bistability, good color performance, roll-to-roll fabricating capability, and multiple writing methods (electrical, thermal, physical, and optical addressing methods). They have many applications such as e-reader, writing board, and e-wallpaper. This section will discuss the research and development activities in the last few years, especially in ITRI.

The development results and products of CLC display (CLCD) for e-paper applications are summarized in Figure 10.21. The first monochrome (black/green) e-reader was launched in 2003

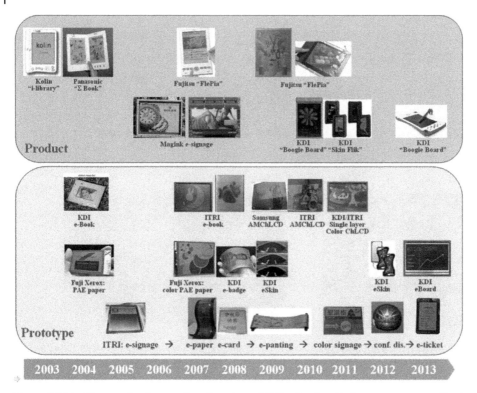

Figure. 10.21 The trends of research and product of CLC reflective display on e-paper applications (summarized by ITRI, 2013).

by Kolin. Panasonic announced double-sided blue/white "Σ Book" in 2004. The first color e-reader "FlePia" was introduced by Fujitsu [46]. The FlePia was fabricated by stacking three separate CLC layers, on plastic substrates, of red, green, and blue colors to show colorful images [47]. Following the implementation of adhesive photo spacers, the light leakage observed with bead spacers is drastically reduced to show higher contrast ratio. Also, fine tuning by a rubbing process further improves the alignment of liquid crystal for superior image quality [48]. The index matching of optical adhesives for each layer effectively reduces the light loss and reflectance between interfaces of substrates. Later, a fast driving scheme was employed to increase the refresh rate. By 2011, the electro-optical performance of FlePia had achieved the specification for 8-inch VGA, 158 ppi, 35% reflectivity, 8.8:1 contrast ratio, 262 k colors, 17% NTSC, and a refresh time of 0.7 seconds. Because of its remarkable performance, FlePia was widely considered for applications such as readers, textbooks, restaurant menu, and advertisement signage.

For large size application of outdoor signage, Magink fabricates color e-signage with similar structure for stacking three CLCDs on glass [49]. Utilizing their proprietary tiling technology, the company produced public signage with size greater than 2 meters. By adopting segmented driving method, the tiled color signage achieved the remarkable contrast ratio of 50:1, 64 gray scales, and color performance of 34% NTSC. These specifications reached almost the highest limits of CLCD and became comparable to those of printing magazines and advertising paper. The tiled unit module is designed by 3 × 3 pieces of 17 × 17 cm CLCD with pixel size of 9 × 9 mm

The specially implemented segmented driving method addresses CLCD between planar state and homeotropic state to demonstrate quick response and video rate possibility.

An interesting and noteworthy handheld product, named "Boogie Board," was introduced by Kent Display Inc. (KDI) in 2010, as a writing board to replace the usage of paper [40]. It can be rewritten on thousands of times by pen or finger, and be refreshed by pushing a button with integrated driving circuit. It is manufactured by a roll-to-roll process, of two plastic substrates with transparent conductive coatings, sandwiching the polymer stabilized CLC layer. The stable focal conic state of the writing board is black. As pressure is applied by pen or finger, the CLC changes its state into the planar state (white state). Polymer is dispersed in the CLC to control the width of the written line. Narrow lines are obtained then the polymer concentration is higher. The electrical field applied to the CLC to switch it to the focal conic state (dark state) erases the images. The newer writing board model has been integrated with touch sensors and memory for storage of writing information. The data written also can be wirelessly transferred to other hand-held devices such as notebook, tablet, and smartphone.

The bistable CLC technology has been studied for single-layered color e-reader by researchers in ITRI and Samsung [50]. The single-layer full-color CLCD could be driven by either passive-matrix or active-matrix driving methods. These color e-readers have the simplified panel struc-ture from three layers down to one layer, and video rate possibility with response time of 20 ms. Another technology of photo addressing e-paper (PAEP) developed by Fuji Xerox shows color image by integrated OPC (organic photo conductive) layer [51–54]. The OPC layer can induce a voltage in the area of light exposure. A mask between light source and PAEP e-paper could create different voltages on areas of different light exposures to show images. Still another concept of e-skin has been revealed by KDI in the product "Reflex Electronic Skins" for the covering of smartphone [55]. A thermal forming process is developed for stacking three CLC layers with dif-ferent colors on PET substrate [56]. It helps to realize the three-dimensional profiles for various shapes of mobile devices. Later, KDI also revealed a seamless tiling technology for stitching four display panels into a larger size eBoard [57]. Lamination of optical coupling film is introduced to decrease the tiling gap in accomplishing a large display with one meter in diagonal.

The research activities in ITRI stay largely focused on glass based signage, single substrate e-paper (named i2R e-paper) and thermal-forming conformal displays [58–64]. The monochrome e-signage developed in 2005 has been upgraded to color signage with optimization of CLC and absorption layer to show multiple colors. The i2R e-paper applications for e-card, e-painting, and e-ticket have been demonstrated by various addressing methods such as electrical, optical, and thermal writing. The thermal writing method, especially, achieves a resolution of 300 dpi, 16 gray levels on a 3-meter-long display. In addition, a conformal display also has been demon-strated successfully. This colorful conformal display based on i2R e-paper takes advantage of thermal forming technology and photo tuning CLC to show different colors as e-skin for decoration applications [60].

10.9.1 Color e-Book with Single-layered Structure

A novel display with performance of low cost, light weight, low power consumption, and flexible form factor is required for e-book applications. The bistable CLCD has become one of candidates because of its low power consumption, bistability, reflectivity, and flexibility by using plastic sub-strates. The characteristic of CLCs reflecting different wavelengths of visible light is suitable for color display. Another distinctive feature of CLCD is the bistability that exhibits two stable states of the planar state, which reflects light, and the focal conic state, which absorbs light. The application of CLCD made by glass substrates ranges from small size displays for e-book, instrumentation displays, and handheld devices to large area displays for signage. CLCDs are particularly suitable for update-on-demand applications because the image can be retained without any applied power. The characteristic of bistability indicates that power is only needed for refreshing images [17].

Typical full-color CLCDs made by glass, as well as plastic substrates, have the structure of triple stacked layers of primary colors and each color layer is addressed by its own pair of indium-tin oxide (ITO) electrodes. This brings the total number of conducting electrodes and the number of substrates to six [65, 66]. Due to index mismatch and light absorption issues of the electrode material, each electrode contributes to the display reflectivity losses especially because of the fact that light reflected from display passes twice through each electrode. This makes the transmission of electrode material and the number of electrodes critical for the display reflectance. Furthermore, conventional full-color CLCDs can only reflect either right- or left-handed circularly polarized light depending on their composition and structure of LC. Nevertheless, here we reveal the study of how to improve the reflectivity of full color CLCDs, and also present a single-layer full-color CLCD with high reflectivity, which was fabricated by pixelized vacuum filling (PVF) method [67].

The single-layer CLC panel is composed of two substrates with three color CLCs sandwiched between them. The cell structure comprises column pixels for filling red, green, and blue color CLCs that are separated by the striped bank structure. The bank structure sustains the space between two substrates and the cell gap of the display is defined by the height of the bank. An adhesive layer is coated between the top electrode and the bank structure to ensure there is no space between them and to prevent the leakage of the LCs in adjacent pixels. The structure of the CLCD and the flow chart of fabrication processes are shown in Figure 10.22. The red, green, and blue CLCs are filled sequentially into the column pixels by vacuum capillary filling.

This sequential filling method by the distinct structure design is called pixelized vacuum filling (PVF). After the filling of the LCs, the cutting and end sealing processes are also developed to fabricate the single-layered color CLCD.

The single-layered color CLCD is developed in the traditional LCD manufacturing process. The transparent electrodes, ITO on glass substrates, are patterned by photolithography. The negative photoresist is spin-coated on the top substrate and then the bank structure formed by photolithography. The thickness of photoresist determined by the rotation speed of the spin coater can define the cell gap of the structure. The alignment accuracy between ITO electrode and bank structure has to be well controlled to prevent light leakage. On the other substrate, an UV-curable adhesive polymer is coated on top of ITO electrode. Before the coating process of adhesive material, the substrate has to be cleaned thoroughly because the coating results of the thin adhesive layer is very sensitive to the surface property of the substrate. The purpose of the adhesive layer is to prevent the overflow of the LCs from the adjacent column pixels and keeps the cell gap uniform between the two substrates. Furthermore, the adhesive layer also acts as a horizontal alignment layer to provide the horizontal anchoring force to the CLC. The next step is to assemble these two prepared substrates and then to continue the process of LC filling.

The color CLCs are then filled into the cell structure by capillary vacuum filling. The PVF method has been developed for filling the single layered red, green, and blue CLCs, and shown in Figure 10.23. The LC filling efficiency of the pixelized column filling is much better than the conventional planar filling. The reason is that the capillary force of the LC in the designed

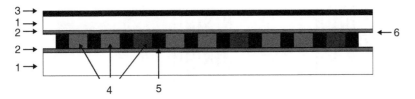

1. Substrate; 2. Electrode; 3. Absorption layer; 4. Cholesteric liquid crystal; 5. Bank; 6. Adhesion layer

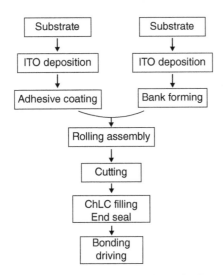

Figure. 10.22 The structure of the color CLCD and the flow chart of fabrication processes.

column structure is much stronger than the force in the traditional filling method. The filling process takes only four hours for the three filling steps of red, green, and blue CLCs. The filling entrance would be sealed by the UV-curable sealant after each step of the pixelized filling. This CLC panel can achieve good bistability of the planar and focal conic states because of the feature of homogeneous aligning surface of adhesive layer on substrate. The high reflectivity of good planar state was induced by the weak homogeneous anchoring force. The CLC panel with good planar state exhibits higher brightness than that with diffuse-like planar state, which was shown in Figure 10.24(a) and (b); furthermore, such weak homogeneous anchoring force condition could stabilize the focal conic texture and perform low reflectance shown in Figure 10.24(c).

For a single layer colored CLCD, the R, G, B sub-pixels are usually aligned in row. As shown in Figure 10.25(a), the common (scan) driving signal will cross the R, G, B sub-pixels simultaneously as addressing the colored CLCD row by row. Because the R-V curves of three colors are different, a column (data) driver IC is required to generate three different voltage levels for three colors, even a PWM driving scheme is applied. This requirement is very difficult and expensive for any commercial PWM column driver IC. To solve the abovementioned issue, we propose a transposed PWM scheme for single-layered colored CLCD. The sub-pixels of our CLC panel were transposed and aligned in column, as shown as Figure 10.25(b). The common (row or scan) or segment (column or data) driving signal is only required to provide unique voltage level as

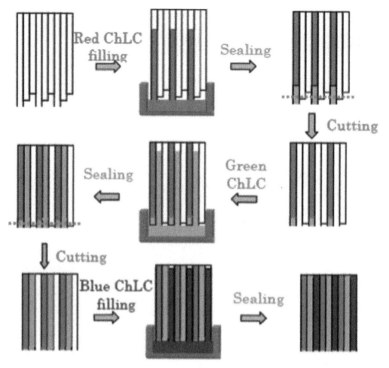

Figure. 10.23 The color single-layer CLCD pixelized vacuum-filling process.

Figure 10.24 (a) the good planar state, (b) the diffuse-like planar state, (c) the focal conic state of the CLC with good planar state.

addressing each row by the proposed transposed PWM unified driving scheme. For clarity, the common driving signals for several consecutive rows are shown in Figure 10.25(c), where VR1, VG1, VB1, etc. are the driving waveform of the consecutive rows in order. Each row was only driven by a single voltage level. The voltage level must be switched between VR, VG, and VB while driving R, G, and B CLC, respectively. The proposed driving scheme greatly simplifies the complexity of system design and significantly reduces the cost of driver IC. Finally, the full color 10-inch bistable CLCD was fabricated (shown in Figure 10.26). The bistability of the display has been verified by checking image performance after turning off the power for a month. Our single-layer full-color CLCD exhibits high reflectivity (~26%) and high resolution (100 ppi) with 4-bit color gray scale.

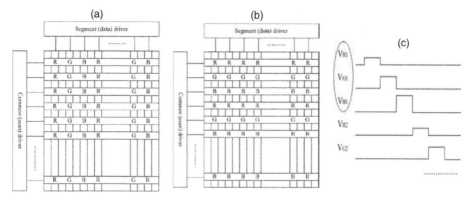

Figure 10.25 (a) Traditional sub-pixels structure for colored CLCD, (b) Transposed sub-pixels structure for the single-layered colored CLCD, (c) Common driving signals for several consecutive rows.

10.9.2 Roll-to Roll E-paper and Applications

Development activity of flexible display has increased rapidly in recent years, for it holds the promise to deliver features such as ruggedness, energy saving, easy to carry, easy to store, and large area display. Encapsulated cholesteric LC technology also has the advantages in bistability, color performance, and is viable for continuous roll-to-roll coating process on flexible substrate. In this section, improvement in the performance of CLCD is reported in reflectance, contrast ratio, and driving voltage. Additionally, a new thermal addressing method is disclosed with system design and superior optical performance. Finally, examples of several new applications in entertainment and interior decorations are described.

Encapsulation of the CLC materials with gelatin, which was initially developed at Eastman Kodak, forms discrete domains and keeps droplets of the CLCs from coalescing. The encapsulated CLCs dispersed in gelatin are coated on a flexible substrate with pre-patterned conductive layer, followed by overlaying the absorption layer and a second electrode. The sandwiched struc-

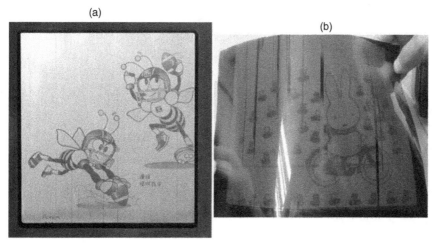

Figure 10.26 (a) The 10-inch single full color CLCD on glass substrates with resolution of 100ppi and (b) on PET plastic substrates with resolution of 40 ppi.

ture of CLCD is shown in Figure 10.27 [56]. The color of this display can be tuned by adjusting LC, chiral dopant formulation, and changing the absorbing material that is made of colored nanoparticles. The optical performance of CLCD is illustrated in Figure 10.28, where the CLCD sample, in the planar state, features yellow color CLC droplets, and a blue absorption layer. When the CLC reflects yellow light, the rest of light propagates to the absorption layer. In the absorption layer, the green and red light will be absorbed, only blue light can pass the absorption layer and be reflected by the second (silver) electrode. Therefore, the reflected, additive yellow and blue (which results in white color), will be seen in the planar state. On the other hand, in the focal conic state, the encapsulated CLC is almost transparent, and only the blue light is reflected. As shown in Figure 10.28. The white and blue color could be shown with yellow CLC and dark blue absorption layer. Our CLCD is fabricated by a roll-to-roll process with process flow shown in Figure 10.29. A pulsed laser is used to pattern the ITO on the PET film. The LC and absorption layer solutions are coated on the PET/ITO film after the laser patterning process. The second electrode, a silver paste, is screen printed on the absorption layer. The entire process, thus, is accomplished in a roll-to-roll manner.

In order to reduce driving voltage, a conducting polymer (Baytron P) is added in the absorption layer. By doing this, the resistivity of the absorption layer is decreased, and the driving voltage is reduced as well. The R-V (reflectance versus voltage) curve of different testing pads shows that driving voltage of the panel with normal absorption layer is about 160 V, and decreases to below 80 V by adding the conducting polymer in absorption layer (Figure 10.30(a)). To enhance the reflectance, the substrate with high transmission (>80%) was tested. The results showed that the reflectance, when using a substrate (part name: 0C300i) with higher transmission, is, indeed,

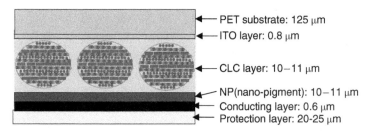

Figure 10.27 Schematic diagram of flexible CLC display.

Figure 10.28 The blue/white CLC display.

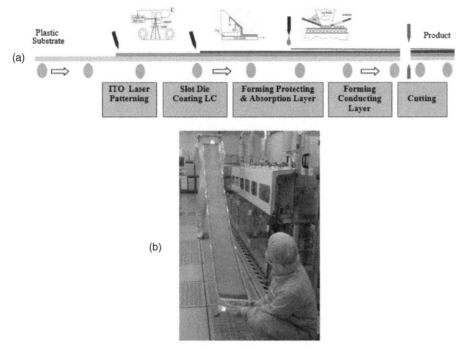

Figure 10.29 (a) The roll-to-roll process flow. (b) A panel made by this process.

greater than that with a lower transmission substrate (part name: 0C100) Figure 10.30(b). In the coating process, the gelatin binder helps to keep the encapsulated CLC maintain its droplet shape. However, the gelatin takes up space in the CLC layer as well, so that the effective LC layer thickness is reduced and the reflectance is affected. To improve the effective LC thickness without increasing the total LC coating thickness, we thus experimented with reducing the gelatin content in the LC layer. When the gelatin ratio in LC layer is reduced from 38% (LC: gelatin = 8:5) to 15% (LC: gelatin = 11:2), the reflectance is improved from 15% to 27%. The panel with lower gelatin ratio shows better performance.

It has been reported in the literatures that the stable states (planar/focal conic) of the CLC can be switched either thermally, or electrically [68–70]. In our laboratory, a thermal print head

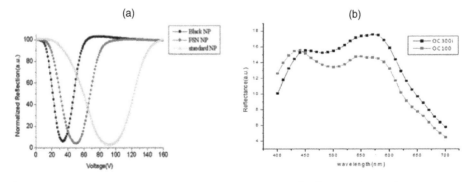

Figure 10.30 (a) Results of reducing voltage. (b) Spectrum of different substrates.

(TPH) is used to address the CLCD. The thermal print head is a commercially available component and widely used in office thermal printers and fax machines. With the auxiliary voltage, the dark state reflectance is dropped to below 2% and the contrast ratio is improved to become comparable to that of electrical driving. The contrast ratio of image is significantly improved three times higher by adding the auxiliary AC voltage. Figure 10.31 shows the thermal writer system integrated with wireless controlled unit and the 24 x100 cm rollable CLCD with resolution of 300 dpi. An ultra-long 24 x 300 cm blue/white CLCD is successfully demonstrated by using thermal addressing method (Figure 10.31). Compared to the electrical addressing, the thermal addressing scheme has the advantages of high resolution, low system cost, and almost no limitation of panel length [59, 71]. The e-painting application of Chinese landscape paintings is showcased by TPH addressing scheme, which can refresh the flexible CLCD longer than 3 meters (Figure 10.32). It can show fine details in the paintings because of its high resolution (>300 dpi) and true eight gray levels capability. The re-writing test reveals that the contrast ratio can be above 4:1 after re-writing more than 260 times. The improvement has been accomplished based

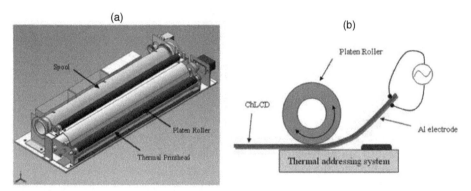

Figure 10.31   (a) Structure of thermal printer module. (b) Addressing mechanism.

Figure 10.32   Wireless controlled thermal writing system on Chinese painting, 24 x 100 cm black/white CLCD (left), and the 24 x 300 cm blue/white CLCD (right).

on the optimization of the protection layer over silver layer, stronger panel structure, and lower temperature activated liquid crystal.

The flexible CLCD revealed here has many potential applications [55, 57, 58, 72]. The soft clock was designed by a leading industrial product designing company, as shown in Figure 10.33(a). It shows the combination of art content and functional device, with an electrical segmented driving system. The e-toy product concept, which was demonstrated in FINETECH Exhibition in 2010, in Japan, combines flexible display and pressure sensor to have the driving system integrated inside (Figure 10.33(b)). This interactive entertainment application could be another example for numerous new market opportunities of flexible displays.

For the advertisement market, CLC e-signage can be built on both segmented and passive matrix-driven panels. The segmented driven panel can be fast refreshed (<0.2 seconds) to attract customers' attention. The passive matrix panel can detail product name and sale information. The color e-signage (40 x 40 cm) shown in Figure 10.33 (c) is made by tiling four A4 size panels with system and battery integrated.

For the i2R e-paper technology, two new process steps have been successfully implemented in the i2R process flow, marked with yellow blocks in Figure 10.34. The first new process step is the screen printing of peel-able material to form the mask for the following coating process. After the coating (LC, NP, and Gel layers) and sputtering (Al layer) process, the peel-able layer would be cut and peeled off to form the patterns of i2R panel as designed card size. The other new process step is the replacement of Ag by Al by sputtering method. Because the second electrode made by Ag constitutes the major part of whole material cost,

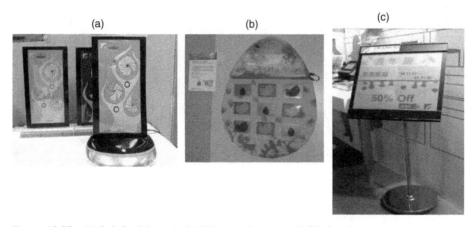

Figure 10.33 (a) Soft flexible e-clock, (b) interactive e-toy, (c) Tiled e-signage.

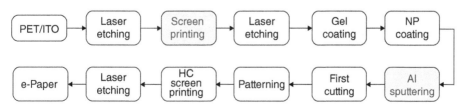

Figure 10.34 The new process flow for i2R e-paper try run line.

replacing Ag by Al could save 90% material costs. The thickness of Al electrode is about 0.06 μm, which is much thinner than that of Ag electrode, which is 20 μm. The thinner electrode can increase heat transfer from thermal printing head to LC layer. The printing quality has been improved to show fine line and picture like bar code or two-dimensional code (Figure 10.35). The line width could be written to as narrow as 82 μm for the resolution of 300 dpi. To enhance the long-term wearability, a protection layer on top of Al electrode is required. The electrode without protection layer is easily damaged from surface contact of thermal printing head and Al electrode during writing. The hardness of protection layer is more than 2H to prevent scratching caused by thermal head on surface of Al electrode. By using the protection layer, the durability of the i2R e-paper is improved tremendously. The reusable and rewritable i2R e-paper could be applied in different circumstances, such as temporary identification card, public transportation ticket, and entrance permits. As shown in Figure 10.36, the temporary identification card integrated with RFID system has been successfully demonstrated for the application of visitor's badge for security purposes. By using i2R e-paper, bus, train, airplane, and boat tickets could be recycled and reused more than 500 times. The usage of printing paper can be drastically decreased. Because the e-paper is flexible, robust, easy to carry, and has freedom of design (Figure 10.37), there has been interest shown for various applications in education, entertainment, transportation, security, membership cards, and so on. Taking the example of tickets for the high-speed train in China, more than 1580 lines, according to official statistics, were used for serving more than 1.33

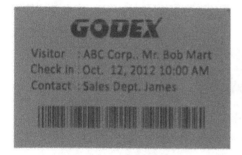 

Figure 10.35 The image quality improved for the applications of bar code (left) and two-dimensional code reader (right).

Figure 10.36 The applications of e-badge with picture (left), visitor card with RF ID (middle), and signage with different colors (right two).

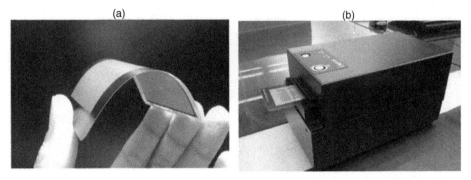

Figure 10.37 (a) The characteristics of flexibility, slim thickness (160 um) i2R e-paper, and (b) conventional thermal printer writing possibility.

million passengers per day at the end of 2012. For the entertainment application of cinema ticket, the market trend in Northern America shows the average number of tickets being sold could be as huge as 1.4 billion per year. There are also other entry tickets for different sporting and entertainment events such as baseball games, basketball games, theater, concert halls, and theme parks. The huge market size of door tickets could be expected to become an attractive market for e-papers.

To meet the need of display on irregularly shaped surfaces, some research groups have suggested the idea of the electronic skin [73, 74]. In our previous studies [60, 75], we proposed two approaches to fabricate the so-called conformal display. One is by three-dimensional coating (spray or dipping) and the other is by thermal forming. Due to the coating uniformity on irregularly shaped surfaces it is difficult to meet the requirement for display, and most effort has since been directed to the thermal forming method. The basic approach is fabricating a flat display by conventional planar processes, such as slot die coating, and then transforming the flat display into a three-dimensional shape display by thermoforming, as shown in Figure 10.38. The encapsulated CLCD with simple structure on a single substrate [38] is modified to develop the thermal formable CLCD, as shown in Figure 10.38. The plastic substrate, amorphous polyethylene terephthalate (APET), is selected for the permissible low process temperature. The brittle ITO electrode is substituted by stretchable PEDOT/PSS, which is a conducting polymer material with better flexibility. The gelatin in the encapsulated CLC layer has been modified with the plasticizer to improve its stretchability. For color display, photo-tunable CLC with modified structure of CLC is used to show different colors depending on the dosage of UV light exposure. The process flow of the thermal formable CLCD is similar to that of i2R e-paper [67]. To prevent the CLC layer from being damaged in the following water-based PEDOT patterning process, a hardener vapor is introduced. The purpose is to make the gelatin binder insoluble via the cross-linking reaction. To improve the thickness uniformity after thermal forming, pre-blow and patterned heating (local area heating) have been implemented. The results, shown in Figure 10.38, meet the target of 40% elongation, with patterned electrodes for segment driving. This thermal formed CLCD could be used on irregularly shaped surfaces, such as the surface of a smartphone, consumer electronic device, dashboard, and the "face" of a robot.

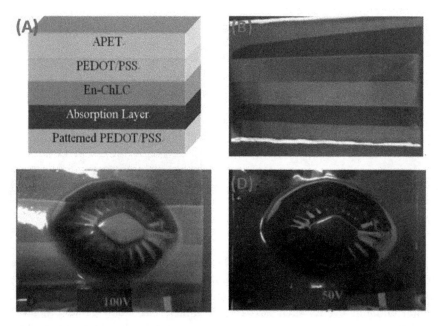

Figure 10.38 (A) and (B) The color photo-addressing encapsulated CLCD panel, (C) the colorful conformal display displayed the planar state (100 V), and (D) focal conic state (50 V).

10.10 Conclusion

CLCs exhibit Bragg selective reflection. Single-layer CLC can reflect nearly 50% unpolarized natural ambient light. CLC reflective displays have high reflection and good readability under room light conditions. They do not need power-hungry backlight, a feature making them energy efficient. They do not need polarizers, which is the first reason that they are compatible with plastic substrates with non-uniform birefringence.

CLCs exhibit two stable states in the absence of applied voltage. They can be used to make multiplexed displays on passive matrix. They do not need active matrix that has to be made under a high temperature, which is the second reason that they are compatible with plastic substrates. They can display static images at 0 V, another feature making them energy efficient.

The third reason CLCs are compatible with plastic substrates is that they can be encapsulated without losing their reflectivity and bistability. They can be used to make flexible displays by roll-to-roll process.

Bistable CLC reflective displays can be addressed by various external stimuli such as voltage, light, heat, and mechanical press, which greatly expand their applications in many areas. Although they are not suitable for high-contrast video applications, they are excellent candidates for e-book, e-signage, e-ticket, e-wallpaper, and instrument display panel, because of their merits of low power consumption, low manufacturing cost, light weight, and mechanical ruggedness.

References

1 De Gennes, P.G. and Prost, J. (1993). *The Physics of Liquid Crystals* (2nd ed.). Clarendon Press.
2 Chandrasekhar, S. (1997). *Liquid Crystals* (2nd ed.).Cambridge University Press.

3 Blinov, L.M. and Chigrinov, V.G. (1994). *Electrooptical Effects in Liquid Crystal Materials*.Springer-Verlag.

4 Yang, D.-K. and Wu, S.T. (2015). *Fundamental of Liquid Crystal Devices* (2nd ed.) Wiley.

5 Reinitzer, F. (1888). Beiträge zur Kenntniss des Cholesterins. *Monatsh. Chem.* 9: 421–441.

6 Wu, S.-T. and Yang, D.-K. (2001). *Reflective Liquid Crystal Displays*. John Wiley & Sons, Ltd.

7 Berreman, D.W. and Scheffer T.J. (1970). Bragg reflection of light from single-domain cholesteric liquid crystal films. *Phys. Rev. Lett.* 25: 577.

8 Berreman, D.W. and Scheffer T.J. (1970). Reflection and transmission by single-domain cholesteric liquid crystal films: Theory and verification. *Mol. Cryst. Liq. Cryst.* 11: 395.

9 Xu, M., Xu, F.D., and Yang, D.-K. (1998). Effects of cell structure on the reflection of cholesteric liquid crystal display. *J. Appl. Phys.* 83: 1938.

10 Gruebel, W., Wolff, U., and Kruger, H. (1973). Electric field induced texture changes in certain nematic/cholesteric liquid crystal mixtures. *Mol. Cryst. Liq. Cryst.* 24: 103–111.

11 Yang, D.-K., Huang X.Y., and Zhu, Y.-M. (1996). Bistable cholesteric reflective displays: Material and Drive Schemes. *Annual Review of Materials Science.* 27: 117–222.

12 Coates, D. (2016). *Cholesteric Reflective Displays in Handbook of Visual Display Technology* (ed. J. Chen et al.) Springer International Publishing Switzerland.

13 Doane, J.W. and Kahn, A. (2005). *Cholesteric Liquid Crystals for Flexible Displays in Flexible Flat Panel Displays* (ed. G.P. Crawford). John Wiley & Sons, Ltd.

14 Yang, D.-K., Doane, J.W., Yaniv, Z., and Glasser, J. (1994). Cholesteric reflective display: Drive scheme and contrast. *Appl. Phys. Lett.* 65: 1905–1907.

15 Yang D.-K., West, J.L., Chien, L.C., and Doane, J.W. (1994). Control of the reflectivity and bistability in displays based on cholesteric liquid crystals. *J. Appl. Phys.* 76: 1331–1333.

16 Lu, Z.-J., St. John, W.D., Huang, X.-Y., Yang, D.-K., and Doane, J.W. (1995). Surface Modified reflective cholesteric displays. *Proc. SID Intnl. Symp.*, XXVI: 172–175.

17 Yang, D.-K. and Doane, J.W. (1992). Cholesteric liquid crystal/polymer gel dispersions: Reflective displays. *Proc. SID Intnl. Symp.*, XXIII: 759–761.

18 Doane, J.W., Yang, D.K., and Yaniv, Z. (1992). Front-lit flat panel display from polymer stabilized cholesteric textures. *Proc. Japan Display*, 92: 73–76.

19 Davis, D., Kahn, A., Huang, X.-Y., and Doane, J.W. (1998). Eight-color high-resolution reflective cholesteric LCDs. *SID Intl. Symp. Digest Tech. Papers.* XXIX: 901.

20 Davis D., Hoke, K., Khan, C.A., Jones, Huang, X.Y., and Doane, J.W. (1997). Multiple color high resolution reflective cholesteric liquid crystal displays. *Proc. Intnl. Display Research Conf.*, 242.

21 Huang, X.-Y., Khan, A.A., Davis, D.J., Podojil, G.M., Jones, C.M., Miller, N.M., and Doane, J.W. (1999). Full color reflective cholesteric liquid crystal display. *SPIE Proceeding*, 3635: 120–126.

22 Hashimoto, K., Okada, M., Nishiguchi, K., Masazumi, N., Yamakawa E., and Taniguchi, T. (1998). Reflective color display using cholesteric liquid crystals. *Journal of SID* 6: 239–242.

23 Chien L.-C., Muller, U., Nabor, M.-F., and Doane, J.W. (1995). Multicolor reflective cholesteric displays. *SID Intl. Symp. Digest Tech. Papers.* XXVI: 169.

24 Braganza, C., Bowser, M., Krinock, J., Marhefka, D., Dysert, K., Montbach, E., Khan, A., Doane, J.W., Chin, C., Cheng, K., Tsai, Y., Liao, Y., Liang, C.C., Shiu, J., and Chen, J. (2011). Single layer full color cholesteric display. *SID Intl. Symp. Digest Tech. Papers.* XXXII: 396.

25 Garner, S.M., Wu, K.-W., Liao, Y.C., Shiu, J.W., Tsai, Y.S., Chen, K.T., Lai, Y.C., Lai, -C.-C., Lee, Y.-Z., Lin, J.C., Li, X., and Cimo, P. (2013). Cholesteric liquid crystal display with flexible glass substrates. *Journal of Display Technology*, 9: 644–650.

26 Tamaoki, N. (2001). Cholesteric liquid crystals for color information technology. *Adv. Mater.* 13: 1135–1147.

27 Lavrentovich, O.D. and Yang, D.-K. (1998). Cholesteric cellular patterns with electric-field -controlled line tension, *Phys. Rev. E* 57, Rapid Communications:, R6269–6272.

28 Helfrich, W. (1970). Deformation of cholesteric liquid crystals with low threshold voltage *Appl. Phys. Lett.* 17: 531–532.

29 Hurault J.P. (1973). Static distortions of a cholesteric planar structure induced by magnetic or AC electric fields. *J. Chem. Phys.* 59: 2068–2075.

30 Scheffer, T.J. (1972). Electric and magnetic field investigations of the periodic gridlike deformation of a cholesteric liquid crystal. *Phys. Rev. Lett.* 28: 593–596.

31 Yu, M., Yang, H., and Yang, D.-K. (2017). Stabilized electrically induced Helfrich deformation and enhanced color tuning in cholesteric liquid crystals. *Soft Matter.* 13: 8728.

32 Yang, D.-K. and Lu, Z.-J. (1995). Switching mechanism of bistable cholesteric reflective displays. *Proc. SID Intnl. Symp.* XXVI: 351–354.

33 Kawachi, M., Kogure, O., Yoshi, S., and Kate, Y. (1975). Field-induced nematic-cholesteric relaxation in small angle wedge. *Japan. J of Appl. Phys.* 14: 1063–1064.

34 Kawachi, M. and Hogure, O. (1977). Hysteresis behavior of texture in the field-induced nematic-cholesteric relaxation, Japan. *J. of Appl. Phys.* 16: 1673–1678.

35 Meina, Y., Zhou, X., Jiang, J., Yang, H., and Yang, D.-K. (2016). Matched elastic constants for a perfect helical planar state and a fast switching time in chiral nematic liquid crystals. *Soft Matter.* 12: 4483.

36 Watson, P., Anderson, J.E., Sergan, V., and Bos, P.J. (1999). The transition mechanism of the transient planar to planar director configuration change in cholesteric liquid crystal displays. *Liq. Cryst.* 26: 1307.

37 Huang, X.-Y., Yang D.-K., Bos, P. and Doane, J.W. (1996). Dynamic drive for bistable reflective cholesteric displays: A rapid addressing scheme. *Proc. SID Intnl. Symp.* XXVII: 347–350.

38 Huang, X.-Y. Yang, D.-K., Bos, P.J., and Doane, J.W. (1995). Dynamic drive for bistable cholesteric displays: A rapid addressing scheme. *J. SID.* 3: 165–168

39 Mi, X.-D., Silbermann, J.M., Rangel, J.I., and Stephenson, S.W. (2009). Cholesteric liquid crystal display system. US Patent U57522141B2, April 21.

40 Chen, J. Shiu, J.-W. Chiu, W.-W. Tsai, C.-C., and Huang, C.-Y. (2011). Roll-to-roll flexible display for e-paper applications. *SID Intl Symp. Digest Tech. Papers.* XXXXII: 107–110.

41 Geng, J., Zhang, D.L., Ma, A., Shi, L., Cao, H., and Yang, H. (2006). Electrically addressed and thermally erased cholesteric cells. *Appl. Phus. Lett.* 89: 081130.

42 Doane, J.W., Yang, D.-K., and Chien, L.-C. (2000). Pressure sensitive liquid crystalline light modulating device and material. US Patent 6,104,448, August 15.

43 Green, A. (2009). Roll-to-roll, flexible displays ... are we there yet? USDC 8th Flexible Electronics and Display Conference: 4.3.

44 Yang, D.-K., Lu, Z.J., Chien, L.C., and Doane, J.W. (2003). Bistable polymer dispersed cholesteric reflective display. *SID Intl. Symp. Digest Tech. Papers.* XXXIV: 959–961.

45 McCollough, G.T., Rankin, C.M., and Weiner, M.L. (2005). Roll-to-Roll manufacturing considerations for flexible, cholesteric liquid crystal (ChLC) display media. *SID Intl. Symp. Digest Tech. Papers* XXXVI:64–67.

46 Kurosaki, Y., Kiyota, Y., Ikeda, K., Tadaki, S., Tomita J., and Yoshihara, T. (2009). Improvement of Reflectance and Contrast Ratio of Low-Power-Driving, Bendable, Color Electronic Paper Using Ch-LCs. *SID Intl. Symp. Digest Tech. Papers.* XXXX: 764–767.

47 Khan, A. (2005). Advances in flexible bistable cholesteric displays. USDC 4th Flexible Electronics and Display Conference, 2–4.

48 Kato, T., Kurosaki, Y., Kiyota, Y., Tomita, J., and Yoshihara, T. (2010). Application and effects of orientation control technology in electronic paper using cholesteric liquid crystals. *SID Intl. Symp. Digest Tech. Papers.* XXXXI: 568–571.

49 Coates, D. (2008). Recent advances in tiled cholesteric billboard displays. *SID DIGEST.* XXXIX: 799–802.

50 Samsung's Exhibition in SID (2009).

51 Hiji, N., Kakinuma, T., Araki, M., Hikichi, T., Kobayashi, H., and Yamamoto, S. (2005). Cholesteric liquid crystal micro-capsules with a perpendicular alignment shell for photo-addressable electronic paper. *SID Intl Symp. Digest Tech. Papers.* XXXVI: 1560–1563.

52 Harada, H., Kobayashi, M., Gomyo, Y. Okano and Arisawa, H. (2004). A new color imaging method for photo-addressable electronic paper. IDW, EP1-1: 1703–1706.

53 Harada, H., Gomyo, Y., Okano, Gan. T., urano, C., Yamaguchi, T., Uesaka, H., and Arisawa, H. (2007). Full color A6-size photo-addressable electronic paper. IDW, EP2-2: 281–284.

54 Sato, M., Hiji, T.I. Tomoda, K., Yamamoto, S. and Baba, K. High resolution electronic paper based on LED print head scanning exposure. *SID Intl. Symp. Digest Tech. Papers.* XXXIX: 923–926.

55 http://www.kentdisplays.com/products.html.

56 Montbach, E., Pishnyak, O., Lightfoot, M., Miller, N., Khan, A., and Doane, J.W. (2009). Flexible electronic skin display. *SID Intl. Symp. Digest Tech. Papers.* XXXX: 16–19.

57 Montbach, E., Krinock J., Liu, L., Ernst, T., Braganza, C., Khan, A., and Doane, J.W. (2013). Large area, seamlessly tiled, flexible eBoard. *SID Intl. Symp. Digest Tech. Papers.* XXXXIX: 1250–1253.

58 Chen, J. and Liu, C.T. (2013). Technology advances in flexible displays and substrate. *IEEE Access.* 1 (1): 1–9.

59 Shiu J.-W., Chiu, W.-W., Tsai, C.-C., Huang, C.-Y., and Chen, J. (2010). Recent advances in flexible displays for e-paper application. Invited Paper, IDW. FLX1-1: 447–450.

60 Shiu, J.W., Lee, K.C., Liang, C.C., Liao, Y.C., Tsai, C.C., and Chen, J. (2009). A rugged display: Recent results of flexible cholesteric liquid-crystal displays. *Journal of SID.* 17 (10): 811–820.

61 Shiu, J.W. (2009). E-paper status and future opportunity. Korean Display Conference.

62 Shiu J.W. (2009). The e-paper applications of cholesteric liquid crystal display (ChLCD) technology. Display Taiwan 2009 Business & Technology Forum

63 Lee, K.C., Tsai, C.C., Shiu, J.W., Lee, C.C., Chang, Y.Y., Liang, J., Huang, C.Y., Huang, W.H., and Hsieh, C.H. (2009).Large area bistable cholesteric liquid crystal display and the driving system. Proc. IDMC/ 3DCA/ Asia Display.

64 Yang M.H., Tsai, C.-C., Chen, W.-T., Chang, R.-L., Chin, C.-L., and Chen, J. (2013). A thermo-formable liquid crystal display. *SID Intl Symp. Digest Tech. Papers.* XXXXIX: 1261–1263.

65 West, J.L. and Bodnar, V. (1999). *Optimization of Stacks of Reflective Cholesteric Films for Full Color Displays*, Proc. 5th Asian Symp. on Inf. Display, 29.

66 Hashimoto, K., Okada, M., Nishguchi, K., Masazumi, N., Yamakawa, E., and Taniguchi, T. (1998). Reflective color display using cholesteric liquid crystals. *Journal of SID* 6: 239–242.

67 Liao, Y.C., Yang, J.C., Shiu, J.W., Tsai, Y.S., et al. (2008). High performance full color cholesteric liquid crystal display with dual stacking structure. *SID Int. Symp. Dig. Tech. Papers* XXXIX: 300–303.

68 Melchior, H., Kahn, F.J., Maydan, D., and Fraser, D.B. (1972). Thermally addressed electrically erased high-resolution liquid-crystal light valves. *Applied Physics Letters* 21 (8): 392.

69 Geng, J., Dong, C., Zhang, L., Ma, A., Shi, L., Cao, H., and Yanga, H. (2006). Electrically addressed and thermally erased cholesteric cells. *Applied Physics Letters* 89 (8): 081130.

70 Yamaguchi, R., Ookawara, H., Ishigame, M., and Sato, S. (1993). Thermally addressed and erasable displays by polymer-dispersed nematic liquid crystals with memory properties. *Journal of the Society for Information Display* 1 (3): 347–352.

71 Liu, P.W., Tsai, C.C., Lee, C.H., Huang, W.H., Chen, C.W., Wang, C.Y., and Lee, K.C. (2009). Large area black/white bistable cholesteric liquid crystal display and thermal-addressing system. IDW FMC2-1.

72 Chen, J. (2009). Recent development of flexible display technology. Eurodisplay Workshop Program.

73 Green, A.M., Montbach, E., Miller, N., Davis, D., Khan, A., Schneider, T., and Doane, J.W. (2008). Energy efficient flexible reflex displays. *IDRC* 08: 55–58.

74 Koch, T. (2009). Roll-to-roll manufacturing of electronic skins. *SID Intl Symp. Digest Tech. Papers* XXXX: 738–741.

75 Liu, C.A., Wu, C.Y., Lin, M.C., Li, C.H., Yang, M.H., and Tsai, C.C. (2012). *Fabrication of Three-dimensional Volumetric Display Dased on Encapsulated Cholesteric Liquid Crystals by Dip Coating Method. ICFPE.*

11

Electronic Paper

Guofu Zhou, Alex Henzen, and Dong Yuan

Electronic Paper Display Institute, South China Normal University

11.1 Introduction

Electronic paper and e-paper are electronic information displays that mimic the appearance of ordinary ink on paper [1]. Unlike conventional flat panel displays that emit light, electronic-paper displays modulate reflected light. This makes their appearance independent of environmental illumination levels, and provides easy sunlight readability. It also arguably makes the displays more suitable for reading text, since it requires no eye accommodation. Also, many electronic-paper technologies hold static images indefinitely without power, making them suitable for low-power battery or solar-powered applications.

Examples of applications using electronic paper include electronic shelf labels and advertising in retail shops as well as timetables at bus stations, electronic billboards, smartphone displays, and e-reading devices. With the continuous improvement of electronic-paper technology, its application scope is expanding continuously while the applicable market grows. According to NPD Display Search's study, 366 million electronic-paper devices were sold in 2013, with a market revenue of $3.6 billion. By 2018, the number of electronic-paper devices will reach 1.8 billion, and the electronic-paper market will increase to almost $10 billion (Figure 11.1).

The main technologies that could be coined as of electronic-paper displays are electrophoretic display (EPD), electrowetting display (EWD), electrochromic display (ECD), and cholesteric liquid crystal display (CLCD). Some more exotic solutions are photonic-crystal display (PCD), interferometric modulator display (IMOD), and liquidpowder display (LPD). The microcapsule-based EPD, manufactured by E Ink Corporation and to a lesser extent Guangzhou OED Technologies (OED), serves almost 100% of the electronic-paper products market because of its outstanding features like bi-stability, low-power DC-field addressing, high-contrast, high, near-Lambertian reflectance. Also fabrication is relatively easy. However, it still has two main challenges to solve: It is hard to realize full color, and the refresh rate is relatively low.

There has been ongoing competition among end-product manufacturers to provide devices with full-color electronic-paper displays. The Ectaco Jetbook Color was released in 2012 as the first color e-reader, using E Ink's color filter-based "Triton" display. Because of a marked lack of brightness and color saturation, this solution for providing color e-paper did not gain acceptance by users. Other color solutions were proposed, like Fujitsu's "Flepia" display based on three-layer cholesteric LCDs in 2009 [2], and the Qualcomm "Mirasol" (IMOD)-based "Kyobo" e-reader in

Flexible Flat Panel Displays, Second Edition. Edited by Darran R. Cairns, Dirk J. Broer, and Gregory P. Crawford.
© 2023 John Wiley & Sons Ltd. Published 2023 by John Wiley & Sons Ltd.

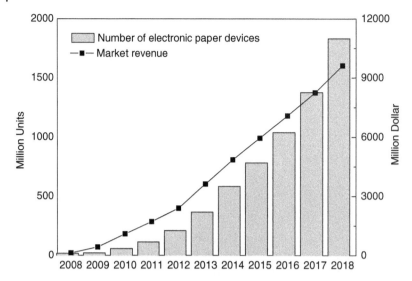

Figure 11.1 2008–2018 global electronic-paper market capacity and development trend. *Source:* NPD Display Search.

Korea in 2012, but none could face the competition with the paper-white background of gray-scale E Ink or the colorfulness of backlit LCD.

For applications other than e-readers, E Ink has been producing another color e ink technology named "Prism" in early 2015. It is capable of displaying an additional color, like red or yellow, using a colored particle combined with the usual black and white, which found its way to a large number of in-store and shelf-label advertising and pricing applications. (main players are Pricer, Displaydata, and SES-imagotag). Then, in 2017, E ink announced another color EPD called Advanced Color ePaper (ACeP), using cyan, magenta, yellow, and white particles to realize full color, as shown in Figure 11.2(a) and (b). However, accurately controlling four different particles is challenging, so the refresh time of the display is expected to be at least several seconds once the technology commercializes. Nevertheless, E Ink believes this may be the breakthrough required to satisfy the users of e-readers, although video applications remain firmly out of reach.

Even for single-color EPD, its low refresh rate, which is several hundred milliseconds, prevents producers from implementing sophisticated interactive applications (using fast-moving menus, mouse pointers, or scrolling) like those common for standard mobile devices. Electronic-paper display technologies with video refresh rate are rare, but typically include electrowetting displays and micro-electromechanical systems (MEMS) displays. Also reflective LCD could be included in the list. The refresh rate of electrowetting display and LPD is several milliseconds, whereas the refresh rate of MEMS display pixels can reach 10 microseconds. Among these three technologies, the electrowetting display is the only one still under development for commercial applications.

Electronic paper is an excellent candidate to be made flexible. The concept of "paper" makes it almost mandatory. Many demonstrations have already been produced, of which some excerpts are shown in Figure 11.3.

Flexibility of displays comprises three different "flexibility demands": the first is the display unit itself (sometimes referred to as the "front plane"); the second is the active area electronics required to generate the pattern of text and images on the display units, (usually referred to as the "backplane"); and finally, the external drive electronics (e.g. display source and gate drivers, controllers, micropro-

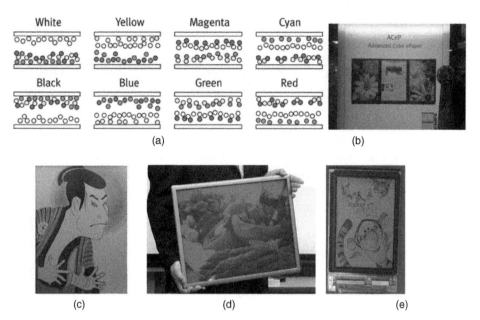

Figure 11.2 Demos of color e-paper. (a) Working principle of ACeP. (b) Demo of ACeP. (c) Demo of CLC color e-paper. (d) Demo of QR-LPD color e-paper. (e) Demo of EWD color e-paper.

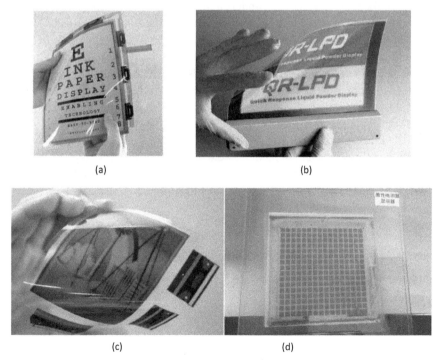

Figure 11.3 Demos of flexible e-paper: (a) EPD, (b) LPD, (c) CLCD, and (d) EWD.

cessors). Most of the front-plane technologies of electronic paper are flexible by design, although in some cases precautions must be taken (like cholesteric LCD and reflective LCD, where the optical structure may be disturbed by liquid crystal flow). Only MEMS displays are so strongly integrated with the "backplane" technology that their flexibility is dependent on the backplane. Therefore, the bottleneck for production of flexible electronic-paper displays mainly lies with the manufacturing of flexible backplanes and making external electronics compatible with a flexing substrate.

Even so, none of the existing e-paper technologies is sufficiently "paper like", meaning the golden combination of thinness, flexibility, infinite bi-stability, zero power, sunlight readability, high resolution, diffuse white appearance, bright colors, and good contrast, as well as low cost. Continuous development will bring this target closer, but the properties of a simple thing like a sheet of paper are surprisingly difficult to mimic with an electronic device.

The rest of the chapter will describe the most important technologies in more detail.

11.2 Electrophoretic Display

11.2.1 Development History and Working Principle

The function of an electrophoretic display (EPD) is based on rearranging charged pigment particles by means of an electric field. An EPD was first presented by Ota et al. in 1973 [3]. Their device was based on colored pigment particles dispersed in a differently colored liquid. The dispersion was encapsulated in a glass cavity in which the movement of particles could be controlled by an external electric field. In the following decades, understanding of the mechanism gradually progressed. The major breakthrough was made in 1998 by Jacobson et al. who reported a novel material and design overcoming the most critical shortcoming of the EPD: lateral movement of particles [4]. The original sample consisted of microcapsules containing white particles and a black liquid. A more mature version of the electrophoretic display, manufactured and sold by E Ink Corporation, uses a dispersion of black and white pigments in a transparent liquid, encapsulated in a microcapsule. Since the white particles in this display provide a diffuse, near-Lambertian reflection, the display looks distinctly paper like, contributing to its popularity as "electronic paper." Since commercial EPDs are mainly based on E ink panels using either microcapsules, or a slightly different but similar encapsulation method, we will focus on this technology.

The working principle of the microcapsule-based EPD is demonstrated in Figure 11.4 [5]. The microcapsules contain black and white particles in a clear carrier liquid. A dispersant is used to

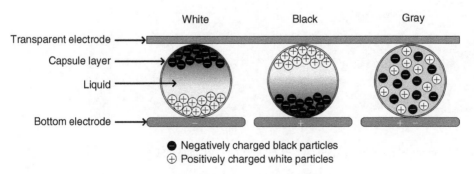

Figure 11.4 Schematic of EPD structure and working principle.

prevent particle agglomeration or precipitation, and a charge control agent provides the appropriate charge to the particles, resulting in opposite charges for the black and white particles. The microcapsules are then mixed into a binder and laminated between two parallel substrates, each provided with electrodes, and separated by a gap of several tens of micrometers. When a voltage is applied across the electrodes, the oppositely charged particles move toward and eventually deposit on opposite electrodes, thus showing a white appearance on one side of the laminate and a black appearance on the other. If the voltage is reversed, the particles each move toward the opposite electrode, thus reversing the appearance of the sides. Gray levels can be realized by halting the reversal process, leaving black and white particles intermixed.

Electrophoresis describes the motion of a charged body under the action of an applied electric field. For display applications, this body usually is a small (tens of nanometers to a few micrometers) particle, dispersed in a liquid. The electrophoretic velocity (U) of the particle is described by the Helmholtz-Smoluchowski equation

$$U = \frac{\varepsilon \xi_{EP} E_x}{\mu},$$
(11.1)

where ε is the dielectric constant of the liquid, ξ_{EP} is the zeta potential of the particle, E_x is the applied electrical field, and μ is the mobility of the particle. The electrophoretic zeta potential (ξ_{EP}) is a property of the charged particle and the suspending liquid (see Figure 11.5).

Key factors affecting an EPD's function are the charge density and size of the white and black particles, the viscosity and dielectric properties of the transparent medium, the size of the microcapsules, and the thickness and dielectric properties of the microcapsule shell. In addition, the electrical behavior of the lamination adhesive plays an important role. In order to secure the dispersion stability, sedimentation needs to be prevented by minimizing the density difference between particles and the dispersing liquid.

EPDs exhibit intrinsic bi-stability, low-power DC-field addressing, and have demonstrated high contrast and reflectivity. Because of these outstanding features, nowadays, microcapsule-based electronic-paper displays occupy almost 90% of the electronic-paper products market.

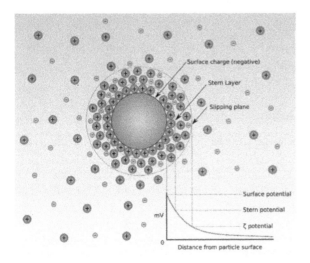

Figure 11.5 Zeta potential. *Source:* Wikipedia.

Examples of commercial applications of EPDs include high-resolution active-matrix displays that are or were used in the Amazon Kindle, Barnes & Noble Nook, Sony Librie, Sony Reader, Kobo eReader, and iRex iLiad e-readers.

Most of these displays are constructed using an electrophoretic imaging film manufactured by E Ink Corporation. Similar imaging films were developed by Sipix, which used an embossed encapsulation structure referred to as "microcup." Sipix was, after being part of AUO from 2009, acquired by E Ink in 2012. Therefore, E Ink now also uses this technology to build their imaging film.

The reflectance and resolution of electrophoretic microencapsulated imaging films is quite good, but the dual particle systems are monochromatic in nature. To create color, E Ink initially joined forces with the Japanese company Toppan Printing, a major producer of color filters for LCD. With Toppan's custom-made color filters, E Ink have made several color EPD films using the imaging film as a black and white shutter. However, these color displays are more expensive than standard EPD and have significantly lower reflectance.

Another drawback of electrophoretic systems is the low migration speed of the particles, leading to a low refresh rate, making EPDs less suitable for displaying animation or video. Nevertheless, some low-end video display is possible, and will allow a reasonable level of animated user interface.

11.2.2 Materials

The main components of the display panel, supporting the working principle and device structure, are the colored particles/pigments, the microcapsule shell, and the (insulating) solvent. The charge control agents and stabilizers also play an important role.

11.2.2.1 Colored Particles/Pigments

Colored nanometer to micrometer sized particles are the key materials for realizing EPDs. The basic requirements for a pigment are:

- specific density, which must be comparable to the suspending solvent in order to help avoid sedimentation
- low solubility in the solvent for long lifetime; non-aggressive solvents
- high refractive index for all particles except black, to enhance light scattering for bright colors
- high specific absorbance of the colored particles to achieve high contrast
- easily chargeable or naturally charged particle surface
- good (environmental) stability
- good manufacturability and purification routes.

When encapsulated into cavities or microcapsules, the particles should not adhere to the cavity or capsule surface.

Many materials have been investigated for EPD applications. The most studied inorganic materials include TiO_2, carbon black, SiO_2, Al_2O_3, chromium yellow, chromium red, and iron red. Examples of organic particles that have been investigated are, among others, toluidine reds, phthalocyanine blue, phthalocyanine green, and highly conjugated oligomers/polymers (like poly-acetylenes). In general, the tens of nanometers to several micrometers sized particles/pigments are dispersed in a carrier liquid. For a long period of time, carbon black and titanium dioxide were used for the black and white particles in EPD devices because of their high white-ness/blackness and easy availability. However, since carbon black is conductive, it negatively influences the electrophoretic properties of the dispersion. Therefore, the carbon black pigment can be coated with a polymer shell [6], as can any other pigment or particle that needs improve-ment of the dispersing or charging properties. The shell materials are typically long-chain organic materials with the ability to form secondary bonds such as materials with alkoxy group, acetyl group, or halogens. If the conductivity of carbon black remains problematic, or a core-shell struc-ture is undesirable, a dark-colored metal oxide can be used in its place to reach the requirements (e.g. copper chromite).

11.2.2.2 Capsule Shell Materials

The EPD device basically consists of microcapsules or pixels enclosed by pixel walls (sometimes referred to as "microcups"). The shell/pixel wall of the capsules turned out to be the key material for this technology, since it takes care of encapsulation of the colored and white particles, spa-tially separating them to prevent lateral migration of the particles, while at the same time only weakly interacting with the particles. The material needs to be transparent, mechanically stable but flexible, have a low conductivity, and be compatible with the materials it encapsulates. For these reasons, materials used are organic polymers like polyamine, polyurethane, polysulfones, polyethylene acid, cellulose, gelatin, and gum arabic. The microcapsule fabrication method is based on the chosen materials. Typical examples are in-situ polymerization of methanol and urea to form urea-formaldehyde resin [7], and composite coagulation of gelatin and gum arabic to form a composite film [8].

11.2.2.3 Suspending Medium (Mobile Phase)

Inside the microcapsules of the EPD device, the colored particles are suspended in the liquid medium. To fulfill the requirements of EPDs, the medium needs to have good insulation prop-erties, low resistance to particle transportation (i.e. low viscosity), thermal and electrical stability, similar density as the particles, and must be environmentally friendly. These requirements can be satisfied by using single or formulated organic solvents such as alkanes, aromatic hydrocar-bons, (cyclo)aliphatic hydrocarbons, and siloxanes. To tune the density, a high-density chlorinated solvent like tetrachloroethylene and a low-density hydrocarbon can be mixed.

11.2.2.4 Charge Control Agents

As seen from Equation 11.1, the electrophoretic velocity is proportional to the zeta potential of the particle. Hence, for a fast display response, the particles should have a high charge density. In a non-aqueous medium, the charges could be obtained via the following technologies: particle surface dissociation in solvent, surfactant dissociation, absorption of ions, or polarization by friction. Typical charge-control agents include sulfates (vitriol), sulfonates, metallic soaps, organic acid amides, amines, organophosphates, phosphate esters, polymers, copolymers, and graft polymers.

11.2.2.5 Stabilizers

A microcapsule contains a system with oppositely charged particles dispersed in an insu-lating liquid. The oppositely charged particles naturally attract each other because of the electrostatic force. Besides, two other degradation modes exist in the suspension: agglomeration, which is caused by an insufficient repulsive barrier between particles, and clustering, which is caused by fluid motion within the microcapsule. All forms of instabilities are detrimental to the life of the display device. A stabilizer (dispersant) is therefore necessary in this system. Polymer coating is an often-used method to bring steric repulsion among particles. In the non-aqueous suspension medium, low molecular weight surfactants cannot stabilize the particles like in aqueous solution. Long molecular chain dispersants, showing affinity to particles and affinity to the solvent have long been in use in the petrochemical industry. The particle affinity part of the dispersant links to the particles via ionic pairing, hydrogen bond, and/or van der Waals force. The solvent affinity part can be solvated to form a polymer chain to become a steric stabilizer among particles. Often-used steric stabilizers or polymer-coating stabilizers are poly(vinyl alcohol) (PVA), polybutylene succinimide diethyl triamine (Chevron OLOA370), polyalkenyl amine (Oronite OGA 472), and Dioctyl sodium sulfosuccinate (Cytec Aerosol OT)

11.2.3 Device Fabrication

The key to microcapsule fabrication is to choose the encapsulation process and the capsule wall material. Important parameters for the performance of the microcapsules are particle size distribution, wall thickness, mechanical strength, particle diffusion constants, electrical and optical properties, and chemical compatibility of the medium and the wall material with the particles. The capsule wall, among other relevant properties, should have a certain level of conductivity, and be colorless and transparent. Moreover, the microcapsules must be monodisperse in size. The formation of microcapsules and the process of microencapsulation may encompass different approaches based on chemical, physical, or physical chemistry processes. The chemical processes used for microcapsule fabrication include interfacial polymerization, in-situ polymerization, and suspension crosslinking method. Among these methods, it is more suitable to use interfacial polymerization and in-situ polymerization for industrialization.

Interfacial polymerization is a condensation polymerization at the interface of a multiphase oil/water system. It is a heterogeneous reaction. In the process of microencapsulation by interfacial polymerization, the capsule wall is formed by the polymerization at least two monomers, which must contain an oil-soluble monomer and a water-soluble monomer. The oil-soluble monomer is located at the water–oil interface of the core material droplet. The water-soluble monomer is located at the same interface in the water phase, and the two monomers react on the surface of the core material to form the polymer film.

In-situ polymerization is similar to interfacial polymerization. The difference is that instead of adding reactive monomers into the core materials and the suspended medium, here the reactive monomers and an initiator are all added into the dispersed phase or continuous phase. Ergo, the monomers and initiator are all located inside or outside the core material droplets. In the microencapsulation system, the monomers are soluble in a single phase, and the polymer is insoluble in the whole system. Upon the in-situ polymerization reaction at the surface of the core material droplets, a solid polymer capsule shell is formed that covers the core material droplet surface.

After the polymerization process, the microcapsules are selected for size (as monodisperse as possible) and dispersed into a binder and coated onto a transparent sheet with an indium tin oxide (ITO) electrode. This coating process is critical, since the coating thickness has to be tuned

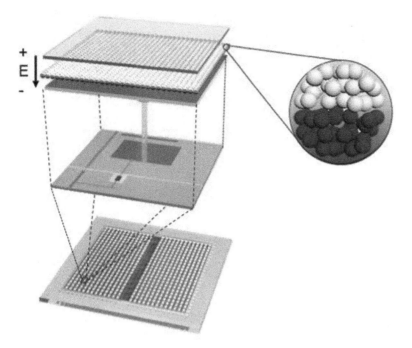

Figure 11.6 Exploded view of an EPD device.

very carefully to produce a monolayer of capsules on the substrate, without interstitial errors (missing capsules or extra capsules). This sheet in turn is laminated onto the plate containing the drive electrodes using a conductive lamination adhesive and sealed, forming the EPD device [9], as shown in Figure 11.6. The lamination process itself is relatively simple, requiring moderate homogenous roller pressure at elevated temperature.

After lamination, the EPD's humidity needs to be stabilized, since the electrical properties of the display change with changing humidity. The laminated polyethylene terephthalate (PET) film's moisture barrier properties are not sufficient to establish this, so additional foils with inorganic barrier films are laminated on top of the microcapsule film, and the edge of the applied barrier film is sealed by an epoxy fillet.

11.2.4 Flexible EPD

The front plane of an EPD is a continuous layer of microcapsules in a binder laminated onto a substrate, any kind of substrate can be used, including rigid or flexible. So, a flexible EPD front plane can be easily constructed with a flexible substrate.

Over the years, a large number of flexible EPD's have been demonstrated, by companies like Polymer Vision Plastic logic, LG displays, E ink, OED, to name a few (see Figure 11.7), but these prototypes were never brought to mass production because of the limited reliability of the flexible backplanes used. More recently, devices containing flexible e-paper displays appeared in the market, such as the iotaphone (with bent, rather than flexible design) and Sony DPT-RP1 drawing tablet, where the device has a certain degree of flexibility.

Figure 11.7 Flexible EPD demonstrated by: (a) E ink, (b) LG Display, (c) OED, (d) Yotaphone, and (e) Sony tablet.

11.3 Electrowetting Displays

11.3.1 Development History and Working Principle

The principle of an electrowetting display (EWD) is based on controlling the contact angle of a hydrophilic liquid on a hydrophobic surface by means of an electric field. Electrowetting for display applications was first proposed by Beni et al. in 1981, working directly on a conductive surface [10, 11]. The reflective-display technology based on electrowetting on a dielectric (EWOD) was first realized and published in 2003 by Hayes and Feenstra at Philips Research Labs [12].

The schematic drawing of an EWD is shown in Figures 11.7 and 11.8. The switching element is formed by a hydrophobic dielectric on an electrode, with a colored, hydrophobic liquid in contact with the dielectric. The structure is submerged in a clear hydrophilic liquid, e.g. water. With no voltage applied, the hydrophobic liquid forms a flat film between the water and the hydrophobic (water-repellent) dielectric, resulting in a colored pixel. When a voltage is applied between the electrode and the water, the interfacial tension between the water and the coating changes. As a result, the flat oil film is no longer stable, causing the water to move the oil aside. This makes the pixel (partly) transparent, or, if a reflective white surface is behind the switchable element, a white pixel.

The mechanism of the replacement of the oil film by water is well understood, and the electrowetting system can be described by the Young–Lipmann Equation (11.2)

$$\gamma_{1g}\cos, = \gamma_{sg} - \gamma_{sl} + \frac{1}{2}\frac{\varepsilon_0\varepsilon_r}{d}V^2 \tag{11.2}$$

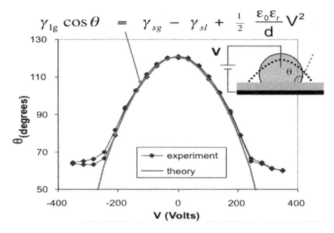

$$\gamma_{lg} \cos \theta = \gamma_{sg} - \gamma_{sl} + \frac{1}{2} \frac{\varepsilon_0 \varepsilon_r}{d} V^2$$

γ_{sl} – The surface tension between the electrolyte and the conductor at zero electric field
γ_{sg} – The surface tension between the conductor and the external ambient
γ_{lg} – The surface tension between the electrolyte and the external ambient
θ – The macroscopic contact angle between the electrolyte and the dielectric
$\varepsilon_r \varepsilon_0/d$ – The capacitance of the interface, for a uniform dielectric of thickness d and permittivity ε_r
V – The effective applied voltage, integral of the electric field from the electrolyte to the conductor.

To obtain a certain contact angle, the applied voltage is determined by the properties of materials (dielectric constant and fluidic interfacial tension) and the thickness of the dielectric material.

The EWD has shown its potential for high quality reflective mode information displays for using ambient light visibility, quick response (<2 ms switching speed has been reached) enabling

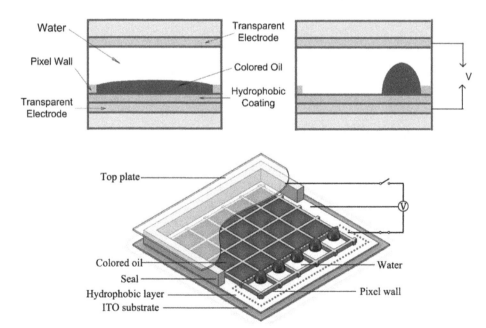

Figure 11.8 Schematic drawing of EWDs' working principle and device structure. (a) In an EWD pixel, without applied voltage, a homogeneous oil film spreads over the pixel area showing the color of the dyed oil (left), with an applied DC voltage.

Figure 11.9 An example of three-dimensional-structured pixel electrodes, providing bi-stability. (*Source:* SPIE)

video display, good optical performance (>50% white-state reflectance), and full color. The development of fluidic and soft display materials provides prospects for flexible displays in the future. Extensive work has been done to improve the display stability, lifetime, brightness, contrast, and color gamut.

The technology of EWDs is developing rapidly. A number of organizations have been working on EWD development and industrialization, including Liquavista (Amazon holding), Advanced Display Technology (Pforzheim University, Figure 11.9) [13], Industrial Technology Research institute (Taiwan), Gamma Dynamics (University of Cincinnati), and South China Normal University. Reduction of material costs and improvements in process efficiency are key to mass production and commercialization.

11.3.2 Materials

Key materials for EWDs are the absorbing hydrophobic liquid (dyed "oil"), the transparent hydrophilic liquid, the dielectric with hydrophobic surface, and the pixel wall material and its design.

11.3.2.1 Absorbing (Dyed) Hydrophobic Liquid

The selection of the absorbing hydrophobic liquid (often referred to as "oil") is critical for the electrowetting display application. The dye needs to provide good electrowetting performance (i.e. not interfere with the electrowetting behavior of the pure hydrophobic liquid), have a good solubility in the hydrophobic liquid and be insoluble in the hydrophilic liquid. Moreover, it should have a high absorbance, and good UV-stability. The two liquid phases should coexist for the lifetime of the display, also under motion and shock loads. Lower N-alkanes or silicone oil [14] are typically chosen as the hydrophobic liquid for interfacial tension, viscosity and dielectric properties. Commercially available dyes soluble in apolar solvents ("solvent dyes") include Oil Blue N, Sudan Red and Sudan Blue 673 (available from BASF), but these materials are susceptible to bleaching. Other dyes like Solvent Blue 98, Solvent Red 164 (both among others by Rohm & Haas) and a yellow solvent dye originating from Keystone [15,16] are more UV-stable. Where needed, black dyes could be formulated from appropriate blend of dyes. Liquavista published on an UV-resistant black dye using purple, magenta, blue and cyan dyes, and Samsung recently published on the dye series of SK1-SK4, specifically for electrowetting displays.

(2) Dielectric and hydrophobic materials.

Another key material in the electrowetting display design is the dielectric. As seen from the Young-Lipmann's equation, the hydrophobicity and the dielectric constant of the insulator determine the driving voltage and the reversibility of the electrowetting process. In previous years, single-layer organic hydrophobic insulators have been investigated intensively, including Teflon AF (DuPont), Fluoropel 1601 V (DuPont), Cytop (Asahi) and Hyflon AD (Solvay). These materials have good solubility in a low surface tension fluorocarbons and these solutions can easily be coated on substrates using a variety of techniques, like spin coating, screen printing or dip coating, thus forming a film with thickness of 0.1–1 μm. However, because of the relatively low dielectric constant of these fluorinated polymers (about 2.0), most of the applied field drops across this dielectric, so a high voltage needs to be applied to achieve the necessary electric field at the surface of the dielectric.

Since the electrowetting effect only requires a field, the insulating properties of the dielectric determine the parasitic current through the dielectric. One of the problems of a single-layer dielectric is the presence of pinholes. Therefore, dual or multiple layers of dielectric material combined with a hydrophobic coating have been proposed and investigated. Typical insulating layers are SiO_2, Si_3N_4, SiOC (silicon-oxide-carbon), or ONO (oxide-nitride-oxide); the hydrophobic layer could be either a thin layer of fluorinated polymer or a vapor deposited hydrophobic organic layer, e.g. Parylene, which forms a conformal pinhole-free film. This strategy solves the pin-hole problem; however, especially the inorganic insulators need to be fabricated under vacuum at high temperature, making the device fabrication more complicated.

(3) Pixel wall.

Another important material in an electrowetting display is the pixel wall which physically separates the colored hydrophobic liquids in order for them to be switched with single pixel resolution. The material should have a higher wettability (lower water contact angle) than the hydrophobic pixel dielectric to hold the oil film inside the pixel (the hydrophilic layer thus forms a barrier against the transfer of hydrophobic liquid into the neighboring pixels). A material that is excellently suited for this purpose is a photo-patternable and highly crosslinked resin known as SU-8, and it is no surprise this material or a derivative can be found in most prototype devices. Recently, other inert materials like polyimide have also been investigated for materials compatibility and ease of fabrication.

(4) Hydrophylic Liquid.

As seen from Figure 11.7, the hydrophylic liquid fills the gap between the oil and the top plate (electrode), acting as the common electrode as well as the dyed oil replacing medium. The conductivity, viscosity and interfacial tension of the conductive liquid needs to be optimized to fit the dual-liquid electrowetting system. Salts like NaCl, $NaSO_4$, Na_2CO_3 have been added to tune the conductivity, and surfactants like sodium dodecylsulfate, Tween20 or Tween80 are added to reduce the interfacial energy. However, small molecules like the alkali metal ions carry a major charge for electrolysis and breakdown of the insulator layer. Small ions also can easily penetrate the pores and pinholes or defects of the insulator, which is detrimental to the devices. Therefore, polyelectrolytes and new designed surfactants have been investigated to improve the device stability. In addition, ionic liquids as well as other polar, non-aqueous liquids have been considered for use in EWDs [14,17].

11.3.3 Device Fabrication

The process flow for manufacturing an electrowetting display is shown in Figure 11.10 [18]. For the bottom substrate any type of substrate can be used, ranging from structured ITO-coated glass for segmented displays and active-matrix substrates for high-resolution, pixelated displays to polymeric substrates for flexible displays.

On the substrate, a sub-micrometer-thick amorphous fluoropolymer layer or a stack of a barrier layer and a fluoropolymer is coated. Photolithographic walls form the pixel structure, which can be filled rapidly by simply dosing across the surface. Depending on the filling method, the height of the pixel walls plays an essential role in determining the amount of oil that self-assembles inside the pixels. This height, determined during a standard photolithographic process, is very uniform across the surface, resulting in a uniform electro-optic response. After the liquids have been applied, the display is closed using an ITO-coated cover substrate to provide the electrical contact with the water.

Electrowetting display processing consists of standard technologies, nearly all of which are used in existing LCD manufacturing facilities. The only exception to this is the filling and coupling, which is different from the filling and coupling process that is used for LCDs. This means that for current players in the industry, electrowetting displays offer a great opportunity to commercialize strongly improved displays with a relatively low investment.

The pixel structure of an electrowetting display is relatively simple and can be fabricated by mainly solution processing, making it very suitable for the application of printed display technology. In prototypes, the insulator and pixel wall materials are usually applied by spin coating, which not only leads to serious waste of material, but also requires additional etching to generate the necessary patterns. For pilot manufacturing, and eventually mass production, printing or patch/slit coating processes have been successfully introduced.

EWD oil filling efficiency and uniformity are still the key challenges on the way to mass production. Originally, the filling process took place submerged in the polar liquid, and the non-polar liquid was filled using the capillary interaction with the hydrophobic surface, applying the

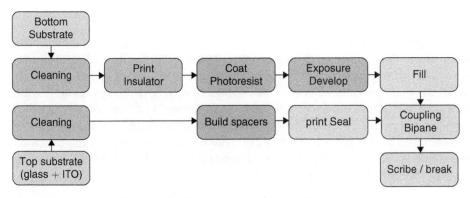

Figure 11.10 Process flow for manufacturing electrowetting displays.

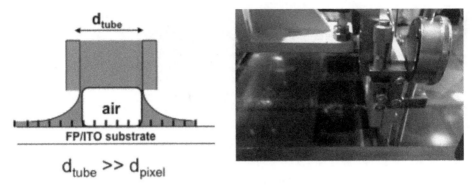

Figure 11.11 Raster filling using three-phase interface.

oil with a needle, one pixel at a time. Later, filling took place by creating a three-phase interface at the display surface, allowing the oil to enter the pixel area, while being covered by the hydrophilic phase (see Figure 11.11). This way of filling is faster than the pixel-by-pixel method, but very sensitive to disturbances of the capillary interface (vibration, contamination). Therefore, achieving a truly homogenous filling remains elusive. An improvement could be found in the use of inkjet printing [19]. Since the deposition of the required liquid quantity by inkjet is intrinsically accurate (usually better than 1%), this opens the way to well controlled oil filness and thus switching voltage. An additional benefit may be found in the possibility to produce single-layer multicolor displays, although this method is not suitable for full color.

11.3.4 Flexible EWD

In 2010, electrowetting operation on paper substrates was shown to function and have switching behavior comparable to electrowetting on glass substrates [20]. This has increased the interest in flexible electrowetting displays. In You's work [21], the operation of electrowetting structures on several types of flexible substrates (paper, plastic, and metal) has been demonstrated, indicating the feasibility of using these substrates as a cheap and flexible option for electrowetting-based e-paper displays.

The mechanical properties of EWD make them suitable for flexible displays. There are no components that will either crack or break during flexing. Also, the imaging plane (oil) is not sensitive to flexural forces, and the water layer covering it is isotropic under any condition. The only component requiring consideration is the glass plate (to be replaced by a flexible substrate) and the ITO electrode, which may crack because of stretching.

You et al. [21] made a flexible electrowetting display consisting of a 45 × 21-pixel array with 300 μm × 900 μm pixel area. The array is fully functional in both OFF and ON states while statically curved as well as during flexing. South China Normal University have also fabricated flexible electrowetting devices based on a PEN film with ITO electrodes, as shown in Figure 11.12. Hyflon (supplied by Solvay) was selected as the hydrophobic insulator due to its low processing temperature ($< 170°C$), and a positive photoresist was used to manufacture the pixel grid. The results were encouraging and showed characteristics comparable to glass-based displays. The driving voltage of the flexible device was 15 V. The response time for the device was less than 12 ms while the bending radius was 10 cm. The display transmission was only about 28%, because of the relatively low drive voltage chosen.

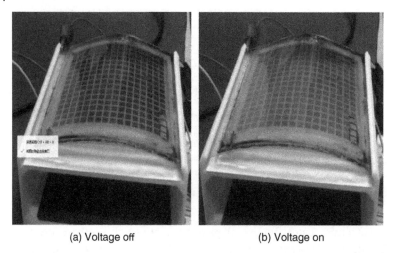

(a) Voltage off   (b) Voltage on

Figure 11.12   EWD operation on flexible substrate (PEN): (a) voltage off; (b) voltage on.

11.4   Other E-paper Display Technologies and Feasibility of Flexibility

11.4.1   PCD

PCD technology is based on the electrical actuation of photonic crystals. The working principle is show in Figure 11.13. Photonic crystals, characterized by a periodic modulation of the refractive index, exhibit exceptionally bright and brilliant reflected colors arising from Bragg reflection. These materials display non-bleachable structural color, and reflect a narrow wavelength band which can be tuned throughout the entire visible spectrum by expansion and contraction of the photonic-crystal lattice [22-24].

Electrical actuation of a photonic crystal film was made possible by its incorporation into a sealed thin-layer electrochemical cell. The device consists of the photonic-crystal composite coated onto a working electrode, a suitable seal adhesive, and a counter-electrode. The cell is filled with an organic solvent-based liquid electrolyte and sealed with epoxy. It is known that electric responsive polymers in solution as well as in supported films display reversible electro-chemical oxidation and reduction, with the partial electronic delocalization along the polymer backbone leading to a continuously tunable degree of oxidation, inducing changes in volume. By virtue of their continuously tunable state of oxidation and volume change associated, the photonic-crystal films display voltage-dependent continuous shifts in reflected colors.

Electrically tunable photonic crystals can provide electronic displays with unique properties. The effect is highly saturated color in high light environments, reflecting a narrow band of wavelengths, has bi-stability, a low operation voltage, and a continuous range of color can be accessed without the need for color filters or other optical elements. It is also possible to integrate a photonic crystal on a flexible substrate and thus make a flexible display. The fabrication and driving mode of a PCD is simple. Each pixel of the PCD can show its own color by electrical actuation. However, the photonic crystal material itself is complicated, and the stability and range of controllability is quite limited. Currently, the switching speed is still low and the color is not bright enough in normal lighting environment.

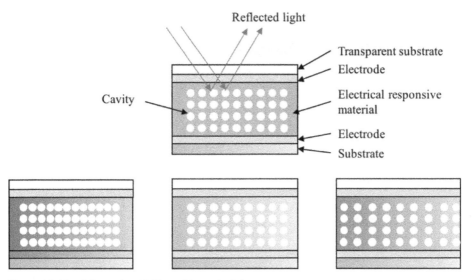

Figure 11.13 Schematic of PCD working principle.

11.4.2 LPD

The "liquid powder" display technology has been developed by Bridgestone, and was branded as "quick response liquid powder display" (QR-LPD) [25]. The working principle is similar to EPD, as it is based on manipulation of black and white particles as illustrated in Figure 11.14 **Reference source not found.** The difference is that the particles are not suspended in a liquid, but move in an inert gas. The positively charged black and negatively charged white particles are enclosed in sealed pixels between front- and back-electrodes. Switching can be done by applying an electric field between front- and back-electrodes, making the particles move to the opposite electrode. The QR-LPD display has one significant advantage over other EPDs: The "air gap" provides a threshold voltage, allowing the display to be driven as a "passive matrix" display, switching the panel line by line, without the need for a thin-film transistor backplane.

Creating gray levels is difficult since it requires carefully controlling the way particles "jump" to the opposite electrode. This level of control has been demonstrated, but repeatability remains uncertain. The system is fully bi-stable, and any image is retained indefinitely after removing the voltage. The gaseous medium makes the particles move very fast, leading to a fast primary response time (0.2 ms). However, Bridgestone continued to have problems with particles sticking to the electrodes, and the necessity to pulse multiple times with high voltage to make a majority of the particles transfer to the opposite surface, leading to much longer refresh times. It is not certain this problem, and the issues with gray levels, can be solved. These, and other less prominent issues, led Bridgestone to decide to discontinue the display effort in 2015.

The front panels of LPDs can be made thin and flexible, and can be fabricated by a roll-to-roll process, which is simple and proceeds at low temperature. This means that flexible plastic substrate could possibly be used that makes flexible display devices. However, the aforementioned problems will still be present.

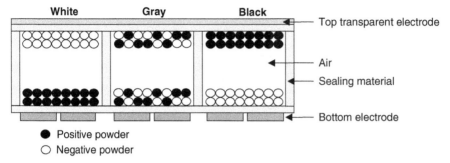

Figure 11.14 QR-LPD working principle.

11.5 Cholesteric (Chiral Nematic) LCDs

One of the first technologies showing an alternative for reflective LCD was the cholesteric display. The optical modulation occurs by selectively switching between the cholesteric state, exhibiting Bragg reflection, and the focal conic state, which has no significant reflection. The change from cholesteric to focal conic can be achieved by applying a moderate voltage, breaking down the ordered cholesteric state to disordered helix fragments, too short to provide Bragg reflection. The route back to the cholesteric state is done by applying a high voltage, bringing the display in a homeotropic state. If this high voltage is removed, the cholesteric state reforms.

The market has been controlled by Kent Displays Inc. (KDI), who have control over most of the patent portfolio.

The technology has so far found limited use, in large-area advertising (full color video, E-Magin, later Magink), as shelf labels (blue–white or green–black, Varitronix/BOE), and as a passive erasable writing screen ("boogie board," KDI).

The flexibility of the technology has been demonstrated by KDI, showing full-color displays with the flexibility of fabrics.

11.6 Electrochromic Displays

Electrochromic materials show a change of color as induced by either an electron transfer (redox) process or a sufficiently high electrochemical potential [26]. An electrochromic display (ECD) principle is demonstrated in Figure 11.15 [27].

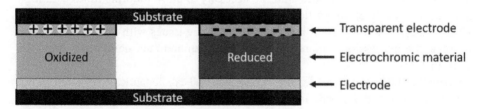

Figure 11.15 ECD principle.

Figure 11.16 Examples of displays by NTera and Siemens.

Many materials can be electrochromially changed: well-known inorganic materials are tungsten oxides, as demonstrated by NTera and Siemens in 2005 (Figure 11.16). The difficulty of these oxides is that there are few if any metal oxides that form a good black. For this reason, organic pigments are now preferred to accurately reproduce colors, as was demonstrated by Ricoh, showing three "primary" pigments (cyan, magenta, and yellow) that can be switched to transparent (Figure 11.18).

An ECD device is essentially a rechargeable battery in which the electrochromic electrode is separated by a suitable solid or liquid electrolyte from a charge balancing counter-electrode. In an ECD device, the electrochromic material is able to reversibly change its color when it is placed in a different electronic state. So, by absorbing an electron (the material is reduced) or by ejecting one (the material is oxidized), the material is able to change its color. The simplest way of making a working cell is to sandwich the electrochromic material between two electrodes, the electrochromic materials are deposited onto the electrode via different methods (see Figure 11.17).

The appearance of an ECD is that of a painted surface, i.e. it is easily readable from any angle. The state of an electrochromic material is determined by the injected charge, which means the

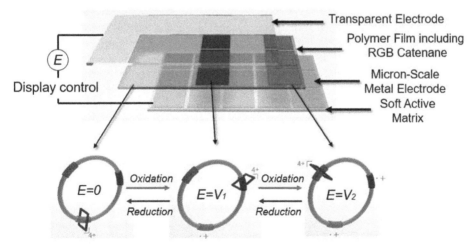

Figure 11.17 Example of a hypothetical multicolor (RGB) electrochromic material ([2] catenane) and its proposed usage in a display device.

Figure 11.18   Image of a 3-layer ECD by Ricoh.

ECD device is bi-stable. Moreover, the ECD device is simple and easy to fabricate, having the advantage of large-area fabrication on flexible substrates.

11.7   MEMS Displays

MEMS display – also called IMOD (trademarked mirasol) [28] – is a technology that can create various colors via interference of reflected light. The principle is shown in Figure 11.19. The basic elements of an IMOD display are microscopic mirror membranes that act as switchable Fabry-Perot resonators. One state is reflecting light due to constructive interference, the other state

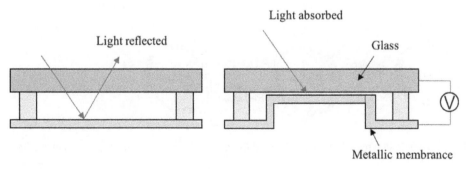

Figure 11.19   Schematic of IMOD working principle.

transmits light. Each of these elements reflects only one exact wavelength of light when turned off, and appears black when the field is on [29].

IMOD Pixels are constructed from small cavities containing the aforementioned mirror membranes that can move in relation to a semi-transparent thin-film stack. The pixels are typically several tens of microns across. The gap between the two mirror membranes determines the reflected color.

This display can hypothetically show multiple colors in one pixel. Multiple-color displays are created by using subpixels, each designed to reflect a specific different color. Multiple elements of each color are generally used to both give more combinations of displayable color (by mixing the reflected colors) and balance the overall brightness of the pixel. Meanwhile, this display exhibiting extremely (microsecond) fast switching. However, according to its sensitivity to the sub-micron cavity, the fabrication is difficult, resulting in low production yields and therefore high cost. Its manufacturing process and working principle also make it hard to construct on flexible substrates.

References

1 Heikenfeld, J., Drzaic, P., Yeo, J.S. et al. (2012). Review paper: A critical review of the present and future prospects for electronic paper. *J. Soc. Inf. Display* 19(2): 129–156.

2 Melissa Perenson. (2010) Fujitsu FLEPia e-reader makes a U.S. appearance. PC World.

3 Ota, I., Ohnishi, J., and Yoshiyam, M. (1973). Electrophoretic image display (Epid) panel. *P. IEEE* 61(7): 832–836.

4 Comiskey, B., Albert, J.D., Yoshizawa, H., and Jacobson, J. (1998). An electrophoretic ink for all-printed reflective electronic displays. *Nature* 394(6690): 253–255.

5 Shui, L., Hayes, R.A., Jin, M. et al. (2014). Microfluidics for electronic paper-like displays. *Lab. Chip.* 14(14): 2374–2384.

6 Werts, M.P.L., Badila, M., Brochon, C., Hebraud, A., and Hadziioannou, G. (2008). Titanium dioxide-polymer core-shell particles dispersions as electronic inks for electrophoretic displays. *Chem. Mater.* 20(4): 1292–1298.

7 Li, J., Wang, S.J., Liu, H.Y., You, L., and Wang, S.K. (2012). Preparation of poly(urea-formaldehyde) microcapsules containing sulfur by in situ polymerization. *Asian J. Chem.* 24(1): 93–100.

8 Wang, D.W. and Zhao, X.P. (2009). Microencapsulated electric ink using gelatin/gum arabic. *J. Microencapsul.* 26(1): 37–45.

9 Rogers, J.A. and Bao, Z. (2010). Printed plastic electronics and paperlike displays. *J. Polym. Sci. A Polym. Chem.* 40(20): 3327–3334.

10 Beni, G. and Hackwood, S. (1981). Electrowetting displays. *Appl. Phys. Lett.* 38(4): 207–209.

11 Beni, G. and Tenan, M.A. (1981). Dynamics of electrowetting displays. *J. Appl. Phys.* 52(10): 6011–6015.

12 Hayes, R.A. and Feenstra, B.J. (2003). Video-speed electronic paper based on electrowetting. *Nature* 425(6956): 383–385.

13 Blankenbach, K., Schmoll, A., Bitman, A., Bartels, F., and Jerosch, D. (2008). Novel highly reflective and bistable electrowetting displays. *J. Soc. Inf. Display* 16(2): 237–244.

14 Staicu, A. and Mugele, F. (2006). Electrowetting-induced oil film entrapment and instability. *Phys. Rev. Lett.* 97(16): 167801.

15 Sun, B., Zhou, K., Lao, Y., Heikenfeld, J., and Cheng, W. (2007). Scalable fabrication of electrowetting displays with self-assembled oil dosing. *Appl. Phys. Lett.* 91(1): 011106.

16 You, H. and Steckl, A.J. (2010). Three-color electrowetting display device for electronic paper. *Appl. Phys. Lett.* 97(2): 023514.

17 Chevalliot, S., Heikenfeld, J., Clapp, L., Milarcik, A., and Vilner, S. (2011). Analysis of nonaqueous electrowetting fluids for displays. *J. Disp. Technol.* 7(12): 649–656.

18 Chen, J.L., Cranton, W., and Fihn, M. (eds.) (2012). *Handbook of Visual Display Technology*. Springer: Berlin, Germany.

19 Ku, Y.S., Kuo, S.W., Huang, Y.S. et al. (2011). Single-layered multi-color electrowetting display by using ink-jet-printing technology and fluid-motion prediction with simulation. *J. Soc. Inf. Display* 19(7): 488–495.

20 Kim, D.Y. and Steckl, A.J. (2010). Electrowetting on paper for electronic paper display. *ACS Appl. Mater. Interfaces* 2: 3318.

21 You, H. and Steckl, A.J. (2012). Electrowetting on flexible substrates. *J. Adhesion Sci. Technol.* 26: 1931–1939.

22 Arsenault, A.C., Miguez, H., Kitaev, V., and Ozin, G.A. (2003). Towards photonic ink (P-Ink): A polychrome, fast response metallopolymer gel photonic crystal device. *Macromol. Symp.* 196: 63–69.

23 Arsenault, A.C., Miguez, H., Kitaev, V., Ozin, G.A., and Manners, I. (2003). A polychromic, fast response metallopolymer gel photonic crystal with solvent and redox tunability: A step towards photonic ink (P-Ink). *Adv. Mater.* 15(6): 503–507.

24 Arsenault, A.C., Puzzo, D.P., Manners, I., and Ozin, G.A. (2007). Photonic-crystal full-colour displays. *Nat. Photon.* 1(8): 468–472.

25 Sakurai, R., Ohno, S., Kita, S.I., Masuda, Y., and Hattori, R. (2006). Color and flexible electronic paper display using QR-LPD technology. *Sid. Int. Symp. Dig. Tec.* 37(1): 1922–1925.

26 Deb, S.K. (1969). A novel electrophotographic system. *Appl. Opt.* 8(S1): 192–195.

27 Ikeda, T. and Stoddart, J.F. (2008). Electrochromic materials using mechanically interlocked molecules. *Sci. Technol. Adv. Mat.* 9(1): 014104.

28 https://en.wikipedia.org/wiki/Interferometric_modulator_display#cite_note-imod-1

29 Miles, M.W. (1997). A new reflective FPD technology using interferometric modulation. *J. Soc. Info. Display* 5(4): 379–382.

12

Encapsulation of Flexible Displays: Background, Status, and Perspective

Lorenza Moro[1] and Robert Jan Visser[2]

[1] *Vice President CTO Group*
[2] *Applied Materials*

12.1 Introduction

Flexible organic light-emitting diode (OLED) displays have become a mainstream technology in high-end mobile consumer applications in the last few years. Introduction of active-matrix top-emitting OLED (TE-AMOLED) as rigid displays fabricated on the glass substrate and encapsulated with glass was done by Samsung in 2010 (Samsung Electronics, Suwon, South Korea). The first commercial AMOLED display on plastic foil and thin-film encapsulation (TFE) with a curved appearance was sold in limited numbers in 2013. Since then, thin AMOLED displays with curved edges have become popular as high-end consumer products and are commercialized by major brands. At this moment (2021) flexible OLED constitutes about 55% of the market, with market share growing year on year. While Samsung Display Corporation (SDC) is still, at the time of writing, the major supplier (70% in 2021), massive investments, mainly in China, suggest that in the next few years flexible displays will become the dominant technology for mobile displays.

By making possible a radical differentiation from the incumbent liquid crystal display (LCD) technology, TFE has been the key technology for the success of OLED displays. Figure 12.1 compares the structure of LCD (Figure 12.1 [left]) and OLED displays (Figure 12.1 [right]). It can immediately be seen how LCDs, by requiring more elements (LED backlight, light diffusers, color filters), are intrinsically thicker and more rigid. The OLED structure presented in Figure 12.1 (right) is slightly deceiving because it does not explicitly show how the OLED film is composed by multiple evaporated layers (six to seven or more). These layers have various functional roles in the device: hole injection, holes transport, emitting layers, electron transport, and injection layers, plus anode and cathode, and others depending on the sophistication of the device. However, all the layers are very thin, 10–100 nm, and the device is directly emissive (no backlight), making fabrication easy and the overall display thickness in the order of 1 micron.

In this chapter we will give a historical and technical background on the TFE evolution applied to OLED, describe the technically challenges overcome in its development, and describe current status and present and future challenges for the next generation of flexible OLED displays.

Flexible Flat Panel Displays, Second Edition. Edited by Darran R. Cairns, Dirk J. Broer, and Gregory P. Crawford.
© 2023 John Wiley & Sons Ltd. Published 2023 by John Wiley & Sons Ltd.

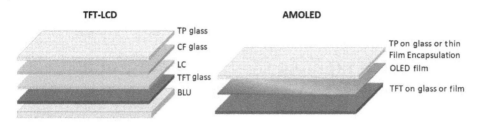

Figure 12.1 Comparison of LCD (left) and OLED (right) display structures.

12.2 Background

After OLED devices were proposed more than 30 years ago by Ching W. Tang and Steven Van Slyke [1], for many years the use of this technology for consumer electronics application was debated. The deployment of AMOLED displays was made possible by improvements in device lifetime, reliability, emission intensity, and color purity of the organic materials used in the OLEDs. Further improvement came by evolution of evaporation and masking tools and protocols when depositing the active OLED materials. At this moment two major applications are on the market: red–green–blue (RGB) displays for middle- and high-end smartphones on rigid glass substrates or flexible polyimide (PI) substrates, and high-end OLED TV with superior contrast and image quality. The OLED TVs currently commercialized [2] encompass white OLED back-light and color filters. Recently, Samsung has proposed a new structure with blue OLED back-light and quantum dot (QD) color filter.

Introduction on the market of small OLED displays for phones started in the early 2000s as monochrome displays and continued as RGB smartphone displays of increasing size for the next 10 years. However, the large-scale mass production deployment of AMOLED displays started when TFE made possible the commercial deployment of new functionalities by allowing the fabrication of flexible displays on plastic foils with a total thickness of a couple of hundred microns. Thinner displays have reduced thickness and weight of the final mobile devices, making possible high-end larger phones sold at premium price. In the following generations the flexibility of the substrate has allowed bending back of the portion of the display supporting the lids to external contacts. "All Screen," "Infinite Display," and "Near Bezel-less" displays have been deployed in stylish handsets where many of the accessory components (cameras, buttons, speakers, etc.) have been integrated in the area of the displays itself. Fully foldable displays for foldable phones or tablets were presented at a developer conference at the end of 2018 and several models and brands entered the market in 2020 and 2021 [3, 4].

In addition, curved displays for automotive application have been presented by major brands of luxury cars for implementation on future models [5].

Commercial deployment of OLED displays requires high-yield and robust encapsulation technology because OLED materials and devices are extremely sensitive to moisture and oxygen contamination. In addition, encapsulation is a back-end process, so any failure will affect the final yield damaging *quasi-finished* devices. Different encapsulation strategies for OLED displays are discussed elsewhere [6–8]. The possibility of flexible OLED displays on plastic foil and encapsulated with thin films has been proposed early in the history of OLED devices, but the debate on its commercial feasibility has been very animated and long-lasting. Despite millions of displays on the market, it is still a common subject of slides at technical conferences and numerous technical papers in the literature address the problem as "unresolved."

Indeed, the TFE critical technical requirements in high-end mobile display applications are extremely demanding. OLEDs are extremely sensitive to moisture and oxygen and barriers making water vapor transmission rates (WVTRs) lower than 10^{-6} g/(m²-day) are required to avoid short-term degradation of devices by black spot formation; oxygen transmission rates lower than 10^{-4} cm³/(m²-day) prevent middle/long-term degradation of light emission efficiency by oxygen. In display applications, point defects cannot be tolerated. Point defects are usually created by particles introduced during processing. A high number of particles are introduced by the evaporation process of the OLED active layers using shadow masks. To achieve the required resolution for mobile displays (540–580 PPI), fine metal masks are fabricated with special and proprietary technologies and used in strict proximity (~10 μm) to the display glass. Advanced tools and strict procedures mitigate – but do not avoid – particle contamination on the display glass. Figure 12.2 illustrates the dependence of defects density from display diagonal [6]. Even a single small defect on a display can ruin it; therefore added defect limitations like those required in semiconductor fabrication are necessary to ensure displays the size of a tablet are defect free. In order to have a good yield of smartphone displays, on a G4, G5, or HG6 [9] larger mother glass, the number of defects should be extremely small. Indeed, high yields have been reached for smartphone displays, i.e. less than 6-inch diagonal, but it is still a limitation for deployment of larger displays.

Inorganic layers are an effective barrier to moisture and oxygen diffusion. The bottom encapsulation of AMOLED flexible displays is done by depositing a PECVD inorganic layer on the polyimide (PI) substrate layer die-coated on glass before the fabrication of the thin-film transistor (TFT) backplane. Because PI can withstand process temperature up to 400°C, standard PECVD technology can be used to form dense inorganic layers. However, in top encapsulation of OLED displays these inorganic layers must be processed at low temperature to be compatible with the

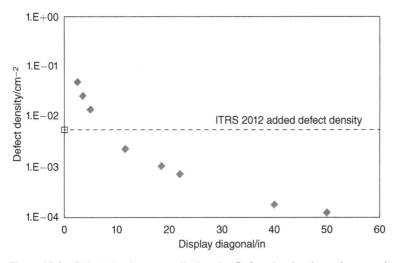

Figure 12.2 Defect density versus display size. Defect density above the curve leads to more than a defect per display. As a comparison the maximum acceptable added defect density for IC devices from the 2012 ITST Roadmap is also shown. Reproduced with permission from [7].

OLED low temperature constraints (80–100°C). Low temperature deposition results in poorly conformal, low-density, and defect-prone inorganic layers. In addition, during plasma-aided inorganic deposition processes, the plasma energy has to be relatively low to avoid degradation by ultra-violet (UV) and neutral or charged particles irradiation. Stress induced by the layer on the OLED device should be low to avoid delamination at the OLED interface with the anode.

Additional functional requirement for top emission displays is transparency of the layers that must be higher than 95%. Therefore, ceramic layers of oxides, nitrides, oxynitrides, or oxycarbides of Al, Si, transition metals have been proposed and implemented using different depositions techniques. Different vacuum deposition techniques achieve comparable barrier performance at different thickness: PECVD/1 µm, Sputtering/100 nm, Atomic layer deposition (ALD) 10 nm. At such a layer thickness the intrinsic barrier properties of these layers is enough to guarantee very long device lifetimes. It is defects (cracks, particles, pinholes) that limit the performance. In general, flexibility requirements limit the thickness of ceramic inorganic layers (max ~ 1 µm). A comparison of inorganic deposition techniques is presented in Section 12.3.2.

Early work on the permeation of single inorganic layers on plastic surfaces showed that with inorganic film thickness above 50 nm there was no improvement by making the oxide thicker but rather degradation at higher thickness because of stress issues [10]. The combination of harsh requirements and strong limitations make impossible a single inorganic barrier layer approach to OLED encapsulation.

Encapsulation by a multilayer approach in which barrier layers of oxide/nitride are alternated with thicker polymer layers [11] is an effective way to address these challenges (Figure 12.3). The ML-TFE technology was proposed to the industry in the early 2000 by Pacific North West National Labs and Vitex (San Jose, California). After early demonstration by the combined team that the concept worked [12, 13], a lot more work needed to be done before the new concept could meet all the requirements and scalability could be addressed [14, 15]. It took more than 10 years before the first scale-up to mass production of a commercial display on plastic substrate and TFE encapsulation was made by Samsung in 2014. Since then, it has become the technology standard for the

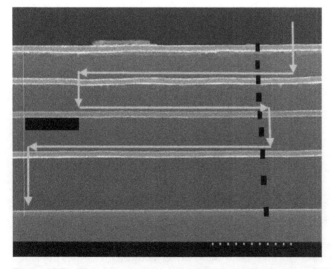

Figure 12.3 Pictorial description of tortuous path for moisture diffusion through defects in the inorganic layer of a multilayer barrier structure. Reproduced with permission from [6].

industry. Figure 12.4 presents a schematic history of the evolution of multilayer TFE for OLED displays. In it, parallel tracks illustrate the evolution of structure, materials and deposition methods, and companies' involvement at different technology stages.

In the early R&D stage multiple dyads (1 dyad: 1 couple of organic/inorganic layer) had to be used due to the high number of extrinsic defects introduced by the OLED production and the early stage tool development. Inorganic layers were deposited by reactive sputtering and polymer films were deposited by vacuum evaporation of monomer blends. By working on materials, processes, and tool improvements, as well as working on an integrated OLED fabrication/encapsulation tool, Vitex was able to demonstrate by 2004 that the number of dyads could be drastically reduced to one or two, making the technology interesting for commercial applications. Several cluster tools were sold for the fabrication of demonstrators on rigid substrate or metal foil and microdisplays. Development of TFE continued in Samsung and eventually brought to the acquisition by Samsung of the Vitex technology in 2010. The first commercial release of a phone with a flexible display fabricated on polyimide (PI) substrate was the Round phone released by Samsung in Korea in 2014.The TFE stack was composed of SiN_x layers deposited by PECVD and a polymer layer deposited by vacuum evaporation. The evolution from sputtering to PECVD has been driven by the availability of commercial PECVD tools developed for display TFE production and by the PECVD advantage to allow frequent in-situ cleaning of chambers and masks necessary to minimize process-added particles while maintaining high tool uptime. The large-scale mass production of display on PI started with the introduction of the high-end smartphone of Galaxy Edge series with a curved display at one or two sides. The TFE structure for such displays is based on 1.5 dyads composed by a stack PECVD-SiN_x/IJ-P polymer/PECVD-SiN_x. Vacuum deposition of polymer was substituted by IJ printing that by allowing selective deposition of polymer without use of masks and reduces costs (material and mask) and contamination. In following generations Samsung introduced direct fabrication of the touchscreen on the TFE structure further reducing the thickness of the displays and pushing forward the integration of the full display integration.

The 1.5 dyads TFE structure PECVD-SiN_x/IJ-P polymer/PECVD-SiN_x has been recently adopted by most of the new suppliers of flexible OLED displays in China, Taiwan, and Japan. The same structure is implemented in the OLED display fabricated by Samsung deployed by Apple in some i-Phone models released after 2017.

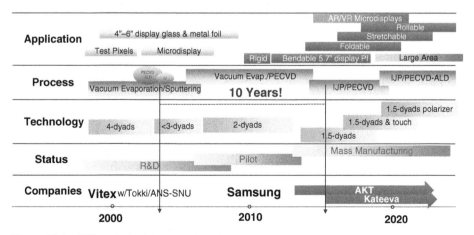

Figure 12.4 TFE evolution history and roadmap.

12.3 Multilayer TFE Technology

12.3.1 Multilayer Approach

The top surface of an OLED device is heterogenous comprising of metal, organic, and oxide layer and it is not planar because of the topology created by the pixel-defining-layer (PDL). This is an intrinsic limitation in depositing inorganic barrier film. Indeed, high-quality and dense inorganic films deposited at low temperature and in mild plasma conditions can be obtained only if the substrate surface is very smooth (RMS < 1\nm) so providing ideal nucleation condition to inorganic layers deposited at low temperature and mild plasma conditions. This is demonstrated in Figure 12.5 where the transmission electron microscopy (TEM) cross sections of two films deposited by the same process on substrates with surfaces of different smoothness are compared. The morphology of the structure is reflected in a columnar structure of the film in Figure 12.5 (left), while the surface with roughness of < 0.5\nm allows the deposition of a perfectly amorphous film even at deposition temperature < 80°C (Figure 12.5 (right)).

It has become commonly accepted that the polymer layer in a multilayer structure has the double function of providing a smooth surface and of planarizing particles and topography of the substrate. Figure 12.6 shows the planarization of a 5-μm large particle present on the substrate from previous process steps in an inorganic/polymer/inorganic TFE structure. The deposition of the first inorganic layer has covered the particle leaving some defect at the bottom edge produced by the localized mechanical stresses around the particle resulting in cracking of the barrier film. However, the particle is completely embedded by the polymer layer that provides a (smooth) planar surface for the nucleation of the following inorganic layer. The topology of AMOLED displays is created using the PDL that creates 2-μm-deep wells with slanted walls for the isolation and definition of individual pixels (Figure 12.7a). While the first inorganic layer completely covers the PDL, the barrier quality on the well is inferior. Planarization by polymer deposition (Figure 12.7b) creates a flat surface for the high-quality deposition of the second inorganic layer.

Some level of defectivity is to be expected for any inorganic layer. The presence of multiple inorganic layers creates a so-called "tortuous path" for water vapor or oxygen molecules diffusing

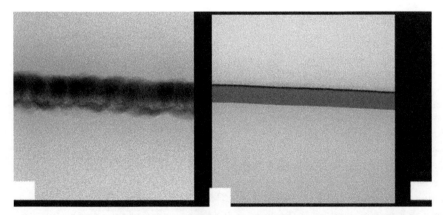

Figure 12.5 Comparison between structure of inorganic film deposited on substrate without planarization layer (left) and with (right). The same deposition process and thickness was used. The film deposited on planarization layer is fully amorphous and dense.

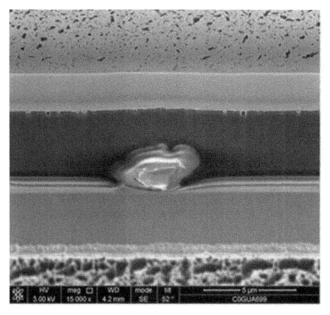

Figure 12.6 Example of particle planarization by polymer film deposited from liquid precursor.

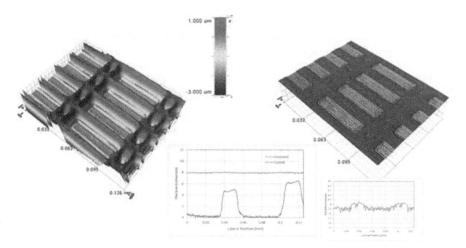

Figure 12.7 Example of PDL substrate (a) the substrate after deposition of planarization layer from liquid precursor (b). Mechanical profilometer profiles are also shown (c) and (d): step high before (bottom line) and after (top line) planarization are respectively 5–6 microns and 200 nanometers (d).

through pinholes in the barrier layers by having to move laterally after penetration through a crack or pinhole to the next available permeation defect in the subsequent inorganic layer rather than traveling vertically through the stack. The barrier performance of the organic itself is low, but the effective thickness through which the molecules must diffuse is defined by the distance between defects.

Graff et al. developed a model for water diffusion in these multilayers [11]. The model shows that the multilayer introduces a lag time for the first water molecules to reach the OLED. The lag

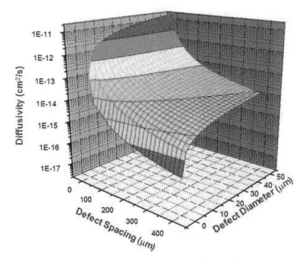

Figure 12.8 Effective diffusivity versus defect size and spacing. Reproduced with permission from [17].

time increases with increasing number of layers, increasing distance between the defects and defect size [16]. The interdependence of the various parameters is shown in Figure 12.8 where the effective diffusivity calculated using the so-called "Crank" diffusion model is plotted versus defect size and spacing [17]. The effective diffusivity initially increases as the pinhole defect size increases for a constant spacing followed by saturation. As the defect spacing is reduced, there is a quadratic increase in diffusivity and the barrier performance is dramatically reduced.

Validity of the model has been experimentally demonstrated by a study of multilayer $Al_2O_3/$ polymer barrier systems with added defects. Diffusion of moisture in the multilayer structure was accelerated exposing the sample at 90% relative humidity conditions at temperature 60°C. Using X-Ray Reflectivity, it has been shown (Figure 12.9) that the moisture permeates from the

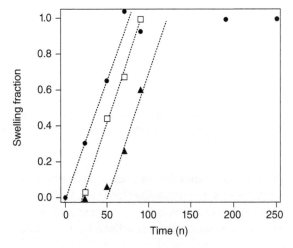

Figure 12.9 Swelling of each polymer layer as measured by X-Ray Reflectivity. The top layer (●) swells first, then the middle layer (□), and finally the bottom layer (▲). The delay time to begin swelling increases as water progresses through the film. Reproduced with permission from [18].

most external polymer layer to the inside. For each layer the lag time to reach the steady-state regime of permeation further increases [18].

Redundancy offered by the structure with multiple inorganic layers mitigates the problem of having films with suboptimal quality and allows the use of thinner films that are more flexible.

Mechanical robustness and flexibility of the barrier structure improves because the thinner individual inorganic layers are more flexible and there is less fragmentation and cracking for brittle ceramic layers sandwiched by compliant organic layers.

Multilayer encapsulation is effective in preventing moisture and oxygen diffusion in the vertical direction. However, saturation of the polymer layer by lateral moisture diffusion must be prevented by proper sealing of the sides of the TFE structure and device itself. This can be achieved by patterning of organic and inorganic layers so that the inorganic layer completely seals the edges of polymer and devices [19]. However, encapsulation edges must be as small as possible to minimize the display bezel.

12.3.2 Inorganic Layer Deposition Techniques

Key requirements of the inorganic layer for being a good barrier are density and low level of added particles. Managing this requirement has driven the choice of deposition technique for inorganic layers in the industry.

Reactive sputtering was initially proposed and demonstrated by Vitex as inorganic barrier layer in the encapsulation stack using a pulsed DC configuration with a planar magnetron cathode [20]. In the proposed configuration a metallic Al target is sputtered at low pression ($<$ 5 millitorr) in an Ar/O_2 atmosphere with O_2 concentration controlled by a feedback loop. The resulting films have a controlled stoichiometry, high density, are deposited at low temperature, and provide intrinsic high barrier performance at thickness $<$ 100 nm. Defects created by particles ejected by the target at the edge of the oxidized area can be a limitation to the overall barrier performance. More recently, this issue has been overcome by switching to rotating cathodes configuration where the targets are continuously and uniformly eroded. Throughput can be enhanced by using AC bias, although this may lead to damage of the substrate by high energy ion/neutral and heating. In roll-to-roll configuration, single layer barriers on planarized plastic substrates have been demonstrated [21, 22]. Main limitations for the application of sputtering in mass production have been the lack of in-situ cleaning and, at the time, of tools specifically designed for display applications. In addition, deployment in mass production was hindered by the need of scanning the substrate that increases the footprint of tools in the cleanroom and therefore their installation/operating cost.

PECVD has become the deposition technique of choice for inorganic layer deposition in TFE applications due to availability of in-situ cleaning capability and of tools equipped with proper handling display glass originally developed for amorphous-Si deposition for TFT fabrication. Whereas many configurations are possible for PECVD, the static showerhead configuration powered by radiofrequency (RF) has been the most successful for its reduced footprint and proved reliability in mass manufacturing. This PECVD technology has been adopted by Samsung and its followers. SiN_x, SiO_yN_y, and SiO_x films deposited by mixture of SiH_4, N_2, NH_3, and N_2O, O_2 with thickness in the range of 1 μm have been used initially. In comparison with sputtered films, thicker films are necessary to achieve acceptable barrier performance. PECVD films deposited at low temperature ($<$ 80°C) and low power are intrinsically less dense because of thermalization of energetic ions occurring at higher deposition pressure (1–2 mbar). Film stoichiometry ranging from SiN_x to SiN_xO_y and SiO_x have been and are considered to meet the requirements of new devices. Film composition and surface chemistry influence wettability and spreading of the monomer formulations used. H-content of the film affecting TFT performances is also lower in PECVD oxides. Incomplete reaction of precursors at low temperature and plasma energy can lead to film hydrolysis and instability upon aging at high temperature and relative humidity producing barrier failure [23, 24]. Faster deposition rates achieved at higher plasma energy can create stressed films with poor adhesion and high absorption in the near UV range (wavelength ≤ 400 nm).

ALD has also been proposed for the formation of low permeation barrier film with layer thickness < 20 nm for enhanced flexibility and foldability [25]. In alternative to time-separated exposure, spatially separated zones for the different reactants have been proposed to increase the deposition rate. Aluminum oxide or transition metal oxides are the most common composition for the film, with nanolaminates of different oxides having demonstrated better barrier performance [26]. Plasma-enhanced ALD (PE-ALD) where the oxidative reaction of the metal-organic precursor is produced by ozone must be used when low temperature is required for completing the reaction of gas precursors and forming stable films. The structure of ALD deposited films is generally amorphous and conformal on surfaces where nucleation occurs. Since the reactions involved in the film formation are purely chemical, no bombardment by energetic particles is involved. While this may be an advantage in terms of damage to the substrate, the negative impact on the final encapsulation structure is that nucleation and adhesion may be problematic on certain surfaces. A flourishing literature exists on the advantages of ALD in OLED encapsulation, but introduction in-pilot or mass-production lines has not occurred, yet. A major obstacle is achieving production-worthy deposition rates. Lack of large size tools with no formation of particles in areas where the gas reactant may get partially mixed and lack of in-situ cleaning chemistry for metal oxide depositions are additional issues hindering industry's acceptance of ALD. ALD films completely covering particles or steep display topologies have been shown as point of strength of the technique. Less discussed is the poor crack resistance under mechanical solicitation. For this reason, combinations composed by thinner PECVD films with very thin ALD layers sealing micro pinholes have been proposed to meet the requirements of high flexibility, mechanical stability, and low permeation [27].

No solution processed layers seem to be able to produce a self-standing barrier layer although some have been demonstrated viable in combination with a thinner inorganic layer deposited by vacuum techniques.

12.3.3 Organic Layer Deposition Techniques

In TFE structure the decoupling organic layers are deposited in liquid form either by polymer vacuum evaporation or by inkjet printing (IJP). Plasma polymerization of hexamethyldisiloxane (HMDSO) has also been proposed, but no implementation other than for R&D has been pursued [28].

The vacuum thermal evaporation (VTE) process was adapted to deposit polymer for OLED TFE application by Vitex system and widely used in TFE R&D tools [29] from a process initially used in roll-to-roll multilayer barrier technology. Vitex, Samsung Display, and SNU Precision successfully scaled the process for 1/4 G5.5 (650 × 750 mm) flexible OLED mass production [30, 31]. The degassed monomer is fed by a syringe pump to an ultrasonic nozzle mounted on top of a thermal evaporator. Monomer liquid is atomized to a fine mist and sprayed over a large area on the interior of an evaporator, and instantly evaporate into monomer vapor. The monomer vapor diffuses through a narrow-slit nozzle, which acting like a linear monomer source and monomer vapor condenses on the shadow-masked substrate as a liquid film as the substrate is moving past the slit nozzle, and the liquid film is subsequently cross-linked to form a polymer film by exposure to an ultraviolet source. Hg lamps were initially used as UV sources, but progress in UV-LED technology led to their substitution with UV-LED lamps making simpler the management of light distribution and thermal load to the substrate.

The advantage of vacuum compatible VTE process is the high deposition rate with high polymer film quality and good uniformity. Polymer process has been studied and can be kept under good control. Polymer film thickness depends on process parameters such as liquid monomer flow rate, evaporator temperature, the linear scanning speed of nozzle or substrate, and substrate temperatures. Substrate temperature affects the condensation rate and monomer re-evaporates rate before curing. The efficiency of polymer deposition can be increased by optimizing monomer formulation and controlling a low substrate temperature.

There are challenges in applying VTE process in TFE application. On the hardware side evaporation and nozzle temperature uniformity must be controlled to prevent polymer or monomer film buildup inside the evaporator and on top of the ultrasonic nozzle due to local hot or cold spot. Ultrasonic nozzle surrounding temperature needs to be controlled and nozzle needs to be serviced periodically to prevent clogging by monomer build up. In addition, requirements related to fluid delivery – i.e. viscosity, vapor pressure, and temperature stability – pose restrictions in the choice of liquid monomers to include in the formulation of a good TFE polymer precursor. In order to use higher molecular weight of the liquid monomer, the higher evaporator temperature is required for flash evaporation process and the more severe the problem of polymerization within the evaporator induced by heating. Monomer material utilization rate is < 20% on pilot tool due to coating on the masks and shields, and some monomer vapor is pumped away by vacuum pumps or trapped by cold surfaces. Polymer film coating inside the chamber, cold traps, and vacuum-pumping lines would need periodic system maintenance and could generate particles if the tool hardware is not designed properly.

IJP deposition of organic layers has been studied by Vitex Systems in the late 2000s as an alternative to vacuum evaporation [32]. The first experimental implementation was by a customized Fujifilm Dimatix printer inside a N_2 glove box equipped with an LED UV curing source. The solvent-free acrylic ink was a modification of the formulation used for vacuum evaporation. The barrier performance of the inkjet polymer TFE process was verified on Ca and OLED samples on Vitex TFE system by replacing the VTE polymer process with IJP process.

The successful demonstration by Vitex led to the subsequent development in Samsung for deployment in mass production OLED fabrication. Since then, IJP deposition has become the incumbent technology for deposition of the organic decoupling layers due to the intrinsic advantage offered by direct patterning of the coated areas.

Drop formation and jetting from the printer-head nozzle is the first step in the deposition process. Ink surface tension, viscosity, and wettability to the printer-head plate influence these process steps and their reproducibility. The ejected droplets land on the substrate forming a grid whose density for a given droplet size, and patterning, depends on the firing sequence and its synchronization with printer-head and substrate movements. For any given drop volume defined by the printer-head choice, the density of droplets on the substrate defines the thickness of the film. On the surface the droplets spread and merge, forming a liquid film. The time allowed for this process is called leveling time and in general for current formulations is of the order of minutes (0.5–3). Film quality in terms of roughness, short- and long-range thickness variation, and general appearance depend on fluid and surface properties. For fluids, surface tension and viscosity are critical because they define the formation and evolution of the drop after ejection. For substrates, surface energy, its possible contamination, topography, and roughness control the formation of a continuous and flat film. In addition, substrate conditions related to handling (e.g. electrostatic charge, temperature gradients, particles) and environmental condition (air/N_2 flow) can change the long-range uniformity of the film. The thickness of IJP organic film ranges between 6 μm and 12 μm in current commercial products. Thin films with thickness < 2 μm have been demonstrated with good planarization on standard PDL configurations [33]. Thick films in the range 20–30 μm can also be deposited by IJP using proper printer-heads.

One advantage of IJP coating is that it allows direct patterning of the coating. This is important because at present the main application of TFE is for OLED displays for smartphones that are fabricated by patterning individual subpanels on a large mother-glass. An area larger than the emitting area of the display but smaller than the area coated by the inorganic films is coated with polymer leaving the external contacts uncoated. The shape of the pattern edges depends on the relationship between fluid and surface properties, the distribution of the droplets in the area close to the edge, and the presence of physical dams that confine the spread of the ink itself. Location and slope of the film edge are important because they define the TFE edge-seal quality and width and therefore its ability to prevent moisture and oxygen side permeation.

After droplets spread and merge, the liquid film is cross-linked in a solid film by UV exposure. Currently, solid-state LED lights with a narrow wavelength distribution around 385 nm are used. Many challenges had to be overcome to use and scale up IJP coating of the organic layer in TFE.

Kateeva, a leading supplier of inkjet equipment for OLED mass production has developed such technology in close collaboration with Samsung [34, 35]. Kateeva is now the leading supplier of TFE IJP tools to the OLED industry with a substantial share of the market in Korea, China, and Taiwan.

One important differentiation of Kateeva's tools is that coating occurs in an inert atmosphere. This is required because, at this process step, OLED devices are protected only by an inorganic layer that is, as discussed earlier, prone to defects that through moisture and oxygen permeation could quickly evolve into blackspots on the OLED device. In addition, any moisture and contamination from the environment could be trapped in the liquid or cross-linked organic film during processing (jetting, leveling, and curing). Such impurities trapped inside the barrier after the subsequent deposition of the inorganic layer(s) would slowly diffuse to the OLED through pinholes and degrade them. Optimization of enclosure volume, N_2 flow, and overall design must be considered to minimize particle addition and decrease recovery time after maintenance.

Long- and short-range film thickness uniformity is an important parameter in TFE encapsulation of top emission displays because it may lead to a luminance non-uniformity of the light emitted (or reflected) by the display. In this context, thickness nonuniformity is named "mura" from the Japanese word meaning unevenness, irregularity. In IJP printing mura can be created by using a regular grid of ink-drop patterns that reproduce nozzle-to-nozzle variations in the printerheads. Mitigations of printing mura can be done by a statistical approach using randomization in nozzle firing or by using more sophisticated algorithms as Kateeva does using its proprietary "Smart Mixing" approach by which ultrafast drop metrology measures the printer-head in real time and software calculates and selects specific nozzle-mixing combinations to enable mura-free printing [36].

A different cause of thickness nonuniformity is "chuck mura," produced by the interaction between the chuck and the substrate. Thermal gradient affecting spreading and merging of drops are created by the contact between substrate and chuck. Uncontrolled variations of the distance printer-head-substrate affect the droplet landing position and therefore the distribution of the ink on the substrate. Kateeva's printers employ proprietary floating glass technology that by using pressure and vacuum holes prevents contact and precisely controls the relative printer-head-substrate position.

Non-compensated electrostatic charges may also create defects in the liquid ink distribution so using ionizers able to compensate tribological charges is advisable.

High reliability and printer speed are necessary to meet throughput and uptime. The printer-heads used to apply the organic material are based on piezo technology. When a voltage is applied to the piezo head, it triggers actuators that jet out ink through nozzles in precisely measured droplets and with great speed and accuracy. With this technology droplets with volume from few to tens of picoliters can be ejected at frequency of few tens KHz. The industrial grade printer-heads offered by Konica Minolta and Fujifilm have a lifetime of billions of print cycles and would last many months in a 24/7 production environment. Multiple printer-heads on each printer allow them to meet 1–2-minute takt time for one sheet of G6H-size glass (925 × 1500 mm).

The inkjet systems are flexible, accommodating multiple display products on the production line without need of physical polymer layer mask. Kateeva have developed and support flexible software in which arbitrary user shapes can be easily imported [33].

The polymer layer edge profile can be tuned by optimization of ink formulation and printing algorithms. Additionally, physical dam structure around the perimeters of the OLED active area can be used to confine the polymer layer.

The organic layers are deposited in liquid form using proprietary blends of monomers and cured by 385-nm UV radiation. Formulations based on multifunctional carbon- and silicon-based acrylates and methacrylates have been used. Epoxy formulations have also been proposed as well as cross-linked organopolysiloxane and inorganic/organic hybrid monomer (ORMOCER ®, PHPS) [37–39].

Viscosity and surface tension of the formulation should be compatible with printing process and easily wet the surface of the inorganic layer to the coated to optimize film formation and uniform planarization of substrates with topography.

A critical characteristic is the ability of the ink to form a uniform film with a smooth surface (RMS, 1 nm). The fluid should be able to wet a variety of materials so to cover particles on the substrate introduced by different processes, e.g. organic and inorganic materials.

The purity of the component and their full reactivity is very important. In fact, after curing the polymer layers should not contain residual low molecular fragments that, remaining trapped in the TFE structure, can diffuse out into the OLED device, damaging its performance [40]. The polymer layers need to be plasma resistant [41] to avoid formation of low molecular compounds by damage produced by the deposition of the next layer using plasma (sputtering physical vapor deposition, PVD, or sputtering plasma-enhanced chemical vapor deposition, PECVD) [42].

TFE is a key component of flexible devices in the form of bendable, flexible, foldable, rollable, and, in the near future conformal/stretchable. Mechanical flexibility and stability of the stack are crucial. Adhesion at the two interfaces (inorganic/organic and organic/inorganic) and layer cohesion need to be preserved after mechanical stress at room temperature and after high temperature/high humidity. Theoretical modeling of the TFE structure and experimental results have demonstrated that for the inorganic layers in the stack, cohesion strength is improved, and crack formation inhibited by the constrain exercised by the polymeric layers.

12.4 Current Technology Implementation

The initial skepticism in the display community on the viability of OLED technology for display commercial applications has been won by the adoption of the technology in high-end smartphones, although after 10-plus years of development and improvement in material performance, tool design, and reliability process development. Initially deployed on rigid substrates, a substantial number is now produced on flexible PI substrates. It has been reported that in 2021 the number smartphone panels produced will be 1561 M units. Of those about 40% are OLED displays and 55% (345 M units!) are on flexible PI substrate [43]. While production-line yields and panel reliability have been proven, cost is still a limitation for widespread adoption. For comparable display size and resolution OLED displays for smartphones are still 15–25% higher than LCD displays. This gap is expected to be reduced and closed in the next year or two with the coming to stable production of the multiple flexible OLED lines in China owned by Chinese Companies (currently 95% of smartphone OLED panels are produced in Korea, mostly by Samsung SDC).

Standard PECVD processes using deposition temperatures in the range 350–400°C are used to form a barrier layer on the PI substrates. To minimize the impact of coating defects in the PI layer of defects induced during the final step of laser lift-off of the finished display from the carrier glass, some companies use a multilayer structure even at the bottom [44].

The TFE structure SiN_xO_y/poly/ SiN_xO_y is mostly used. Companies tune the inorganic film composition differently, or even differently for each product, to match requirements of wettability by the ink, H-content that may affect TFT performance, and mechanical and optical performance of the stack. PECVD is the dominant deposition technology with tools supplied mostly by Applied Materials (Santa Clara, CA) [45] and, with a minor market share, Jusung (Gyeonggi-do, South Korea) [46]. The major supplier of acrylic formulation for the deposition is Samsung SDI, but many other chemical companies are trying to get into the market with alternative acrylic and/or epoxy formulations. The immediate evolution of the technology has seen efforts in the reduction of edge width and the inclusion in the displays area and stack of sensing and actuation capabilities. One example of this is the integration on top of the external SiN_x of the touch sensor as shown in Figure 12.10, implemented by Samsung on the Galaxy S8 (5.77-inch display) and in more recent models [47]. This is achieved by fabricating the touch-screen panel by integrated processing. Integration of the phone of other sensors (e.g. multiple cameras, fingerprint sensors, etc.) and the attractiveness of locating them inside the display area

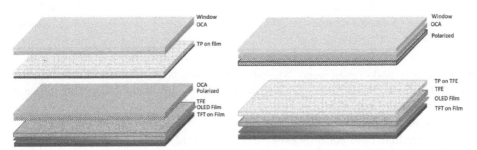

Figure 12.10 Comparison of display layer structure with touch panel (TP): traditional TP on film and TP integrated on TFE [47].

Figure 12.11 Mask-less deposition process flow. Adapted from [48].

require having vias and recessed areas cutting through the TFE structure, further challenging inorganic and organic coating technologies. Reduction of bezel width requires reduction of the shadowing created by masks used in the inorganic deposition and narrowing of the edge area where inorganic layers meet, sandwiching and isolating the polymer inside them.

Mask-less deposition of SiN_x has been proposed with the double advantage of reducing the bezel-less and reducing cost of mask and time of maintenance [48]. The process flow for a layer TFE stack is described in Figure 12.11. A final polymer layer is deposited on top of the second SiN_x layer, followed by an etching step to remove SiN_x film from the external contact area. The function of the last polymer layer is to be a mask to protect the TFE structure in the display area. A final step to remove the last polymer layer is optional and will depend on the quality of the film.

Although millions of high-quality flexible AMOLED displays have been commercialized, the flexibility of the display has not been exploited other than for creating elegant and sometimes functional curved edges with fix curvature, few mm in radius. Full flexibility in foldable displays has been demonstrated by all major OLED manufactures. The first foldable commercial displays were presented in November 2018 by Samsung [49] and Royole [3]. The Royole's FlexPai has been commercially available in China on a limited production since January 2019 while the Samsung phone was launched in September 2019. Figure 12.12 depicts Royole's FlexPai (Figure 12.12a) and the prototype shown by Samsung (Figure 12.12b). Throughout 2020 and 2021, more models were released by Royole and Samsung, as well as other brands (Huawei, Motorola, Xiaomi). Google, Oppo, Honor, and Vivo have also announced models for late 2021/early 2022.

For such displays the structure is still possibly a three-layer structure, with a reduced thickness of the inorganic PECVD layers. Common thinking and presentations [44, 50, 51] suggest that the thickness of the polymer layer should be reduced to reduce the minimum critical radius for folding of the structure with no cracking of the inorganic barrier layers. Whereas this is correct considering the TFE structure by itself, it may not be necessary or even deleterious looking at the complete display/phone structure where many other layers are present. From the point of view of mechanical robustness upon folding, the entire TFE structure should be in or close to the neutral plan to minimize stresses. In this case the thickness of the organic layer may not matter much because it is only a fraction of the total thickness. On the other hand, localized defects if not well planarized by the organic film may become crucial for their effect in concentrating mechanical stress and starting cracks in the inorganic layers upon mechanical solicitation (i.e. folding). For this reason, organic layers may need to be thicker than in rigid structures. For foldable displays the use of inorganic layers deposited by ALD is often indicated as a solution. However, the inherent fragility of few-nanometer-thick layers makes it unsuitable for robust encapsulation. Thinner PECVD layers combined with a very thin ALD layer is the solution proposed by others [27]. On the other hand, improved PECVD films with graded or variable density may also be able to provide enough barrier

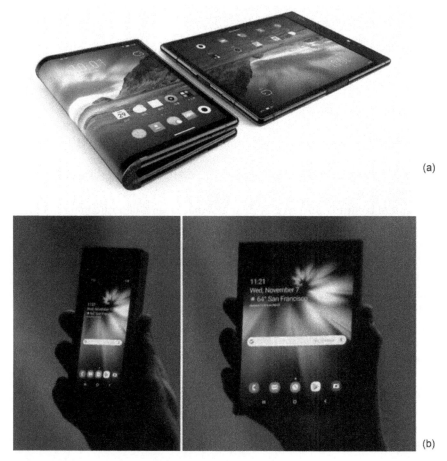

(a)

(b)

Figure 12.12 Royole's FlexPai (a) and Samsung foldable e (b).

performance while maintaining flexibility. One example of such structure constituted by alternating layers of high and low density is a SiN_x film, as shown in Figure 12.13. The TEM picture, and the inset in it, clearly shows that the 600-nm film is composed of multiple layers. Each sublayer, about 12–15-nm thick in the example shown, has different density. Using this novel experimental linear PECVD source [52, 53], schematically shown in Figure 12.14, the film is deposited by scanning the substrate under two electrodes which two RF frequency are applied while the reactant gases are injected in-between. In the injection zone the higher RF excites and decomposes the precursors, depositing a larger amount of low-density material that constitutes a first type of sublayer. More material is deposited in the area under the electrodes. Here the lower RF applied between the electrodes and the substrate itself densifies it, forming a second type of sublayer with higher density [54]. Multiple linear sources could be stacked side-by-side to increase the deposition rate. Changing deposition conditions and scanning speed, the relative thickness of the sublayer can be changed to allow good control of density and stress in the final film; and by controlling them producing films with low WVTR, low stress, and high transparency. Similar inorganic multilayer structures could be created by conventional showerhead sources.

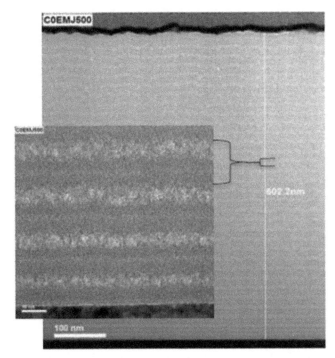

Figure 12.13 TEM cross section of SiN$_x$ film with high/low-density multilayer structure. Reproduced with permission from [53].

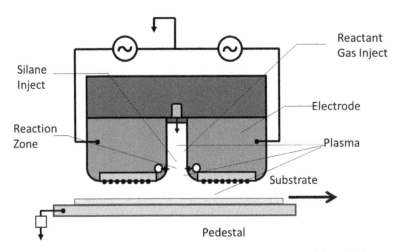

Figure 12.14 Novel experimental linear PECVD source. Adapted from [53].

12.5 Future Developments

Stretchable displays are the new horizon in the development of displays. Looking to the deployment of stretchable displays, the science fiction dream of high-end display directly attached to people skin may be not so close, but two applications are emerging:

- Three-dimensional (3D) conformal high-end AMOLED displays may find applications as automobile displays or other applications where 3D curved surfaces are available. These 3D-freeform diverse shapes, as Samsung calls them [55], including dome or convex/concave surface, and stretchability are necessary in both orthogonal directions. The backplane for this type of display must be low-temperature poly-silicon (LTPS) to meet the required display performance.
- Low-end skin-adhering displays associated with a sensor used for immediate and simple communication of data to the user [56]. Most of these displays may be deployed for medical applications. These displays are single driving or passive-matrix displays and have no or less requirement for the backplane.

For both types of stretchable displays, multiple technical challenges remain. They are related to the choice and reliability of materials, not least the reliability of TFE to maintain the required lifetime of the devices. Samsung demonstrated a stretchable 9.1-inch full-color display at SID Week 2017 [55]. To ensure the required stretchability maintaining the TFT performance, a backplane fabricated with standard technology on PI was etched in the area outside the TFT, connecting the matrix of TFT with stretchable interconnections (see the process flow depicted in Figure 12.15a). The product was a matrix of interconnected islands (Figure 12.15b) with 5% stretchability on which the OLED devices were then evaporated. While the demonstrator shown at the show was changing the shape from flat, the associated paper discloses a low-temperature thermoforming process for its use as a conformal 3D display. In this approach TFE must be implemented at the pixel or pixel-set level where the two inorganic layers uniformly coat a set of pixels and the interlayer polymer planarize the PDL structure. Barrier performance can be improved by deposition of inorganic layer(s) after thermoforming using a non-directional coating method such as ALD.

It must also be mentioned that the presence of relatively thick polymeric surface coating further improves the encapsulation reliability. The "island" approach is the one adopted and suggested by many authors [57, 58]. Corrugated surfaces on stretchable substrates have also been proposed [59–61]. While these approaches work at the demonstration level, more R&D and innovation are required before mass manufacturing become reality.

For displays associated with sensors that may be deployed for medical application or other short-term sensing application, mechanical performance and thickness of all displays are the focus of the current research. Reliability and lifetime expectations are more limited, driven by the overall performance of the materials utilized (flexible conductors, organic TFT, etc.). The application itself enables us to distinguish between a (relatively) long shelf-life that can be achieved by proper packaging and operational lifetime that can be relatively short and obtained with limited use of inorganic layers, mostly ALD, and quite thick, but very flexible polymeric layers that slow down moisture and oxygen diffusion to the OLED. The polymer most used now for these applications is polydimethylsiloxane, also because of its known compatibility with human use [62].

Large "paperwall" or "rollable" TV is another emerging field of application for flexible displays requiring TFE. Currently, mass-production OLED TV are bottom emission AMOLED.

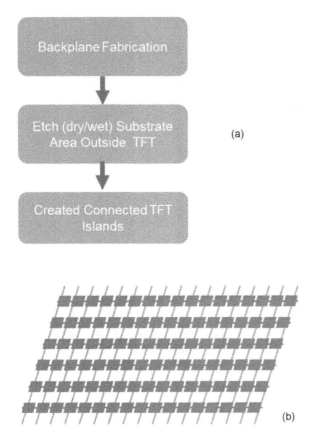

Figure 12.15 Samsung stretchable backplane fabrication: process flow (a) and schematic structure (b). Reproduced with permission from [55].

Encapsulation is generally done by glass or metal foil on top of a passivation inorganic layer. LG in its line of "Wall-paper" OLED TV has demonstrated and started to commercialize 65-inch and 77-inch TV panels thinner than 1 mm, which enables them to make TV sets as thin as 5 mm and lighter than 13 kg. Major manufacturer companies are actively investigating the deployment of TFE for large high-resolution (4 K, 8 K) and transparent displays. For example, LG demonstrated an 88-inch 8-K OLED panel at IFA 2018 [63].

Not many examples of flexible of flexible large-screen TVs have been presented at this time. Significant ones are from LG: Crystal Sound TV and Crystal Sound Speakers, a 65-inch rollable TV, and a 77-inch transparent flexible display. The Crystal Sound OLED was shown first at CES 2018 and a few other following shows in that year (e.g. SID Display Week, IMID 2018). The Crystal Sound OLED is an OLED panel that contains an embedded sound system. The TV has "exciters" –low-power RF transmitters – located behind the thin OLED panel that produces the sounds by vibrating the display since the display does not require backlights [64]. The same technology has been demonstrated by Luflex, the LG OLED lighting company, as a standalone multifunctional product that emits light and sound simultaneously [65].

At CES 2018, LG presented a 4-K 65-inch rollable OLED TV [66] complete with a base that holds the TV when it is rolled away (Figure 12.16). Rolling into and out of the base was demon-

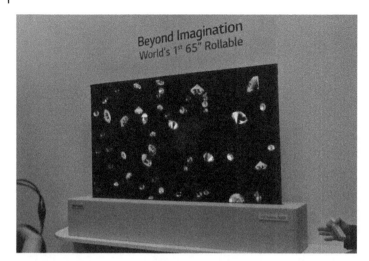

Figure 12.16 LGD Rollable TV. Ken Werner, Nutmeg Consultants.

strated, showing that if the display was rolled into the housing to present a reduced area (21:9 area for instance) the electronics automatically presented a similarly scaled image to the screen. The rollable OLED was fabricated on 80-micrometer-thick flexible glass, but fabrication on a plastic substrate is planned to allow the screen to be rolled more tightly. This was a prototype display. The commercial TV set introducing the technology was launched in Sout Korea in Fall 2020 and internationally in Fall 2021.

An impressive 77-inch transparent display was shown at SID 2018 and IMID 2018 (Figure 12.17). The display was developed during the 10-year government Transparent Flexible Display National Project under the supervision of the Ministry of Trade, Industry, and Energy and of the Korea Institute of Industrial Technology. The display features a resolution of 4 K/UHD (3840 × 2160), an 80-mm radius of curvature and 40% transparency [67, 68]. The fabrication process is illustrated in Figure 12.17. Two carrier glasses are used, one for the fabrication of the white

Figure 12.17 LGD Rollable transparent TV.

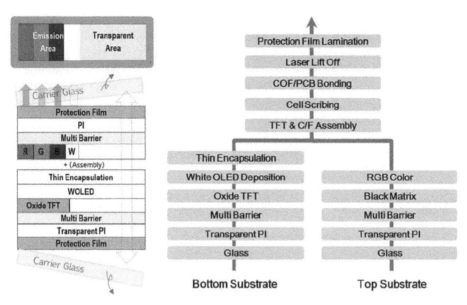

Figure 12.18 Fabrication process flow of LGD rollable transparent TV. Reproduced with permission from [67].

OLED on TFT, the other for the fabrication of the color filter. The two panels are then laminated, the cell is scribed, the driving electronic assembled, and finally the assembled display is detached from the glass by lift-off and laminated with protective films.

The encapsulation structure is complex, and it has been studied and optimized for low WVTR, high transparency, and robustness toward defects. The fabrication process includes TFE and barrier film lamination (Figure 12.18) [64]. A PECVD multilayer barrier is deposited at high temperature on both the PI films coated by die – coating on the carrier glasses. Multiple layers of SiN_x/SiO_x are deposited. Thickness and refractive index of the layers are optimized to reach a flat (no interference) transmission curve with transparency > 93.5%, at wavelengths > 450 nm. Barrier film stress and particles have also been optimized to allow the defect-less laser lift-off process, aided by the presence of a sacrificial layer at the interface. The WVTR of the barrier has been measured by the tritiated water test and is < 3.0×10^{-7} g/m^2/day. After oxide-TFT and 4 K TE white-OLED fabrication, a TFE structure is coated on the device before lamination. The total transparency of the display is > 40% and its rollable radius 80 mm. While clearly this display is still a prototype, the technical achievement is extraordinary and has demonstrate the possibility of new large-scale and high-quality flexible displays.

While existing TFE materials and deposition methods are suitable for TFE of OLED, the biggest technological challenge is tool improvements to further reduce particle addition.

12.6 Conclusions

TFE has enabled large-scale commercialization of flexible OLED displays by allowing innovative form factors for the displays and therefore making possible the differentiation of OLED technology from other incumbent technologies. Mass production of hundreds of millions of OLED displays is established for smartphone-size displays on plastic substrates. The established TFE technology is multilayer barrier that manages defects in the single layers and with particles that are being generated in the OLED production. Low-temperature PECVD and IJP are the coating methods used in the cost-effective deposition of the inorganic and organic layers, respectively.

Recently, multilayer barrier technology has been extended to foldable OLED devices, which entered into mass production and commercialization at the beginning of 2019.

Modification of the multilayer TFE technology integrated with barrier films in a hybrid encapsulation scheme is expected to be the upcoming evolution of TFE for application to stretchable displays and large-size flexible TV.

Acknowledgments

The authors recognize all Vitex team members and partners for the development of the multilayer TFE technology. They recognize the hardworking Samsung Display U Project team in commercializing the TFE technology, the work of Samsung Cheil (now SDI) in on polymer, and the Kateeva's contribution to the technology deployment. The authors thank the Applied Materials TFE team for exciting discussions and work on PECVD technology for TFE.

References

1 Tang, C.W. and Slyke, S.V. (1998). Organic electroluminescent diodes. *Appl. Phys. Lett.* 51: 913. https://doi.org/10.1063/1.98799.

2 LGD, the main manufacturer of OLED IV displays, supplies display to all the OLED TV manufacturers https://www.flatpanelshd.com/news.php?subaction=showfull&id=1541074266. Accessed November 10, 2021.

3 Samsung Developer conference November 7–8, 2018, Moscone West, San Francisco, CA, https://www.youtube.com/watch?v=Sp6RVQUQ1G4. Accessed November 10, 2021.

4 Royole Corporation Fremont, California http://www.royole.com, October 31, 2018. https://www.prnewswire.com/news-releases/royole-introduces-flexpai-the-worlds-first-commercial-foldable-smartphone-with-a-fully-flexible-display-a-combination-of-mobile-phone-and-tablet-300741383.html. Accessed November 10, 2021.

5 See for examples: https://www.businesswire.com/news/home/20181002006155/en/Samsung's-7-inch-OLED-Display-Selected-Audi-e-tron. Accessed November 18, 2018. Also http://www.businesskorea.co.kr/news/articleView.html?idxno=21156.

6 Moro, L., Boesch, D., and Zeng, X. (2015). OLED encapsulation. In: *Fundamentals of High-Efficiency OLEDS, Basic Science to Manufacturing OF Organic Light-Emitting Diodes* (ed. D.J. Gaspar and E. Polikarpov), 25–68. London: Taylor & Francis

7 Tominetti, S., Gigli, J., Shih, S.H., Su, Y.T., and Jou, J.H. (2018). Seal encapsulation: OLED sealing processes. In: *Handbook of Organic Light-Emitting Diodes* (ed. C. Adachi, R. Hattori, H. Kaji, and T. Tsujimura), 23-2. Tokyo: Springer.

8 Morena, R.M., Bayne, J.F., Westbrook, J.T., Widjaja, S., and Zhang, L. (2018). Frit sealing of OLED displays. In: *Handbook of Organic Light-Emitting Diodes* (ed. C. Adachi, R. Hattori, H. Kaji, and T. Tsujimura), 24-1. Tokyo: Springer.

9 G4, G5, G6 are definition of the motherglass size. G= Generation; H=Half. For details see also https://www.samsungdisplay.com/eng/media/news_view.jsp?publishdt=2018-05-29&brdGb=R

PT&cpage=1&boardidx=666&searchFlag=&searchValue and http://contents.dt.co.kr/ima ges/200711/2007111502011832673002.jpg. Accessed November 10, 2021.

10 Da Silva Sobrinho, A.S., Czeremuszkin, G., Latrèche, M., and Wertheimer, M.R. (1998). Study of defect numbers and distributions in PECVD SiO2 transparent barrier coatings on PET. *MRS Proc.* 544: 245–250.

11 Graff, G.L., Williford, R.E., and Burrows, P.E. (2004). Mechanisms of vapor permeation through multilayer barrier films: Lag time versus equilibrium permeation. *J. Appl. Phys.* 96: 1840–1849.

12 Graff, G.L., Gross, M.E., Hall, M.G., Mast, E.S., Bonham, C.C., Martin, P.M., Martin, P.M., Shi, M.-K., Brown, J.J., Mahon, J., Burrows, P., and Sullivan, M. (2000). Fabrication of OLED device on engineered plastic substrates. *Proc. Annu. Tech. Conf. Proc. Soc. Vac. Coaters. Dever* April 15–20: 397.

13 Weaver, M.S., Hewitt, R.H., Kwong, R.C., Mao, S.Y., Michalski, L.A., Ngo, T., Rajan, K., Rothman, M.A., Silvernail, J.A., Bennet, W.D., Bonham, C., Burrows, P.E., Graff, G.L., Gross, M.E., Hall, M., and Martin, P.M. (2001). Flexible organic light emitting devices. *Proc. SPIE.* 4295: 113.

14 Moro, L., Krajewski, T.A., Rutherford, N.M., Philips, O., Visser, R.J., Gross, M.E., Bennett, W.D., and Graff, G.L. (2004) Process and design of a multilayer thin film encapsulation of passive matrix OLED displays. Proceedings of SPIE-The International Society for Optical Engineering 5214 (Organic Light-Emitting Materials and Devices VII): 83–93.

15 Lin, S., Chu, X., and Rosenblum, M.P. (2010). "Ultra-Barrier Coatings Enabled by Inkjet Printing" 855. In: *American Chemical Society, Division of Polymeric Materials: Science and Engineering (ed. American Chemical Society)*. San Francisco, California, USA: Spring. 21–25 March 2010, PMSE Preprints Volume 102. Washington, DC: American Chemical Society.

16 Graff, G.L., Burrows, P.E., Williford, R.E., and Praino, R.F. (2005). *Barrier Layer Technology for Flexible Displays* (ed. G.P. Crawford). First published: June 15 https://doi. org/10.1002/0470870508.ch4 Cited by: 17 Series Editor(s): Anthony C. Lowe Online ISBN: 9780470870501 Print ISBN: 9780470870488.

17 Moro, L., Visser, R.J. et al. (2018). Barrier film development for flexible OLED. In: *Handbook of Organic Light-Emitting Diodes* (ed. C. Adachi, R. Hattori, H. Kaji, and T. Tsujimura), 24. Tokyo: Springer.

18 Vogt, B.D., Lee, H.-J., Prabhu, V.M., DeLongchamp, D.M., Lin, E.K., Wu, W.L., and Satija, S.K. (2005). X-ray and neutron reflectivity measurements of moisture transport through model multilayered barrier films for flexible displays. *J. Appl. Phys.* 97 (11): 114509.

19 Moro, L., Chu, X., and Hirayama, H. (2006) A mass manufacturing process for Barix encapsulation of OLED displays: A reduced number of dyads, higher throughput and 1.5 mm edge seal, in IMID/IDMC '06 Digest, Daegu, South Korea, August 22–25, 754–758.

20 Visser, R.J., Moro, L. et al. (2018). Thin film encapsulation. In: *Handbook of Organic Light-Emitting Diodes* (ed. C. Adachi, R. Hattori, H. Kaji, and T. Tsujimura), 26-1. Tokyo: Springer.

21 Lorenza Moro Environmental Barriers for Flexible Displays; Invited Presentation at LOPEC 2014, Messe Munich, Germany, May 26–28, 2014

22 Ravi Prasad Transparent Ultra-Barrier Films for OLED Devices Society for Information Displays International Symposium Digest of Technical Papers 48 (1) Los Angeles, CA May 23–25, 2017 195–196

23 Chiang, J.N., Ghanayem, S.G., and Hess, D.W. (1989). Low-temperature hydrolysis (oxidation) of plasma-deposited silicon nitride films. Chem. Mater. 1: 194–198.

24 Visweswaran, B., Harikrishna, S., Mandlik, M.P., Silvernail, J., Ma, R., Sturm, J., and Wagner, S. Predicting the Lifetime of Flexible Permeation Barrier Layers for OLED Displays Society for Information Displays International Symposium Digest of Technical Papers 45 (1), San Diego, CA, June 1–6, 2014 June 2014 111–113.

25 Bulusu, A.S., Wang, C.Y., Dindar, A., Fuentes-Hernandez, C., Kim, H., Cullen, D., Kippelen, B., and Graham, S. (2015). Engineering the mechanical properties of ultrabarrier films grown by atomic layer deposition for the encapsulation of printed electronics. J. Appl. Phys. 118: 085501. https://doi.org/10.1063/1.4928855.

26 Meyer J., Schmidt H., Kowalsky W., Riedl T., and Kahn A. (2010) The origin of low water vapor transmission rates through Al2O3/ZrO2 nanolaminate gas-diffusion barriers grown by atomic layer deposition. Appl Phys Lett 96(24): 243308–243303.

27 Zeng, X., Moro, L., and Boesch, D. (2016). Gas and moisture permeation barriers. US Patent 9,525,155 Published December 20

28 SID 2018 Digest of Technical Papers Los Angeles, CA, May 22-25, 2018, ISSN 0097-996X/18/4702-1103-$1.00 © 2018 SID SID 2018 DIGEST • 1103 Large Area Thin Film Encapsulation from Bendable to Rollable and Foldable; Helinda Nominanda, Wenhao Wu, Jerry R. Chen, Soo Young Choi; AKT Applied Materials Inc., 3101 Scott Blvd., Santa Clara, California, 95054.

29 Graff, G.L., Gross, M.E., Affinito, J.D., Shi, M.K., Hall, M., and Mast, E. (2003) Environmental barrier material for organic light emitting device and method of making U.S. Patent 6,522,067 B1. App 09/427,138. Filing date October 25, 1999; Published February 18, 2003.

30 Samsung Electronics, Suwon, South Korea.

31 SNU Precision Co Ltd Asan, South Korea.

32 Lin, S., Chu, X., and Rosenblum, M.P. Ultra-Barrier Coatings Enabled by Inkjet Printing American Chemical Society, Division of Polymeric Materials: Science and Engineering, Spring 2010, San Francisco, California, USA, 21–25 March 2010, PMSE Preprints Volume 102

33 Ink Jet Application in Thin Film Encapsulation of Flexible, Lorenza Moro, Chris Hauf, June Zhang, and Jeff Hebb in Digest IMID 2018 (International Meeting on Information Display) August 28–31, 2018/BEXCO, Busan, Korea.

34 Kateeva, Inc., Newark, CA, USA. http://kateeva.com. Accessed November 10, 2021.

35 Madigan, C.F., Hauf, C.R., Barkley, L.D., Harjee, N., Vronsky, E., and Van Slyke, S.A. (2014) Advancements in Inkjet Printing for OLED Mass Production. In: SID Symposium Digest of Technical Papers 45: 399–402,San Diego, CA, June 1–6, 2014.

36 Hebb, J., "Inkjet Printing for Manufacturing of Flexible and Large-Size OLEDs" presented at Vacuum society NCCAVS Joint Users Group Technical Symposium "Novel Materials, Processes, and Devices for Future Generation Electronics" San Jose. February 23, 2017 consulted online on October 20, 2018.

37 Amberg-Schwab, S.U.W. Inorganic-organic polymers in combination with vapor deposited inorganic thin layers an approach to ultra barrier films for technical applications. Presented at LOPE-C, International Conference and Exhibition for the Organic and Printed Electronics Industry, Frankfurt, Germany, June 23–25, 2009.

38 Tomonori Kawamura Barrier film and production method thereof US 2012/0153421 A1 Published June 21, 2012.

39 Hiromoto Ii Satoshi Ito Makoto Honda Kiyishi Oishi Issei Suzuki Gas barrier film, process for production of gas barrier film, and electronic device US2013/0115423A1 Published May 9, 2013.

40 Moro, L., Boesch, D., Zeng, X., and Maghsoodi, S. (2013) Barrier Stacks and Methods of Making the Same, US Patent Application 20130330531.

41 Moro, L. and Krajewski, T.L. (2010) Encapsulated devices and method of making. U.S. Patent 7,767,498.

42 Moro, L., Krajewski, T.L., Ramos, T., Rutherford, N.M., Chu, X., Hirayama, H., and Rj, V. (2006) UV curable layers in barrier films for flexible electronics. Presented at RadTech 2006 – UV & EB Technology Expo & Conference, Chicago IL, USA

43 Ross Young SID/DSCC Business Conference: Smartphone display market outlook. Presented at SID Display Weed 2021, Business Conference, 58th International Symposium, Seminar and Exhibition. Virtual event held May 17–21, 2021.

44 SID 2017 Digest of Technical Papers Los Angeles, CA, May 23-26, 2017, The Challenges of Flexible OLED Display Development, Shan-Chen Kao, Liang-Jian Li, Ming-Che Hsieh, Song Zhang, Pao-Ming Tsai, Zhong-Yuan Sun, Da-Wei Wang, BOE Technology Group Co., LTD, Beijing, China, 1034 • SID 2017 DIGEST ISSN 0097-996X/17/4702-1034-$1.00 © 2017 SID.

45 Applied Materials, Inc. Santa Clara, California, USA.

46 Jusung (Gyeonggi-do, South Korea. Gyeonggi-do 1, South Korea. http://www.jseng.com. Accessed November 10, 2021.

47 https://www.displaydaily.com/article/display-daily/so-what-is-y-octa. Accessed November 10, 2021.

48 Steltenpool, M. and Gautero, L. (July 2017). Thin film encapsulation of moisture-sensitive flexible electronic. Organic and Printed Electronics Magazine No 19 | OPE journal. https://www.ope-journal.com/magazine.html. Accessed November 10, 2021.

49 https://www.oled-info.com/etnews-sdc-building-qd-oled-tv-pilot-production-line accessed November 18, 2018.

50 SID 2018 Digest of Technical Papers Los Angeles, CA, May 22-25, 2018, 5.8-inch QHD Flexible AMOLED Display with Enhanced Bendability of LTPS TFTs, Jaeseob Lee, Thanh Tien Nguyen, Joonwoo Bae, Gyoochul Jo, Yongsu Lee, Sunghoon Yang, Hyeyong Chu, Jinoh Kwag, Display Research Center, Samsung Display Co., LTD., Giheung-Gu, Youngin-Si, Gyeonggi-Do, Korea, Contact Author Email: Jaeseob.lee@samsung.com, ISSN 0097-996X/18/4702-0895-$1.00 © 2018 SID SID 2018 DIGEST • 895

51 SID 2017 Digest of Technical Papers Los Angeles, CA, May 23-26, 2017, Challenges and Progress of Small Bending Radius Foldable AMOLED Display Module Technology; Li Lin, Pengle Dang, Kun Hu, Xiaoyu Gao*, Xiuqi Huang, Kunshan New Flat Panel Display Technology Center Co., Ltd, Kunshan, Jiangsu, P.R. China, Kunshan Govisionox Optoelectronics Co., Ltd, Kunshan, Jiangsu, P.R. China, ISSN 0097-996X/17/4701-0445-$1.00 © 2017 SID SID 2017 DIGEST • 44531.4L.

52 Flexible Plasma-Deposited Encapsulation Barrier for OLED Displays; Carl Galewski, Allan Wiesnoski, Xianghui Zeng, Lorenza Moro, and Stephen Savas in Digest IMID 2014 (International Meeting on Information Display) August 26–29, 2014/Exco, Daegu, Korea.

53 Kreis, J., Freiberger, P., Galewski, C., Wiesnoski, A., and Zeng, X. (2016). Scalable Deposition Technology for Barrier Films Yielding Unique Material Property Set, 2016 Flex Conference, 02-03, Monterey CA.

54 Aixtron OptaCap Technology. http://www.euflex.com.tw/en/books/html/?145.html. Accessed October 15, 2018.

55 Hong, J.-H., Shin, J.M., Kim, G.M., Joo, H., and Park, G.S. (2017). In Bom Hwang, Min Woo Kim, Won-Sang Park, Hye Yong Chu, Sungchul Kim 9.1-inch stretchable AMOLED display based on LTPS technology. J. SID 25 (3): 194–199. doi:10.1002/jsid.547.

56 Steinmann, V. and Moro, L. (2018). Encapsulation requirements to enable stable organic ultra-thin and stretchable devices OLED encapsulation. J Mater Res 33 (13): 1925–1936. doi:10.1557/jmr.2018.194.

57 Axisa, F., Vanfleteren, J., and Vervust, T. (2012). Method for manufacturing a stretchable electronic device. US Patent 8,207,473.

58 Lee, C.-H., Hong, J.-H., Park, W.-S., and Baek, J.-I. (2014) Stretchable base plate and stretchable organic light-emitting display device. US Patent 8,860,203.

59 Arora, W.J. and Ghaffari, R. (2013) Extremely stretchable electronics. US Patent 8,389,862.

60 Han, D.-W., Visser, R.J., and Moro, L. (2015) Barrier film composite and display apparatus including the barrier film composite. US Patent 8,987,758.

61 Kim, D.-H., Xiao, J., Song, J., Yonggang Huang, J.A., and Stretchable, R. (2010). Curvilinear electronics based on inorganic materials. Mater. Adv. Mater. 22: 2108–2124.

62 Sekitani, T., Nakajima, H., Maeda, H., Fukushima, T., Aida, T., Hata, K., and Someya, T. (2009). Stretchable active-matrix organic light-emitting diode display using printable elastic conductors, Nature Materials 8 (June). www.nature.com/naturematerials. doi:10.1038/NMAT2459. Accessed November 10, 2021.

63 https://www.oled-info.com/tags/8k-oled accessed October 16, 2018.

64 OLED-Info Posted: January 5, 2017 by Ron Mertens. https://www.oled-info.com/lgs-crystal-sound-oleds-embed-sound-system-within-display-panel. Accessed October 16, 2018.

65 Werner, K. (MARCH/APRIL 2018). Five short display stories from CES 2018. SID Inf. Display 34 (2): 28.

66 Park, C.I., Seong, M., Kim, M.A., Kim, D., Jung, H., Cho, M., Lee, S.H., Lee, H., Min, S., Kim, J., Kim, M., Park, J.-H., Kwon, S., Kim, B., Kim, S.J., Park, W., Yang, J.-Y., and Yoon, S. (2018). Inbyeong Kang World's first large size 77-inch transparent flexible OLED display. J. SID 26: 287. doi:10.1002/jsid.663.

67 Yoon, J., Kwon, H., Lee, M., Yu, Y.-Y., Cheong, N., Min, S., Choi, J., Im, H., Lee, K., Jo, J., Kim, H., Choi, H., Lee, Y., Yoo, C., Kuk, S., Cho, M., Kwon, S., Park, W., Yoon, S., Kang, I., and Yeo, S. (2015). World 1st large size 18-inch flexible OLED display and the key technologies. SID DIGEST 962.

13

Flexible Battery Fundamentals

Nicholas Winch, Darran R. Cairns, and Konstantinos A. Sierros

West Virginia University, Statler College of Engineering, WV

13.1 Introduction

Flexible displays and devices that incorporate them can come in many different forms and sizes but for some applications it would be beneficial if the entire device is flexible. One particular challenge for this is the need for a flexible power source. There has been a significant amount of work on flexible solar panels [1–3], but even with flexible solar panels there will likely be the need for a flexible battery. This raises a number of challenges and we believe that the development of robust, high-performance, flexible batteries will become an important area in the field and provide an enabling technology for a range of new applications for flexible displays. In this chapter some of the history of flexible batteries, individual designs and applications, as well as future perspectives will be discussed.

A battery is a cell or series of cells that contain electrochemically active material that exchanges ions through chemical reactions to move electrons through a circuit and power a device. Batteries consist of multiple components that serve different purposes to the system. These include the anode, cathode, electrolyte (and separator), current collectors, and the encapsulation. A sample cell can be seen depicted in Figure 13.1. In very simple terms, a battery operates as follows. A battery's potential originates from the theoretical difference in standard potential that the cathode and anode materials possess. This potential is due to the thermodynamic relationship between the two materials, and an inherent tendency for materials to go to their lowest energy state in a closed system. Through redox reactions, the anode material is reduced, where a positive ion is formed, and negative electrons are produced. Ions are transported across the electrolyte barrier while electrons circulate and power the load circuit, both reaching the cathode and oxidizing it. Current collectors are conductive materials, most typically aluminum, copper, or stainless steel in commercial devices, which are non-reactive conductors that store and transport electrons from the redox reaction to the outside circuit. They are very important to the device to maintain stability and consistency. Eventually the battery exceeds its amount of ionizable material (its capacity limit), and the discharge ceases. At this point, the difference between primary and secondary batteries becomes relevant. A primary battery is intended for only a single use, while a secondary battery is designed for multiple uses. After depletion, secondary batteries must be recharged in order to operate again, which is accomplished by reversing the flow of current and running the reactions in reverse. Many factors are considered when determining the optimal and practical combinations of materials for high performing batteries, and this is a

Flexible Flat Panel Displays, Second Edition. Edited by Darran R. Cairns, Dirk J. Broer, and Gregory P. Crawford.
© 2023 John Wiley & Sons Ltd. Published 2023 by John Wiley & Sons Ltd.

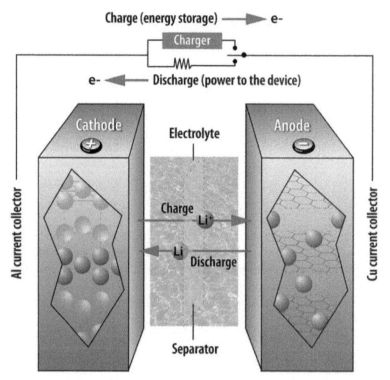

Figure 13.1 Rudimentary illustration of an example lithium-ion battery in charging/discharging state and directionality of particle motion through the circuit. Fair use image.

constantly evolving field of study. A battery is typically a closed system unaffected by external conditions with all of its inherent potential stored inside. It is therefore very important to have a robust outer casing to keep the battery operating smoothly and safely. All the aforementioned information about batteries is intended to be introductory and can be found in more detail in the *Handbook of Batteries* [4].

Flexible batteries are likely to be mostly employed as secondary cells with an emphasis on being recharged safely and consistently. One of the most widely used batteries is the lithium-ion cell, although other ions can be utilized in modern battery technology, including zinc and even potassium [5–14].

It is important that flexible batteries maintain structural integrity during bending and that safety is ensured. There have been a number of high-profile instances of fires or overheating in lithium-ion batteries in consumer devices and it will be important to ensure that flexing or bending of batteries for flexible devices do not cause similar issues.

13.2 Structural and Materials Aspects

A battery for a flexible device has to be inherently deformable, else it would not function. To afford this deformability we can select materials that are flexible, and we can use mechanical design to incorporate flexibility into the battery structure. While some materials are inherently deformable and can be used relatively easily without accommodation, there are often trade-offs in the electrical properties of the material such as poor capacity or power density [5, 6, 8, 11, 14–24]. In contrast, battery materials with high capacity and power density are often brittle and fragile leading to fracture at very low deformations. To mitigate this, it is possible to integrate such brittle materials inside of a structure or framework designed to absorb these stresses and prevent deformation of the structure from affecting the battery negatively. This approach requires more forethought and complex design [25].

13.2.1 Shape

Flexible batteries have many shape constraints compared to other types of battery. They must be thin enough to allow them to deform to a reasonable degree; excessive thickness will result in breakages or user discomfort over time. This limits the use of complex three-dimensional structures that are designed to maximize surface area contact at the interfaces and increase overall power density. This has led to a focus primarily on one-dimensional and two-dimensional type structures to accommodate the role of batteries in flexible devices. This being said, as they are specialized per application, the overall scaling and dimensional ratios can vary greatly from case to case, designed for optimal device performance.

13.2.2 One-dimensional Batteries

One-dimensional battery structures have made significant progress over the past decade. These battery configurations are labeled as one-dimensional due to the length being the dimension that exceeds height and width by orders of magnitude. Similar in macroscopic appearance to a strand of thread, these structures can even be woven into fabric to further integrate with the foundation upon which the flexible device is situated. Since they are thread-like in nature, these batteries are designed with the most flexibility available, allowing large twisting and deformation to occur. This means that the chosen materials almost exclusively need to be inherently flexible to account for their innate lack of any fixed structure. The stresses faced by this type of battery lead them to be radial in nature, following a fixed pattern about the central point along the longitudinal axis. Different forms of one-dimensional battery can be seen depicted in Figure 13.2; these will be discussed in the subsequent paragraphs.

Perhaps the simplest form this can take is that of the coaxial battery. Commonly seen in many production batteries such as AAs, its cross-section appears as several circles fixed about a central point, with either an electrode or a flexible substrate and insulation forming the

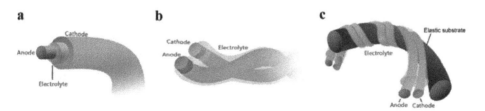

Figure 13.2 (a) Depiction of a coaxial flexible battery. (b) Depiction of a twisted flexible battery. (c) Depiction of a stretchable flexible battery. Reprinted with permission from [26]. Copyright 2018, Royal Society of Chemistry.

central rod, and subsequent components being layered further out from the core, concentrically [12, 14, 17, 23, 24].

Another form that one-dimensional batteries can take is that of twin lengths of electrode separated by an electrolyte and surrounded by an insulating material. This is the one-dimensional adaptation of a planar, layered battery. This form of one-dimensional batteries is normally the easiest to manufacture and assemble, and all parts can be created separately if so desired [1, 27, 28].

Similarly, twisting spiral-type batteries have been created that utilize a polymer electrolyte as both a coating and barrier between the two electrode strands. These twin electrodes spiral evenly about the center of the axis, akin to a DNA helix in appearance. This twisting reduces the maximum tension and compression stresses in the materials. It is more time-consuming and complex to create than the layered structure but can be mechanically more durable [1, 28].

Stretchable variants of these batteries are possible to create, as well. The battery strands are wrapped along the length and secured around an elastic substrate much like the tight coils of a tiny spring. This substrate then serves as the central axis for the system. As the battery length is pulled axially, the substrate carries the majority of the load and deforms, leaving the battery unharmed and power uninterrupted [1, 27, 28].

13.2.3 Two-dimensional Planar Batteries

Planar batteries are the other common form of flexible battery. Made as layers of thin sheets, these batteries offer less variation in design type, but can be shaped to accommodate most form factors. They have less designed bendability and are typically a larger part of the structure of the device they are powering. Batteries created in this fashion are almost exclusively designed to be the singular power source of a device, as opposed to one-dimensional models, where many small units can be connected. In planar form, batteries are far less likely to be made stretchable, as the likelihood of material delamination and cracking are unavoidable by geometric design and cannot be coiled around a flexible substrate as in the one-dimensional example. There are benefits to the planar design, however. They are much easier to manufacture, being able to be assembled easily one layer at a time simply by stacking. This, as well as a uniform surface area from all component interfaces, will increase consistency in the electrochemical reactions in the cells layer by layer, and theoretically offer greater specific capacity than their one-dimensional counterparts [2, 5, 8, 11, 14–16, 19, 20, 22].

Planar flexible batteries can also be created to accommodate ceramics if this is desired. This is achieved by creating a thin layer of material with evenly spaced pores in two dimensions. The active ceramic material is then placed inside those pores, flush with the top and bottom of the layer. In this way the ceramic active material will contact both of the adjacent battery layers with enough consistency to create the necessary channels throughout the battery and the flexible layer bends in response to external loading instead of the ceramic breaking. The smaller the pore size, the less likelihood the ceramic will be bent to any appreciable degree, and the more robust the system becomes. This is most commonly utilized for solid electrolytes, as many designers prefer ceramics over polymers for their superior performance, and this is a creative way to overcome material fragility while retaining ionic conductivity. The ceramic can also be applied to nanofibers that will bend in ways to protect the ceramic while maintaining overall system flexibility [25]. Ideally, it would be beneficial for the electrical performance of the battery to maximize the volume fraction of active material in the battery, but this must be balanced with the structural integrity of the battery, which is impacted by the requirements for device flexibility and expected number of bending cycles.

In instances where only a minor amount of consistent bending is expected, some designers will eschew the flexible substrate entirely, intending for the stresses to be overcome by the neutral plane effect. As any rudimentary exploration into material stresses will show, the greatest forces, either tension or compression, caused by bending occur at the extreme edges of the object in question. If only unidirectional bending can be expected, the centroid of the object will be preserved in a stress-free environment. This tactic can be utilized when bending force is allowable, but axial stresses are prohibited. Again, this is most commonly seen in fragile solid electrolyte material, conveniently the central component of the battery. In contrast to the case for electrolytes where there has been the development of novel materials for flexible batteries, electrodes are often similar to their rigid battery counterparts. Often it is a thin layer of cathodic or anodic paste spread thinly on a carbonic or metallic foil substrate that are inherently malleable. The difference is that the paste must be formulated or treated with special care to ensure it sticks properly to the substrate and does not exhibit cracking.

13.2.4 Solid versus Liquid Electrolyte

When discussing battery design, one of the initial questions should be: Will this battery have a liquid electrolyte? Or is it going to be all-solid-state? Liquid electrolytes are the best ionic conductors on the market today. Solid electrolytes have come a long way in the past two decades, but they cannot yet compete based on profitability or consistency. However, that is not to say that liquid electrolytes have no drawbacks. Most of the materials used for the most efficient liquid electrolytes are flammable, toxic, or both. Pairing this with wearable devices, which have lower mechanical strength and closer contact to consumers, could prove hazardous to consumer safety. Any device designed with a liquid electrolyte introduces risk [2, 8, 17, 20, 22, 24]. This is a primary reason why, especially in the wearable devices field, solid electrolytes are seeing a lot of application. They are generally benign substances that have no flammability or serious health and human contact concerns. Problematically, most of the leading solid electrolytes are brittle ceramic substances, but novel polymer electrolytes are now receiving serious consideration, particularly in the flexible electronics field. With new breakthroughs occurring in materials synthesis all the time, the future is bright [5, 6, 8, 11, 14, 15, 19, 21, 23–25, 27, 29, 30].

13.2.5 Carbon Additives

Many of the leading electrode substances in the battery field do not conduct electricity particularly well; in fact some transfer electrons quite poorly. This would seem like a serious concern, but fortunately an extremely lightweight, facile method exists for boosting electronic conductivity by large margins: the addition of carbon. Carbon comes in many different forms, many of which are inexpensive, and can be combined with the majority of electrode materials. The most common additives are carbonaceous due to their low weight and the high inherent electron conductivity of carbon. The additives differ primarily with their atomic lattice shapes and their intended implantation material [31]. These come in the form of carbon powder, carbon nanofibers, carbon nanotubes (CNTs), and graphene. Carbon powder is the simplest but is an excellent conductor in its own right. This is a result of graphitic carbon's ability to hold a free electron due to its tendency to form only three covalent bonds, leaving space for a fourth electron. When it is mixed into a substance thoroughly, it spreads throughout the bulk, creating electron jumping points dispersed throughout the material [2, 11, 15, 19, 20, 31]. Carbon nanofibers differ from the powder's point-to-point bulk conductivity by creating narrow amorphous carbon chains with occasional graphite clumping throughout a bulk material that allows for more consistent electron flow [5, 8, 15, 23, 31–34]. CNTs are a similar material concept, but more rigid by design. Unlike nanofibers, they are deliberately designed cylindrical lattices of carbon atoms. This serves to add fully formed linear conductive pathways into the electrode material, allowing for even higher levels of conductivity to be achieved. They come in different types, ranging from single cylindrical units to several concentric cylinders, depending on the application and localized conductive channeling required. Single-walled nanotubes can even serve as semiconductors or be non-conductive depending on their chirality, or bend angle in the helix. However, multi-walled nanotubes are exclusively conductive [6, 16, 27, 28, 30, 31, 35, 36]. Graphene is similar in concept to CNTs, but different in that the carbon atoms form in a two-dimensional monolayer, or several layers depending on production quality. Dubbed a modern super-material, it is incredibly strong, flexible, and extremely conductive, allowing for an entire layer of conductivity in a material. Graphene is now being utilized three-dimensionally in foams. This new method is useful for creating extremely hierarchical lattices, and are now being incorporated into flexible technologies, including batteries [18, 31, 36–38].

While all these materials are useful, it is key to understand when to use one over the other. In general, we have listed the carbon additives in order of their recommended usability from bulk to thin layer. The materials also range from least to greatest expense on average; the more specific and atomistically scaled the orientations have to be, the more costly they become. This is why for extremely thin substances, bulk graphite powder may not be the best choice, as it has the least targetable conductivity bonus to offer. However, in a bulky electrode, there is no reason to use many interspersed layers of graphene or nanotubes, as it is difficult to maintain their precise orientations and the expense and time involved preclude it. In flexible electronics, specifically, there can be far less use of thick or otherwise bulky materials; there simply is not enough space. This frees up researchers to utilize more advanced carbon additives in relatively low quantities with an emphasis on electron path creation and orientation. Clearly, these conductive agents will be of paramount importance when it comes to the thin batteries required by wearable displays.

Aside from the electrodes themselves, the current collector is another challenge in flexible batteries. Traditionally, most current collectors have been metals, such as aluminum or copper. These materials do not allow for the highest levels of design flexibility, so research is being conducted utilizing smarter material design. In 2013, Wang et al. [28] showed that CNT films functioned as a suitable replacement material in thin batteries, with data suggesting there was less interfacial resistance between the electrode and the CNT film while also showing superior energy density.

13.3 Examples of Flexible Batteries

We have discussed what makes a flexible battery different from a standard battery and what additional design steps have to be undertaken to achieve best results. Now we will go into some of the key works that have shaped the field in recent years and paved the way for future advances. This is an attempt to illustrate the design variation that flexible batteries show, and what makes them unique or preferable to other forms of battery.

In 2012, Kwon et al. [17] successfully demonstrated a functional flexible lithium-ion battery. This battery was demonstrated to power an LED display, as seen in Figure 13.3, and has led to direct advancements in the field of wearable displays. The flexible battery created by this group was of the one-dimensional variety, in a hollow coaxial configuration. It involved the creation of several tightly spiraled Ni-Sn coated Cu wires to serve as the innermost layer and anode of the battery. This anode was covered by a nanowoven polyethylene terephthalate separator and coated with thin Al wire and a lithium cobalt oxide ($LiCO_2$) cathode slurry. The liquid ($LiPf_6$) electrolyte was inserted into the central gap spacing and was sealed with shrink wrap tubing leaving the Cu and Al wiring tips exposed to connect into the load circuit. The resultant battery resembled a thick piece of wiring that could be bent, twisted, or knotted without succumbing to failure. The hollow core design was actually shown to improve functionality and flexibility of the battery immensely. This is clearly visible in Figure 13.4 and the data speaks for itself. Additionally, it allowed for extremely facile insertion of the liquid electrolyte the battery utilized.

There were several initial concerns with the Ni-Sn coating on the copper wiring. These concerns can be seen in detail in Figure 13.5. Ni-Sn alloys make for good anode material, but it was illustrated that to ensure flexibility the elemental ratio had to be specific to guarantee both performance and durability, otherwise cracking would occur under little stress. The tin is most beneficial to the flexibility of the device, while the nickel benefits the electrical performance and cyclability of the battery, indicating that the materials have a negative correlation of their most beneficial attributes. It is a material that has potential to be replaced with a better option in future studies. It was illustrated that when the Ni composition was brought down to 5.5% the cracking subsided, while maintaining a minimal level of cyclability. The biggest long-term questions stemming from this project are the safety of the battery in consumer hands, the application of a battery designed like this, and the necessary methodology to test the degradation after many cycles and an amount of bending.

One of the earliest fully functional flexible planar Li-ion batteries was created by Hu et al. in 2010 [16]. They utilized a lamination process to fully integrate all battery components onto a single sheet of paper. This battery consisted of freestanding CNT current collectors and $Li_4Ti_5O_{12}$ (LTO) and $LiCoO_2$ (LCO) electrodes. Each current collector and individual electrode were man-

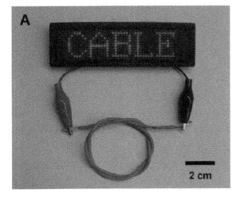

Figure 13.3 Demonstration of functional LED display utilizing cable-like battery. Reprinted with permission from [17]. Copyright 2012, Wiley.

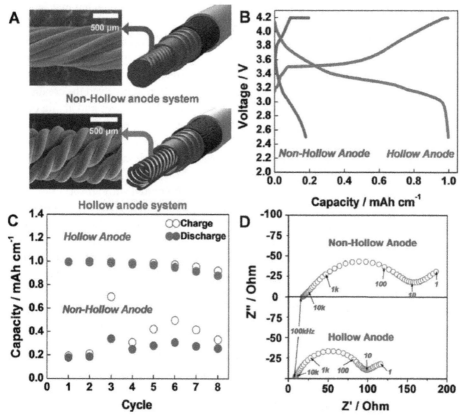

Figure 13.4 Comparison of the hollow and non-hollow anode system, (A) SEM comparison, (B) charge and discharge profile comparison (C) cyclic capacity comparison (D) EIS comparison. Reprinted with permission from [17]. Copyright 2012, Wiley.

ufactured in double-layers by first drying a CNT slurry onto a stainless-steel substrate using blade-coating. Afterwards, the electrode slurry was applied in a similar thickness atop the CNT layer, which was dried and then delicately removed from the substrate and cut into workable components for either electrode. The authors utilized Xerox-type paper as the separator and primary structure of the battery due to its lack of pores and referenced stability to LiPF$_6$, their liquid electrolyte of choice. After layering properly, the battery was assembled and sealed with a thin layer of polymer. This method, shown in detail in Figure 13.6, shows large steps forward in several facets, including cyclability, overall thickness, and manufacturing ease and consistency. Within a certain range of variance, the paper batteries created here were less than 300 microns in height, making them easily thin enough to conform to most relevant objects with its intended usage of electronic paper, interactive labeling, and radiofrequency sensing. The paper cells could be bent without battery failure or material separation and were shown to power an LED when bent 180 degrees (6 mm end to end). Interestingly, the impedance spectroscopy revealed that the paper functioned with lower impedance than the commercial electrolyte–separator combinations on the market at the time. The battery also retained 95% of initial capacity after 300 cycles. This data is shown in Figure 13.7 alongside other relevant flexible battery points of emphasis.

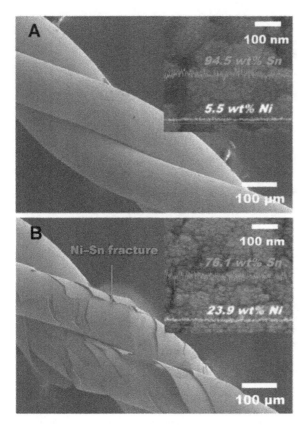

Figure 13.5 Comparison of elemental concentrations in the Ni-Sn coated Cu wire.

The battery is best applied to applications requiring low power, as they illustrate extremely good energy density, but could only power an LED intensely for roughly 10 minutes at the time of this project. This can be offset by connecting multiple of these batteries in parallel, either stacked or adjacently depending on available space, and this will multiply the capacity by the number of batteries connected.

A more recent example of a planar Li-ion flexible battery was created by De et al. in 2017 [15]. They focused on an all-solid-state model with an emphasis on safety and manufacturing ease and used layer-by-layer drop-casting on fabric for wearable devices. This battery utilizes $LiFePO_4$ as a cathode, TiO_2 as the anode, Cu-coated carbon fibers as the cathodic current collector, aluminum foil for the anodic current collector, and lastly a composite polymer of lithium salt and coarse glass fibers as the solid electrolyte. The cathode, anode, and cathodic current collector were all prepared with carbon powder additives for extra electrical conductivity. Due to the assembly nature involving drop casting, inks were created for all materials (excepting the aluminum foil) and were prepared and mixed using polyethylene oxide (PEO) and methanol to an ideal drop-casting consistency. This drop-casting methodology led to excellent interlayer adhesion in the battery and served to create a very homogenous system. An excellent feature of this methodology is that the PEO prevents environmental degradation in ambient conditions.

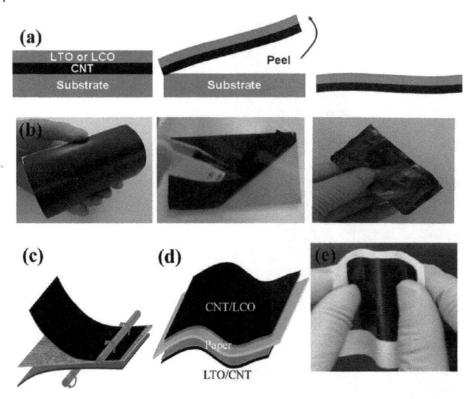

Figure 13.6 Manufacturing process of paper battery. (a) Apply and peel method for anode and cathode. (b) Examine. (c) Roll onto paper separator with electrolyte. (d/e) Bend and check for defects. Reprinted with permission from [16]. Copyright 2010, American Chemical Society.

Ambient manufacture and operation are both allowed and recommended due to this favorable materials selection, massively cutting down on expense. PEO-based composites display flexibility and elasticity; however, they will retain their form, limiting the ability of the fabric to behave as usual and possibly contributing to wearer discomfort. In terms of performance, the battery is capable of producing over 2.5 V and can operate an LED for hours while bending constantly, all while possessing a thickness of around 100 microns. Bending did not appear to negatively affect this battery, showing only minimal discrepancies as the bend angle was adjusted. While bend angle did not impact the battery performance operational voltage increased by almost a full 2 V when the aluminum foil was added. Both comparisons are shown in Figure 13.8. One potential challenge for developing these batteries for commercial applications is degradation of battery performance after multiple bending cycles the comparison of charging curve versus discharge curve across multiple cycles in Figure 13.9 shows more degradation in the discharge profile over time.

Due to their position atop of commercial battery technology, Li-ion technology has naturally garnered the majority of attention in flexible batteries. However, a recent study in 2018 by Ma et al. [33] created a functional Al-air battery using a novel cathode material composed of nitrogen-doped carbon fibers (CF) combined with Fe_3C. It created a battery with a capacity of over

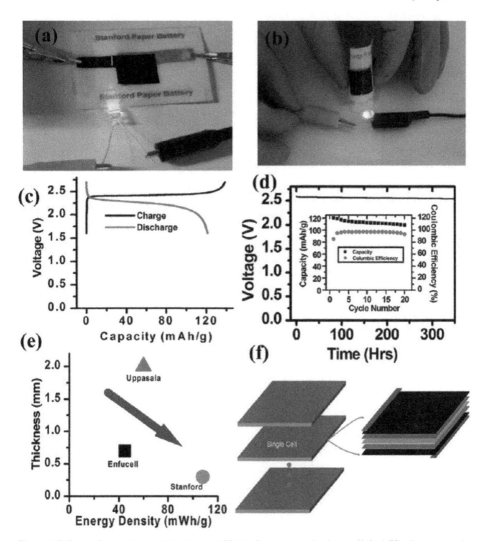

Figure 13.7 (a) Paper battery lighting an LED. (b) Battery continuing to light LED after severe bending. (c) Charge/discharge profile. (d) Voltage retention over time and cyclic capacity and efficiency curves. (e) Comparison graph showcasing this study (Stanford) vs others in regards to energy density and thickness. (f) Stackability demonstrated in both parallel and series. Reprinted with permission from [16]. Copyright 2010, American Chemical Society.

1200 mAh/g. A major problem with Al-air batteries is the slow kinetics of the oxygen diffusion and oxygen reduction. This is a huge hurdle to the success of a high-performance battery that needs to generate electricity quickly. The N-CF-Fe₃C is shown to help catalyze oxygen reactions and promote activity similar to that of Pt or other metal catalysts, which is why it was explored here. This study also attempted to test the merit of a liquid versus solid electrolyte by testing both options using a 6 M KOH solution. It is shown that on average the discharge voltage for the solid-state system was higher by up to 0.15 V. However, this comes with the downside of almost 50% less capacity and runtime. In either case, the study showed improvement compared to

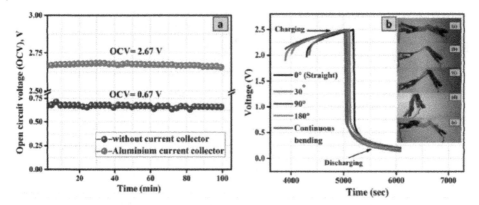

Figure 13.8 (a) Voltage-time curve showcasing importance of current collectors in a battery. (b) Voltage-time curve illustrating consistency of flexible battery throughout bending. Reprinted with permission from [15]. Copyright 2017, American Chemical Society.

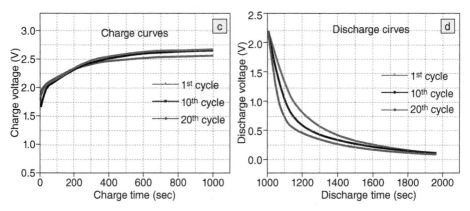

Figure 13.9 (a) Voltage-time graph for charge cycles. (b) Voltage-time graph for discharge cycles. Reprinted with permission from [33]. Copyright 2018, American Chemical Society.

similar style Al-air batteries using different cathodes. When the solid-state variant was inserted into a watchband, it was capable of powering the watch for 22 hours, making it a potentially viable and safe option.

13.4 Future Perspectives

The field of flexible and wearable devices is constantly growing. It is clear that more work will be undertaken in the next few years that will take flexible batteries ever closer to the level that current inflexible technology sits. Also, it cannot be overstated that there are multiple promising pathways to a viable commercial flexible battery. One-dimensional technologies have the potential to be reduced in size sufficiently to become intertwined into clothing enough to go unnoticed. Planar-type batteries could develop into coatings we apply on a wide range of applicable items, similar to a protective coating of energy storage. Custom-sized and custom-shaped batteries could be created to fill vacant spaces in moving parts, perhaps being paired with actuators to assist in energy retrieval and increase efficiency in a system. It should be expected that with the increasing proliferation of additive manufacturing technologies will come the ability to make thinner, lighter, and more consistent battery components that can only serve to aid this developing field. Direct ink writing is an expanding field in flexible battery technology. It is a methodology that dispenses shear-thinning ink blends onto a substrate to create custom-patterned battery components. This technique is useful for creating customizable components with properties that allow for inherent flexibility and tailorable molecular concentrations. The advent of more inherently ductile and beneficial battery components will obviously assist, as well.

References

1 Lee, S.Y., Choi, K.H., Choi, W.S., Kwon, Y.H., Jung, H.R., Shin, H.C., and Kim, J.Y. (2013). Progress in flexible energy storage and conversion systems, with a focus on cable-type lithium-ion batteries. *Energy Environ. Sci.* 6: 2414–2423.

2 Lee, Y.H., Kim, J.S., Noh, J., Lee, I., Kim, H.J., Choi, S., Seo, J., Jeon, S., Kim, T.S., Lee, J.Y., and Choi, J.W. (2013). Wearable textile battery rechargeable by solar energy. *Nano Lett.* 13: 5753–5761.

3 Wang, Y., Bai, S., Cheng, L., Wang, N., Wang, J., Gao, F., and Huang, W. (2016). High-efficiency flexible solar cells based on organometal halide perovskites. *Adv. Mater.* 28 (22): 4532–4540.

4 Linden, D. and Reddy, T.B. (2002). *Handbook of Batteries*, 3rd e. New York: McGraw-Hill.

5 Hiralal, P., Imaizumi, S., Unalan, H.E., Matsumoto, H., Minagawa, M., Rouvala, M., Tanioka, A., and Amaratunga, G.A.J. (2010). Nanomaterial-enhanced all-solid flexible Zinc−Carbon batteries. *ACS Nano* 4 (5): 2730–2734.

6 Li, H., Han, C., Huang, Y., Huang, Y., Zhu, M., Pei, Z., Xue, Q., Wang, Z., Liu, Z., Tang, Z., Wang, Y., Kang, F., Li, B., and Zhi, C. (2018). An extremely safe and wearable solid-state Zinc Ion battery based on a hierarchical structured polymer electrolyte. *Energy Environ. Sci.* 11: 941–951.

7 Li, Y., Fu, J., Zhong, C., Wu, T., Chen, Z., Hu, W., Amine, K., and Lu, J. (2019). Recent advances in flexible Zinc-based rechargeable batteries. *Adv. Energy Mater.* 9: 1802605.

8 Liu, Q., Wang, Y., Dai, L., and Yao, J. (2016). Scalable fabrication of nanoporous carbon fiber films as bifunctional catalytic electrodes for flexible Zn-air batteries. *Adv. Mater.* 28: 3000–3006.

9 Ma, Y., Xie, X., Lv, R., Na, B., Ouyang, J., and Liu, H. (2018). Nanostructured polyaniline–cellulose papers for solid-state flexible aqueous Zn-ion battery. *ACS Sustain. Chem. Eng.* 6: 8697–8703.

10 Tan, P., Chen, B., Xu, H., Zhang, H., Cai, W., Ni, M., Liu, M., and Shao, Z. (2017). Flexible Zn – and Li –air batteries: Recent advances, challenges, and future perspectives. *Energy Environ. Sci.* 10: 2056–2080.

11 Tehrani, Z., Korochkina, T., Govindarajan, S., Thomas, D., O'Mahony, J., Kettle, J., Claypole, T., and Gethin, D. (2015). Ultra-thin flexible screen printed rechargeable polymer battery for wearable electronic applications. *Org. Electron.* 26: 386–394.

12 Wang, Z., Ruan, Z., Liu, Z., Wang, Y., Tang, Z., Li, H., Zhu, M., Hung, T.F., Liu, J., Shi, Z., and Zhi, C. (2018). A flexible rechargeable zinc-ion wire-shaped battery with shape memory function. *J. Mater. Chem. A* 6: 8549–8557.

13 Xiang, P., Chen, X., Xiao, B., and Wang, Z.M. (2019). Highly flexible hydrogen boride monolayers as potassium-ion battery anodes for wearable electronics. *ACS Appl. Mater. Interfaces* 11: 8115–8125.

14 Zamarayeva, A.M., Ostfeld, A.E., Wang, M., Duey, J.K., Deckman, I., Lechêne, B.P., Davies, G., Steingart, D.A., and Arias, A.C. (2017). Flexible and stretchable power sources for wearable electronics. *Sci. Adv.* 3: 1602051.

15 De, B., Yadav, A., Khan, S., and Kar, K.K. (2017). A facile methodology for the development of a printable and flexible all-solid-state rechargeable battery. *ACS Appl. Mater. Interfaces* 9: 19870–19880.

16 Hu, L., Wu, H., Mantia, F.L., Yang, Y., and Cui, Y. (2010). Thin, flexible secondary Li-ion paper batteries. *ACS Nano* 4 (10): 5843–5848.

17 Kwon, Y.H., Woo, S.W., Jung, H.R., Yu, H.K., Kim, K., Oh, B.H., Ahn, S., Lee, S.Y., Song, S.W., Cho, J., Shin, H.C., and Kim, J.Y. (2012). Cable-type flexible lithium ion battery based on hollow multi-helix electrodes. *Adv. Mater.* 24 (38): 5192–5197.

18 Li, N., Chen, Z., Ren, W., Li, F., and Cheng, H.M. (2012). Flexible graphene-based lithium ion batteries with ultrafast charge and discharge rates. *Proc. Natl. Acad. Sci.* 109 (43): 17360–17365.

19 Ostfeld, A.E., Gaikwad, A.M., Khan, Y., and Arias, A.C. (2016). High-performance flexible energy storage and harvesting system for wearable electronics. *Sci. Rep.* 6 (1): 26122.

20 Pu, X., Li, L., Song, H., Du, C., Zhao, Z., Jiang, C., Cao, G., Hu, W., and Wang, Z.L. (2015). A self-charging power unit by integration of a textile triboelectric nanogenerator and a flexible lithium-ion battery for wearable electronics. *Adv. Mater.* 27: 2472–2478.

21 Saunier, J., Alloin, F., Sanchez, J., and Caillon, G. (2003). Thin and flexible lithium-ion batteries: Investigation of polymer electrolytes. *J. Power Sour.* 119-121: 454–459.

22 Tajima, R., Miwa, T., Oguni, T., Hitotsuyanagi, A., Miyake, H., Katagiri, H., Goto, Y., Saito, Y., Goto, J., Kaneyasu, M., Hiroki, M., Takahashi, M., and Yamazaki, S. (2015). Truly wearable display comprised of a flexible battery, flexible display panel, and flexible printed circuit. *J. Soc. Inf. Disp.* 22 (5): 237–244.

23 Yadav, A., De, B., Singh, S.K., Sinha, P., and Kar, K.K. (2019). Facile development strategy of a single carbon-fiber based all-solid-state flexible lithium-ion battery for wearable electronics. *ACS Appl. Mater. Interfaces* 11: 7974–7980.

24 Zhu, Y.H., Yuan, S., Bao, D., Yin, Y.B., Zhong, H.X., Zhang, X.B., Yan, J.M., and Jiang, Q. (2017). Decorating waste cloth via industrial wastewater for tube-type flexible and wearable sodium-ion batteries. *Adv. Mater.* 29: 1603719.

25 Fu, K.K., Gong, Y., Dai, J., Gong, A., Han, X., Yao, Y., Wang, C., Wang, Y., Chen, Y., Yan, C., Li, Y., Wachsman, E.D., and Hu, L. (2016). Flexible, solid-state, ion-conducting membrane with 3D garnet nanofiber networks for lithium batteries. *Proc. Natl. Acad. Sci.* 113 (26): 7094–7099.

26 Sumboja, A., Liu, J., Zheng, W.G., Zong, Y., Zhang, H., and Liu, Z. (2018). Electrochemical energy storage devices for wearable technology: A rationale for materials selection and cell design. *Chem. Soc. Rev.* 47 (15): 5919–5945.

27 Ren, J., Zhang, Y., Bai, W., Chen, X., Zhang, Z., Fang, X., Weng, W., Wang, Y., and Peng, H. (2014). Elastic and wearable wire-shaped Lithium-Ion battery with high electrochemical performance. *Angewandte Chemie International Edition* 53: 7864–7869.

28 Wang, K., Luo, S., Wu, Y., He, X., Zhao, F., Wang, J., Jiang, K., and Fan, S. (2013). Lithium-Ion batteries: Super-aligned carbon nanotube films as current collectors for lightweight and flexible lithium ion batteries. *Adv. Funct. Mater.* 23: 846–853.

29 He, H., Fu, Y., Zhao, T., Gao, X., Xing, L., Zhang, Y., and Xue, X. (2017). All-solid-state flexible self-charging power cell basing on Piezo-electrolyte for harvesting/storing body-motion energy and powering wearable electronics. *Nano Energy* 39: 590–600.

30 Noerochim, L., Wang, J.Z., Chou, S.L., Wexler, D., and Liu, H.K. (2012). Free-standing single-walled carbon nanotube/SnO2 anode paper for flexible Lithium-Ion batteries. Carbon 50: 1289–1297.

31 Rao, J., Liu, N., Zhang, Z., Su, J., Li, L., Xiong, L., and Gao, Y. (2018). All-fiber-based Quasi-solid-state Lithium-Ion battery towards wearable electronic devices with outstanding flexibility and self-healing ability. *Nano Energy* 51: 425–433.

32 Liu, Q.C., Liu, T., Liu, D.P., Li, Z.J., Zhang, X.B., and Zhang, Y. (2016). A flexible and wearable lithium-oxygen battery with record energy density achieved by the interlaced architecture inspired by bamboo slips. *Adv. Mater.* 28: 8413–8418.

33 Ma, Y., Sumboja, A., Zang, W., Yin, S., Wang, S., Pennycook, S.J., Kou, Z., Liu, Z., Li, X., and Wang, J. (2018). Flexible and wearable all-solid-state Al–air battery based on iron carbide encapsulated in electrospun porous carbon nanofibers. *ACS Appl. Mater. Interf.* 11: 1988–1995.

34 Wang, C., Song, Z., Wan, H., Chen, X., Tan, Q., Gan, Y., Liang, P., Zhang, J., Wang, H., Wang, Y., Peng, X., Aken, P.A.V., and Wang, H. (2020). Ni-Co selenide nanowires supported on conductive wearable textile as cathode for flexible battery-supercapacitor hybrid devices. *Chem. Eng. J.* 400: 125955.

35 Chew, S.Y., Ng, S.H., Wang, J., Novák, P., Krumeich, F., Chou, S.L., Chen, J., and Liu, H.K. (2009). Flexible free-standing carbon nanotube films for model lithium-ion batteries. Carbon 47: 2976–2983.

36 Wang, J.Z., Zhong, C., Chou, S.L., and Liu, H.K. (2010). Flexible free-standing graphene-silicon composite film for lithium-ion batteries. *Electrochem. Commun.* 12: 1467–1470.

37 Zhou, G., Li, L., Ma, C., Wang, S., Shi, Y., Koratkar, N., Ren, W., Li, F., and Cheng, H.M. (2015). A graphene foam electrode with high sulfur loading for flexible and high energy Li-S batteries. Nano Energy 11: 356–365.

38 Zhou, G., Li, L., Wang, D.W., Shan, X.Y., Pei, S., Li, F., and Cheng, H.M. (2015). A flexible sulfur-graphene-polypropylene separator integrated electrode for advanced Li-S batteries. *Adv. Mater.* 27: 641–647.

39 Peng, H.J., Huang, J.Q., and Zhang, Q. (2017). A review of flexible lithium–sulfur and analogous alkali metal–chalcogen rechargeable batteries. *Chem. Soc. Rev.* 46: 5237–5288.

40 Fu, K.K., Cheng, J., Li, T., and Hu, L. (2016). Flexible batteries: From mechanics to devices. *ACS Energy Lett.* 1: 1065–1079.

41 Gaikwad, A.M., Arias, A.C., and Steingart, D.A. (2015). Recent progress on printed flexible batteries: Mechanical challenges, printing technologies, and future prospects. *Energy Technol.* 3 (4): 305–328.

42 Zhou, G., Wang, D.W., Li, F., Hou, P.X., Yin, L., Liu, C., Lu, G.Q., Gentle, I.R., and Cheng, H.M. (2012). A flexible nanostructured sulphur–carbon nanotube cathode with high rate performance for Li-S batteries. *Energy Environ. Sci.* 5: 8901–8906.

43 Deng, Z., Jiang, H., Hu, Y., Liu, Y., Zhang, L., Liu, H., and Li, C. (2017). 3D ordered macroporous MoS2@C nanostructure for flexible li-ion batteries. *Adv. Mater.* 29: 1603020.

44 Liu, Z., Mo, F., Li, H., Zhu, M., Wang, Z., Liang, G., and Zhi, C. (2018). Advances in flexible and wearable energy-storage textiles. *Small Methods* 2: 1800124.

45 Gwon, H., Hong, J., Kim, H., Seo, D.H., Jeon, S., and Kang, K. (2014). Recent progress on flexible lithium rechargeable batteries. *Energy Environ. Sci.* 7: 538–551.

46 Zhou, G., Li, F., and Cheng, H.M. (2014). Progress in flexible Lithium batteries and future prospects. *Energy Environ. Sci.* 7: 1307–1338.

47 Mackanic, D.G., Kao, M., and Bao, Z. (2020). Enabling deformable and stretchable batteries. *Adv. Energy Mater.* 2001424.

48 Zhu, Y.H., Yang, X.Y., Liu, T., and Zhang, X.B. (2019). Flexible 1D batteries: Recent progress and prospects. *Adv. Mater.* 32: 1901961.

49 Liu, B., Zhang, J., Wang, X., Chen, G., Chen, D., Zhou, C., and Shen, G. (2012). Hierarchical three-dimensional $ZnCo_2O_4$ nanowire arrays/carbon cloth anodes for a novel class of high-performance flexible lithium-ion batteries. *Nano Lett.* 12: 3005–3011.

50 Hu, Y. and Sun, X. (2014). Flexible rechargeable lithium ion batteries: Advances and challenges in materials and process technologies. *J. Mater. Chem. A* 2: 10712–10738.

14

Flexible and Large-area X-ray Detectors

Gerwin Gelinck

Holst Centre, TNO, Eindhoven, The Netherlands
Eindhoven University of Technology, Eindhoven, The Netherlands

14.1 Introduction

Flat panel x-ray detectors for medical diagnostic applications were commercialized just before the turn of the millennium. Being digital, these x-ray detectors offer a number of benefits over older analog systems. Images are available faster, are easier to share, and can be achieved using less radiation. The lack of a practical means to focus x-rays implies that the shadow x-ray image has a size that is equal to or larger than the object to be imaged. A full-chest x-ray image can require up to a 17×17-inch detector. The heart of these detectors is formed by a two-dimensional array of pixels (Figure 14.1). Each pixel in the array comprises a light (or x-ray) sensing element plus an associated switching transistor. The development of large arrays of hydrogenated amorphous silicon (a-Si:H) thin-film transistors (TFTs) for active-matrix liquid crystal displays (LCDs) made it possible to achieve high signal to noise ratio at low radiation dose and large area. The basic mode of operation and the materials as described in the early pioneering work [1–3] are still used today. Via many incremental technology improvements in the pixel array as well as the other building blocks, the latest generation of detectors reach very high performance in terms of signal to noise ratio, dynamic range, spatial resolution, readout speed, and reliability, and they can be made at reasonable costs at the same time [4]. As a result, flat panel x-ray detectors are widely used in medical applications such as radiography, ultra-low-dose fluoroscopy, and CT-like three-dimensional imaging. They also found their way into many industrial imaging solutions and security industries.

Today's x-ray detectors are produced on glass substrates, making them heavy, difficult to transport, and prone to breakages. This is particularly important for portable radiography applications. Flexible x-ray detectors processed on plastic substrates would be less vulnerable than current glass-based designs. Less protection materials have to be applied to ruggedize the portable x-ray system, saving cost and making it lighter for mobile users to carry. Some degree of mechanical flexibility or bendability can furthermore enable innovative product designs, concepts, and applications. Flexible digital radiography can reduce the discomfort caused by the rigid flat panel sensors in intraoral radiography. Curved detector can result in a smaller three-dimensional imaging x-ray system with better, more uniform image quality [5]. A mechanically flexible x-ray panel is also appealing for integrity analysis of, for instance, pipelines or inspection of aircraft structures.

Flexible Flat Panel Displays, Second Edition. Edited by Darran R. Cairns, Dirk J. Broer, and Gregory P. Crawford.
© 2023 John Wiley & Sons Ltd. Published 2023 by John Wiley & Sons Ltd.

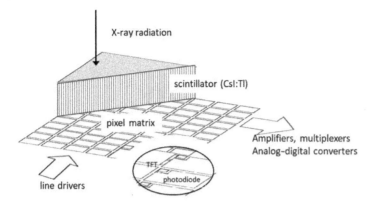

Figure 14.1 Schematic of the two-dimensional image sensor. Here, a scintillator-based (indirect) x-ray detector is shown. A direct-conversion detector consists of a similar stack: TFT pixel array on glass or plastic substrate, thick photoconductor layer on top.

Admittedly, considerable technical development is still necessary before flexible detectors will appear in the diagnostic radiology department, but spectacular breakthroughs are no longer needed. This chapter aims to give an overview of the status and the challenges in this area. Section 14.2 describes basic operation principles of a flat panel detector and the different building blocks. In Section 14.3 we will discuss novel semiconductor materials that can be used as sensing photodiodes. We focus on materials that can be processed from solutions over large area using low-temperature processes, making them attractive from a cost point of view for detectors on plastic as well as glass. Because the flexible TFT technologies will be discussed in detail elsewhere in this book, we will introduce them only briefly in Section 14.4, guided by their requirements in x-ray detectors. Earlier prototypes on flexible TFT arrays for large-area imagers are discussed in Section 14.4.2, followed by a more detailed description of a recent "medical-grade" detector in Section 14.5. Finally, directions for future research are finally outlined in Section 14.6.

14.2 Direct and Indirect Detectors

Flat panel digital x-ray detectors are usually classified as direct conversion or indirect conversion, depending on the process for converting the incident x-rays into charge. The basic components of an indirect-conversion digital x-ray detector include a scintillator phosphor conversion film that converts the incident x-rays into visible photons, a photodiode – usually amorphous silicon – that detects the light emitted from the x-ray phosphor conversion film and converts the light into an electric charge, and a TFT backplane that reads out the photo-charges for each pixel in the backplane individually.

The scintillator is typically composed of high Z-atoms, to capture the x-ray photons efficiently, and doped with phosphor atoms to tune the light emission in the visible spectrum of 500–550 nm. Two materials that are frequently used are gadolinium oxysulfide doped with terbium (GOS:Tb) and thallium-doped cesium iodide (CsI:Tl). The GOS:Tb x-ray conversion film can be a separate detachable sheet laid on top of the array. CsI:Tl is usually directly deposited on the array. This gives better optical coupling efficiency. CsI:Tl crystals have a needle-type structure. The columnar structures effectively behave like an optical waveguide and decrease optical crosstalk [6].

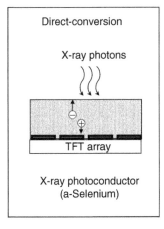
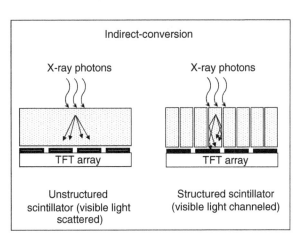

Figure 14.2 Direct-conversion detectors have an x-ray photoconductor such as amorphous selenium that converts x-ray photons into an electric charge directly, with no intermediate stage. Indirect-conversion devices have a scintillator that first converts x-rays into visible light. That light is then converted into an electric charge by using a thin-film photodiode array. In indirect-conversion detectors, the x-ray scintillator can be structured or unstructured. Structured scintillators such as CsI:Tl reduce the spread of visible light. This improves spatial resolution and permits the use of thicker scintillator materials for improved x-ray absorption.

Direct-conversion detectors utilize a photoconductive layer – usually amorphous selenium [7] – that absorb the x-ray and produce an electron–hole pair. When a bias is applied across the photo-conductive layer, the generated charge carriers are pulled toward the electrodes. When an x-ray is absorbed and generates an electron–hole pair, the charge carrier is drawn to the opposing electrode with limited scattering in comparison to indirect detection techniques, see Figure 14.2. As a result, the image resolution of the direct imager is better than that of an indirect detector. However, the amorphous selenium has a smaller capture cross-section in comparison to the CsI:Tl screens used in indirect conversion, thus requiring the patient to be exposed to a higher dosage of x-rays to achieve the same resolution. In addition, the selenium film is usually 0.25–1.0 mm thick, requiring a large voltage on the order of 5–10 kV across the selenium to provide a sufficient electric field to extract the incident x-ray-generated charge carriers. There have been several reviews on this topic: see, for example, [7–9].

In this review we concentrate only on indirect-conversion detectors, and highlight some of the new approaches in the field and the progress made to date with special attention to flexible detectors and detectors made using solution-processed semiconductors.

14.3 Thin-film Photodiode Sensors for Indirect-conversion Detectors

14.3.1 Performance Parameters

The most important figure-of-merit parameters of the photosensors can be summarized as follows:

(1) Responsivity (R) indicates how efficiently a photodetector responds to an optical signal. It is defined as the ratio of photocurrent to incident-light intensity,

$$R = J_{ph} / L \tag{14.1}$$

where J_{ph} is the photocurrent and L is the incident-light intensity. Since R is proportional to the external quantum efficiency (EQE) of the photodetector that evaluates the conversion rate from photons to electrons–hole pairs, it can also be expressed as

$$R = EQE * \lambda q / hc \tag{14.2}$$

where λ is the incident-light wavelength, q is the absolute value of electron charge (1.6×10^{-19} C), h is the Planck constant (6.63×10^{-34} Js), and c is the speed of light (3.00×10^{9} m/s). Efficiency should be quoted at 500–550 nm, as that is the wavelength of the emitted photons of commonly used scintillator materials.

(2) Detectivity (D*) characterizes the weakest level of light that the device can detect. D* is determined by the responsivity and noise of the photodetector. D* (measured in units of Jones, (1 Jones = $1 \, cmHz^{1/2}$/W) is calculated using

$$D* = R / \left(2qJ_d\right)^{1/2} \tag{14.3}$$

where J_d is the dark current density. The use of Equation 3 implies that the dark current is the major contribution to the noise that limits D*.

(3) Dark current (J_d) is the current flowing through the diode per unit area in the absence of light under a given reverse bias voltage. The dark current should be sufficiently small, as it increases recombination losses of photogenerated charge carriers and is a source of electronic noise determining the minimum detectable optical signal. The dark current in most photoconductive semiconductors normally comes from one of two factors: the rate of injection of carriers from the contacts into the photoconductor and the rate of thermal generation of carriers. A small dark current implies that the contacts to the photoconductor should be non-injecting, and the rate of thermal generation of carriers from various defects or states in the bandgap should be negligibly small (i.e. dark conductivity is practically zero). Low dark conductivity generally requires a wide bandgap semiconductor. The acceptable dark current depends on the exact application. Values in the range 1–10 pA/mm^2 or 0.1–1 nA/cm^2 are often quoted [10]. According to simulation [11], an imager array needs EQE of 40–50% with dark currents of less than 0.1 nA/cm^2 to be quantum limited over x-ray exposure range.

(4) Noise equivalent power (NEP). The sensitivity of the array can be expressed in terms of the noise equivalent power, NEP, using

$$NEP = q_N / eF \, d^2 EQE \tag{14.4}$$

where q_N is the image sensor readout noise charge, F is the fill factor of a sensor pixel, and d is the edge dimension of a square pixel.

(5) Response speed. The rise time and fall time of the photodiode response to an optical signal is typically defined as the time between 10% and 90% of the maximum photocurrent. It is related to the charge transport and collection, which means a small electrode spacing (charge carrier path length) benefits fast response, while ensuring sufficient light absorption and small leakage current. The 10–90 time constants describe the behavior of the majority of the photogenerated charges. A minor fraction of the photogenerated charges can, however, show significantly slower fall time due to charge trapping effects. Effectively this leads to a small persistent photosignal

after exposure. Even when this amounts to < 1% after 100 ms, it can be problematic in case of real-time applications such as computed tomography, CT, or fluoroscopy.

(6) Large area. One of the most important requirements for a large area x-ray photoconductor is that it must be capable of being coated to the required thickness over a large area to cover the TFT array, e.g. 24 cm × 30 cm for mammography. The large area coating requirement obviously rules out the use of photo- or x-ray sensitive semiconductors, which are difficult to grow in such large areas, and would require process temperatures incompatible with the TFT backplane and its (plastic) substrate.

14.3.2 Photodiode Materials on Plastic Substrates

14.3.2.1 Amorphous Silicon

Almost all commercial indirect large-area x-ray detectors use amorphous silicon as photoactive materials. The a-Si is grown to a thickness of 1.0–1.2 µm using PE CVD. The typical photodiode fabrication approach is to pattern the n-Si layer of the photodiode at each pixel, and to blanket deposit the i-Si, p-Si, and transparent conductor top metal contact continuously over the entire array. A major advantage of this full fill factor approach is that the entire pixel is covered by the i-Si absorption layer, thus eliminating any "dead" spots in the pixel where photon strikes will not be detected. This full fill factor approach also produces the smallest pixel size (and hence highest resolution) for a given set of design rules.

The a-Si:H photodiodes are typically fabricated using process temperatures of around 300°C. The high temperatures in combination with the relatively thick film thicknesses can cause problems such as strain-induced cracking of the sensor layer due to thermal mismatch, when integrating these a-Si:H diodes on plastic. The built-in stress in the a-Si:H layer of p-i-n diodes can be reduced by lowering the deposition temperatures, albeit with a degraded performance [12]. The defect density increased by a factor of 20 when reducing the deposition temperature to 150°C. By reducing the thickness of the intrinsic layer charge carrier recombination was minimized, and 150°C diodes with an EQE up to ~ 70% and dark current < 1 nA/cm^2 were realized [12]. Since then, other groups have successfully reported detectors on polyethylene naphthalate (PEN) foil [13, 14]. However, if the i-Si deposition process is not tightly controlled, stress-related failure may occur. Moreover, when the i-Si layer is deposited at lower temperatures, it will take hours to deposit a ~ 1 µm-thick film.

The issue of the intrinsic film stress of amorphous silicon was dealt with in another way by Marrs and Raup [41]. They patterned the i-Si layer, thus imparting less mechanical stress on the substrate, and placed the diode next to the TFT. This reduces the area available for detection by approximately 15%. An advantage of the pixelated p-i-n diode approach is that less photons absorbed over one pixel electrode are read out by a neighboring pixel. This reduction in crosstalk between neighboring pixels results in increased contrast and improved resolution. By using another plastic substrate, polyimide, which has coefficient of thermal expansion closer to that of the deposited thin films, and by optimizing the diode passivation layers, they were able to make photodiodes at 275°C with greatly improved electrical performance (a dark current of 0.5 pA/mm^2 and photodiode quantum efficiency of 74%) and mechanical reliability.

14.3.2.2 Organic Semiconductor Materials

Organic semiconductors are very appealing for light detection applications. They combine effective light absorption in the green region of the spectrum with good photogeneration yield, sensitivity, and response time [15]. Since many organic materials can be dissolved in organic solvents, organic thin films can be easily formed using a solution or printing process without the need for vacuum equipment. This makes them specifically attractive for large area imagers. The low process temperature, typically less than 150°C, creates the possibility to use a wide range of plastic substrates instead of glass. Furthermore, the (thermo)mechanical properties of organic semiconductors are compatible with plastic substrates.

As organic semiconductors have low dielectric constants and rather high exciton binding energies (0.1–1.4 eV), the photocarrier yield is too low for applications. When, however, two (dissimilar) organic materials are used then efficient charge transfer can take place at the interface (Figure 14.3). This concept was first demonstrated using organic bilayers. Halls et al. [16] and Yu et al. [17]. demonstrated the concept of bulk heterojunction (BHJ) structure, in which semiconducting polymers as the electron donor and fullerene derivatives as the electron acceptor are mixed in the bulk, phase separated in micron and nanometer scale, and form junctions with great interface area to dissociate excited electron–hole pairs. This work was initially geared toward photovoltaic applications, but it was soon realized that the same concept could be very useful to make organic photodetectors (OPDs) as well. Ng et al. demonstrated a flexible photoimage sensor array using solution-processed organic photodiodes on a flexible a-Si:H matrix backplane [18]. To achieve low dark currents of < 1 nA/cm², very thick layers of 4 μm were used. As a result, the EQE was reduced to 35%. This illustrates their remarkable potential and has spurred further research. In recent reviews [15, 19], it is shown that solution-processed organic photodiodes can show D* detectivity values exceeding 10^{13} Jones, combined with large linear dynamic range of and short

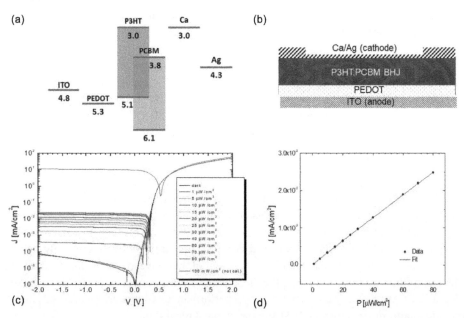

Figure 14.3 Example of bulk heterojunction photodetector: A solution-processed blend of a p-type polymer, poly(3-hexylthiophene (P3HT), and n-type molecule (PCBM) sandwiched between two electrodes. Energy diagram (a) and schematic cross-section (b) of the photodetector. Current-voltage characteristics as a function of different light intensities. In (c) the voltage was swept from −2 to 2 V, and back. In (d) the current density is plotted as a function of light intensity, P. From the fit the responsivity, R, of 0.31 A/W can be extracted.

response time of 0.1 ns. The photoresponse is high (EQE ~ 50% or higher) at low thicknesses (hundreds of nanometers). Dark current densities can be low, with values reported from 10^{-3} mA/cm^2 down to (<) 1×10^{-7} mA/cm^2 at voltages of -2 V or lower. Further work is in progress to add additional layers to reliably lower the dark currents even further.

14.4 TFT Array

14.4.1 Pixel Architecture and Transistor Requirements

The electrical schematic of one sensor pixel is shown in Figure 14.4A. The anodes of all photodiodes are connected to a common electrode, which is connected to an external bias voltage source. The diodes are operated with a reverse bias voltage, of typically a few volts. The cathodes of all photodiodes of each column are connected to a common readout line via a TFT. Each readout line is connected to the input of its assigned readout amplifier (Figure 14.4b). The gates of all

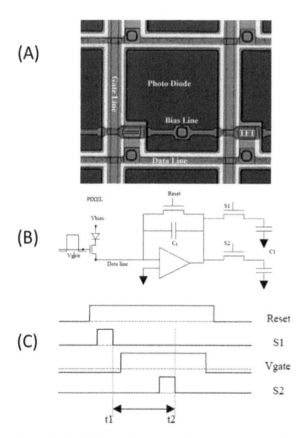

Figure 14.4 (A) Photomicrograph of a single pixel, consisting of TFT, photodiode connected to gate, bias, and data lines. (B) A schematic diagram of a pixel connected to one channel of a charge amplifier and its two sampling circuits. (C) Readout scheme. During one frame time all the rows are sequentially selected by applying a voltage that changes the TFTs from the non-conducting to the conducting state. In this line selection time, the readout TFT transfers the charge from the photodiode capacitance to the data line. Reproduced with permission from [20].

TFTs of each row are connected to a common gate line, which is driven by a dedicated row driver output. The flat panel sensor is scanned one row at a time, in a similar way as active-matrix displays (Figure 14.4c).

Photons absorbed in the active layer in the diode create electron–hole pairs. The electrons and holes are separated by the electric field generated by the reverse bias on the diode and collected at the terminals. The resulting photocurrent discharges the diode capacitance and reduces the reverse bias across the diode. The readout charge is proportional to the number of photons absorbed in the photodiode. External electronics amplify the readout charge from each pixel (Figure 14.4b) and multiplex the data signals from the array into a video signal, representing the two-dimensional image.

During one frame time, all the rows are sequentially selected by applying a voltage that changes the TFTs from the non-conducting to the conducting state. In this line selection time, the readout TFT transfers the charge from the photodiode capacitance to the data line and resets the voltage across the photodiode capacitance to its original value. The short time (typically tens of microseconds) that is available to read out the charge that was accumulated on the photodiode determines the minimum value of the resistance of the transistor when it is switched on on-resistance (R_{on}). During this time, the charge must be transferred from the photodiode to the integrating amplifier, and the output of the amplifier must be scanned. So, all pixels of an entire line are read out simultaneously. The maximum readout speed of the configuration for one line is characterized by the time constant according to $\tau = C_D \times R_{on}$ where C_D denotes the capacitance of the pixel photodiode. An important characteristic of the detector plate with respect to the affordable noise performance is the parasitic capacitance of the readout lines.

For a discharge down to 1% after the third frame, a τ of 10 μs would be sufficient. For a diode capacitance C_D of, say, 2 pF, R_{on} must be less than 5 MΩ to keep the RC time constant below 10 μs. With mobilities of 0.5 cm^2/Vs and gate voltages of ~ 10 V, achieving such channel resistances is no problem for TFTs with dimensions as small as 10×10 μm^2, i.e. relatively small compared to the typical pixel size of 150×150 μm^2. The off-resistance must be more than 10^{12} Ω to keep the charge loss during one frame time (~20 ms) below 1%. As the leakage currents of all pixels at the readout line add to the signal of one pixel that is addressed, an even higher off-resistance is desirable to reduce crosstalk from strongly illuminated regions in the arrays.

Other requirements for the TFT are low noise, in particular low frequency (flicker) noise, and sufficient radiation hardness in the conditions of radiology.

14.4.2 Flexible Transistor Arrays

a-Si:H TFTs arrays are commonly used as the active electronic elements in both indirect- and direct-conversion flat panel detectors [24]. The a-Si:H technology provides good uniformity over a large area, at an adequate performance. The TFT manufacturing process is essentially the same as that used to make flat panel LCDs. This makes the technology readily available at affordable prices. The hydrogenated amorphous silicon layer is deposited using plasma-enhanced chemical vapor deposition, typically at ~ 300°C on glass, and subsequently patterned via photolithography and chemical etching techniques. As a result, a high-resolution array of pixels can be fabricated.

By lowering the deposition temperatures to 150°C, several groups have made flexible a-Si:H TFTs on plastic [25, see also Table 14.1]. Researchers at PARC demonstrated its use in flexible 180×180 pixelated image sensors, using low-temperature a-Si:H [12, see also Figure 14.5] as well

Table 14.1 Specifications of x-ray detector made on flexible plastic substrates with amorphous silicon or soluble OPD sensors using different active-matrix TFT technologies.

Parameter	[12]	[18]	[13]	[21]	[22]	[23]
Panel size	180 × 180, 3.5-in diagonal	180 × 180, 3.5-in diagonal	160 × 180, 3.2 × 3.6 cm	16 × 16, 2 × 1.5 cm	32 × 32	120 × 160
Resolution	75 ppi	75 ppi	127 ppi		127 ppi	200 ppi
Pixel size	340 μm × 340 μm	340 μm × 340 μm	200 μm × 200 μm	~1 × 1 mm	200 μm × 200 μm	126 × 126 μm
Substrate	PEN	PEN	PEN	Polyimide	PEN	PEN
TFT material	a-Si:H	a-Si:H	IGZO	Organic, carbon nanotubes	Organic, pentacene	IGZO
Mobility (cm²/Vs)	0.9	0.9	10–15	17.4	0.2	15
Photosensor material	a-Si:H p-i-n	Organic BHJ	a-Si:H p-i-n	Organic BHJ	Organic BHJ	Organic BHJ
Dark current density (nA/cm²)	<1	<1	0.3	~4200	960	0,1–0,5
EQE	70%	35%	N.A.	40%	60%	25%
Responsivity	–	0.14 A/W	N.A.	0.15 A/W	0.32 A/W	0.2–0.25 A/W
Detectivity (D*, NEP)	NEP 1.2 pW/cm²	NEP 30 pW/cm²	N.A.	N.A.	N.A.	$D* 3 \times 10^{13}$ Jones
Scintillator	–	–	GOS:Tb	GOS:Tb	CsI:Tl	CsI:Tl
Lowest detectable dose	–	–	N.A.	~10 mGy/s	0.27 mGy/s	3 μGy/frame

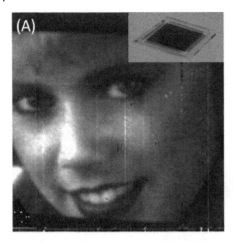 

Figure 14.5 Optical images captured with a flexible, 3.5-inch diagonal a-Si:H TFT sensor arrays fabricated at 150°C, patterned by inkjet printing the patterning resists on PEN. In (A) the TFT backplane is combined with a low-temperature a-Si:H p-i-n diode. Reproduced with permission from [12]. In (B) an organic bulk heterojunction photodiode is used as photosensing element. Reproduced with permission from [18]. (Insets show photographs of the actual arrays).

as organic photosensors [18]. It is noteworthy that in this work patterning was done by inkjet printing wax resists rather than using traditional photolithography to drive down manufacturing cost.

The thus far largest flexible a-Si:H-based x-ray detector was made by a US consortium, led by Arizona State University. Its diagonal size was 10 inches [26]. Instead of manufacturing a single large-area detector, one alternative is to abut several smaller x-ray detectors together. Using conventional arrays on thick glass this approach allows a $4 \times$ larger detector from four separate digital x-ray detectors. With x-ray detectors on thin and flexible plastic substrates, it was shown that it is possible to seamlessly tile more than four digital x-ray detectors that are overlapped similar to placing roofing shingles [27]. This assembly makes it in principle possible to create extremely large digital x-ray imaging arrays for medical imaging applications, such as single-exposure, low-dose, and full-body digital radiography.

Despite their success, amorphous silicon a-Si:H TFTs have some drawbacks, especially when the process temperatures are reduced, and this spurs interest in alternative semiconductor materials. Driven again by the display market trends, i.e. flexible, low-cost displays, ultra-high-definition TV standards, and active-matrix organic light-emitting diode displays (OLEDs), new transistor technologies such as organics and amorphous metal oxide (such as indium gallium zinc oxide, IGZO) are making their way onto the stage. Organic TFTs offer mobilities on par with a-Si:H but processed at much lower temperatures, and potentially at much lower cost. The improved speed capabilities (with mobility values of IGZO TFT multiplied by 10 compared to a-Si:H) represent an opportunity for next-generation flat panel detectors with even faster readout modes and longer exposure windows.

Organic small molecule and polymer-based TFTs can reach mobilities > 5 cm^2/Vs [28, 29, 30]. Mobility values of ~ 0.5 cm^2/Vs are, however, sufficient for most applications, as shown in the previous section, and many organic materials have met this mobility value at very low processing temperatures. A 5×5 cm^2 organic image sensor array was made by laminating a foil with pentacene TFTs with another foil that contains (evaporated) photoconductors [31]. Limited foil-to-foil overlay accuracy and relatively large electrical connections between the two foils resulted in a pixel resolution of only 36 ppi. Nausieda et al. demonstrated a small 4 × 4, 36 ppi active-matrix by direct fabrication of the diodes on top of the TFT backplane [32]. Again, the semiconductor was pentacene deposited by vacuum deposition. Solution processing is, however, preferred for high-volume, low-cost production. Using drop-cast solution-processed nanotube TFTs with mobility of ~20 cm^2/Vs, a low-resolution 18 × 18 array (physical size is 2 cm × 1.5 cm) was made [21]. A GOS:Tb scintillator film is used to convert x-ray photons into green light with an emission peak of ~545 nm, which is then detected by the organic photodiodes in the imager. The imager showed a linear response of pixel photocurrent as a function of incident x-ray dose rate, down to ~10 mGy/s. The lowest detectable intensity is governed by the OPD dark current. With a TFT technology originally developed for flexible electrophoretic displays [33], Gelinck et al. demonstrated integration of soluble organic TFTs and organic photodiodes into high-resolution image arrays on a PEN foil of only 25 μm thickness. Arrays with different pixel densities up to 200 ppi were made, and it was shown that the thin, unpatterned OPD does not lead to significant optical crosstalk between neighboring pixels [22]. Except for the metals, all layers are processed from solution. As the active semiconductor in the TFTs, a soluble precursor of pentacene was applied by spin coating that was subsequently converted to functional pentacene by annealing. TFT mobility was very uniform with an average mobility of ~ 0.2 cm^2/Vs and a very low off current of ~ 0.1 pA. The ratio between the TFT's off- and on-resistance exceeds 10^6. The leakage current influences from the TFTs are negligible. In combination with a CsI:Tl scintillator, static x-ray images were recorded using a typical radiography x-ray spectrum with a mean energy of 40 keV. X-ray images could be recorded at dose levels down to 0.27 mGy/s, i.e. in the range normally used in medical applications. This sensitivity was limited by (the non-uniformity of) the dark current of the photodiodes of ~ 10^{-5} mA/cm^2.

IGZO TFTs outperforms amorphous silicon transistors in terms of charge carrier mobility as well as leakage current. Lujan and Street demonstrated a flat panel flexible x-ray image sensor fabricated with IGZO TFTs [13]. The detector array had a 160 × 180-pixel format with a 200-μm-pixel size for an overall dimension of 3.2 × 3.6 cm. Amorphous silicon p-i-n photodiodes were used in this work, with a maximum process temperature of 170°C, with a leakage current of 3 pA/mm^2 at 5 V bias. This is within a factor 3 of the typical high-temperature conventional devices on glass. OPD/amorphous IGZO TFTs x-ray imagers were recently proposed for next-generation digital breast tomography systems [34]. Simulation results showed that IGZO enabled superior imaging performance to a-Si:H due to a low electronic noise. The possibility to vertically stack the OPD on top of an active TFT pixel circuit – consisting of three TFTs per pixels – rather than side-by-side allows further pixel downscaling. In Section 14.5 we will describe the fabrication and characterization of recent x-ray detectors with IGZO TFTs.

Independent of the TFT material used, the way to fabricate the flexible TFT and detector is by first attaching a thin flexible plastic foil onto a glass substrate. The functional layer stacks are subsequently processed and finally the plastic foil containing the microelectronic devices is delaminated from its support. The adhesion forces should be sufficiently large to withstand all processing steps but low enough to enable controlled mechanical debonding without introducing defects to the finished detector. The rigid support can be reused. This allows the use of standard off-the-shelf patterning and deposition equipment. Typically, a registration better than 2 μm can be achieved, when the plastic substrates are thin (125 μm or less) and have a coefficient of thermal expansion

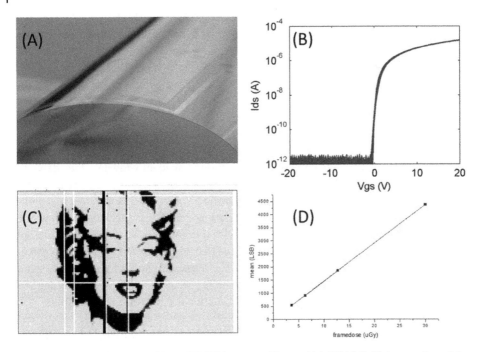

Figure 14.6 (A) Photograph of an IGZO TFT array on 25-μm-thick PEN foil. (B) Impression of the good transistor uniformity; 168 TFT curves are plotted here on top of each other. (C) Optical image taken with the photodetector. (D). Dose response curve.

(CTE) that matches well with that of the glass substrate. The requirements on the allowed CTE mismatch become increasingly more stringent with increasing process temperature.

14.5 Medical-grade Detector

In 2015, Holst Centre presented an x-ray detector array using a solution-processed OPD on top of an IGZO TFT backplane, integrated on a very thin plastic substrate that is capable of medical-grade performance Figure 14.6 [23]. As an indirect-conversion flat panel detector, it combines a CsI:Tl scintillator with an organic photodetector layer and an oxide TFT backplane. Readout electronics was standard. The organic bulk heterojunction photodetector film was slotdie coated directly on the active-matrix backplane. To comply with the maximum process temperature allowed for polyester substrates, a robust and reliable IGZO transistor process was developed using temperatures not exceeding 200°C. Its high mobility and low leakage current aided in improving the sensitivity and robustness of the final detector array. The photodetector exhibited the best performance of any bulk heterojunction photodetector studied to date, with dark current density as low as 1 pA/mm², and high sensitivity of 0.2 A/W in the green wavelength range, i.e. nicely compatible with emitted wavelengths of typical x-ray scintillator materials. The outstanding optical characteristics are retained when these photodiodes are combined in a high-resolution TFT array. Results showed there is no need to fine-pattern the active organic layer and transparent top electrode. So, a number of costly vacuum deposition and lithography steps are eliminated compared to the case of conventional inorganic (amorphous silicon-based) detectors. This greatly simplifies the manufacturing process, opening the door to lower cost. With a scintillator layer on top of the arrays, low-dose x-ray images could be obtaine, i.e. down to 3 μGy per frame, meeting medical specifications.

The use of reactive materials in the cathode implies that the photodetector arrays need to be protected from ingress of water and oxygen. A flexible multilayered encapsulation stack, developed for OLEDs, was employed to ensure a shelf life of more than 10 years. The barrier is optically transparent and is a multilayer stack of two low-temperature plasma-deposited amorphous hydrogenated silicon nitride (a-SiN$_x$:H) films and an organic intermediate layer [35]. Other lifetime tests were performed on the full detectors as well as the separate building blocks. The TFTs and the organic photosensors were subjected to x-ray-hardness tests as well as temperature-cycle tests. All tests were passed without any apparent degradation in the optical properties and without any additional line or pixel defects.

Since this earlier publication, researchers from Holst Centre have produced the first commercial-sized flexible x-ray detectors. Measuring 17 × 17 cm, the new detectors replicate the performance of the smaller prototypes. Together with its partners, Holst Centre also realized a prototype curved x-ray detector that was integrated into a medical cone-beam CT (CBCT) system [5], paving the way for smaller optical and three-dimensional imaging x-ray systems with better, more uniform image quality.

Using solution-based processes, they are suitable for upscaling to even larger sizes. Currently being transferred to production facilities, the technology could reduce costs for rigid detectors and enable lightweight, thin, and robust flexible detectors

14.6 Summary and Outlook

The commercialization of the x-ray image detectors was enabled in part by the successful development of a-Si:H TFT technology for flat panel (liquid crystal) displays. And where today digital x-ray sensors are a-Si:H based and made on large glass substrates it is merely logical that the recent development of flexible displays and their underlying technologies (TFT arrays on plastic, novel high-performance TFT materials such as IGZO, and (printed) organic semiconducting materials) will be evaluated in light of making next-generation glass-based detectors cheaper, extending the portfolio of imagers broader by making plastic-based detectors with reduced weight for portable applications, and eventually detectors that can be tiled, curved, and stacked.

The pixel in Figure 14.4 is simple and the readout scheme is straightforward. With only one TFT per pixel that acts as an addressable switch to transfer the charge to the external circuit, it is compact and amenable to high-resolution imaging. It therefore represents the default pixel configuration in most flat panel x-ray detectors today. It is possible to introduce one or more additional TFTs at each pixel and introduce gain for the sensor at the pixel level. Such TFT arrays are called active-pixel sensors (APSs). The APS performs in-pixel signal amplification providing – ideally – higher immunity to external noise, hence preserving the dynamic range. Such detectors have the potential of providing better image quality at lower dose. More complicated pixel electronics, of course, implies a more complex fabrication process and more stringent requirements on the TFT compared to the passive pixel sensor architecture. APSs with a-Si have been demonstrated, but they never made it into mainstream commercial products, presumably because the limited mobility of a-Si TFTs resulted in too little amplification, which is even further degraded by the poor reliability of the TFTs under prolonged bias stress. This may change in the foreseeable future as new TFT technologies mature that have higher mobility, lower leakage current, and/or less bias-stress effects.

New x-ray sensitive materials and concepts are proposed to replace amorphous selenium in direct-conversion type detectors. Büchele et al., for instance, demonstrated a quasi-direct x-ray detector [36]. Scintillating terbium-doped gadolinium oxysulfide (GOS:Tb) particles of ca. 1-micrometer size were mixed into a thick organic bulk heterojunction film. The x-ray induced light from the GOS:Tb particles is absorbed in the BHJ, minimizing optical crosstalk and enabling a high image resolution.

Solution-processed halide-based lead perovskites also have great potential for direct-conversion detectors. These perovskite compounds have the general formula $[(RNH_3)mMX_n]$, in which the exact composition of metal (M, typically Pb), halide (X = Cl, Br, I), and organic groups (R) control the perovskite structure and determine the physical properties. Their large mobilities and carrier lifetimes in single crystals [37, 38] as well as thin films [39, 40], and the high atomic numbers of Pb, I, and Br make them ideal for direct-conversion x-ray and gamma-ray detection.

References

1 Kobayashi, K., Makida, S., Sato, Y., and Hamano, T. (1993). 640 × 400 pixel a-Si: HTFT driven 2-dimensional image sensor. *SPIE Electron. Imag. Conf. Charge-coupled Dev. Solid State Opt. Sensor.s III* 1900: 40–46.

2 Powell, M.J., French, I.D., Hughes, J.R., Bird, N.C., Davies, O.S., Glasse, C., and Curren, L.E. (1992). Amorphous Silicon Image Sensor Arrays. *Mat. Res. Soc. Proc.* 258 (1): i27.

3 Street, R.A., Nelson, S., Antonuk, L.E., and Perez Mendez, V. (1990). Amorphous silicon sensor arrays for radiation imaging. *Mat. Res. Soc. Proc.* 192: 441–452.

4 Ducourant, T., Wirth, T., Bacher, G., Bosset, B., Vignolle, J.M., Blanchon, D., Betraoui, F., and Rohr, P. (2018). Latest advancements in state-of-the-art aSi-based x-ray flat panel detectors. *SPIE Med. Imag.* 10573: 105735V. doi: 10.1117/12.2291908.

5 Van Breemen, A.J.J.M., Simon, M., Tousignant, O., Shanmugam, S., Van der Steen, J.L., Akkerman, H.B., Kronemeijer, A.J., Ruetten, W., Raaijmakers, R., Alving, L., Jacobs, J. Malinowski, P.E., De Roose, F., and Gelinck, G.H. (2020). Curved X-ray detectors. *npj Flex Electron* 4: 22.

6 Zhao, W., Ristic, G., and Rowlands, J.A. (2004). X-ray imaging performance of structured cesium iodide scintillators. *Med. Phys.* 31: 2594–2605.

7 Kasap, S., Frey, J.B., Belev, G., Tousignant, O., Mani, H., Laperriere, L., Reznik, A., and Rowlands, J.A. (2009). Amorphous selenium and its alloys from early xeroradiography to high resolution x-ray image detectors and ultrasensitive imaging tubes. *Phys. Status Solidi B Basic Solid State Phys.* 246: 1794–1805, and references therein.

8 Kasap, S.O. and Rowlands, J.A. (2002). Direct-conversion flat-panel x-ray image detectors. *IEEE Proc. Circ. Dev. Syst.* 149: 85–96.

9 Overdick, M., Baumer, C., Engel, K.J., Fink, J., Herrmann, C., Kruger, H., Simon, M., Steadman, R., and Zeitler, G. (2009). Status of direct conversion detectors for medical imaging with X-rays. *IEEE Trans. Nucl. Sci.* 56: 1800–1809.

10 Colbeth, R.E., Allen, M.J., Day, D.J., Gilblom, D.L., Harris, R.A., Job, I.D., Klausmeier-Brown, M.E., Pavkovich, J.M., Seppi, E.J., Shapiro, E.G., Wright, M.D., and Yu, J.M. (1998). Flat-panel imaging system for fluoroscopy applications. *SPIE Proc. Medical Imaging: Physics of Medical Imaging* 3336: 376.

11 Blakesley, J.C. and Speller, R. (2008). Modeling the imaging performance of prototype organic X-ray imagers. *Med. Phys.* 35: 225–239.

12 Ng, T.N., Lujan, R.A., Sambandan, S., Street, R.A., Limb, S., and Wong, W.S. (2007). Low temperature a-Si: Hphotodiodes and flexible image sensor arrays patterned by digital lithography. *Appl. Phys. Lett.* 91: 063505.

13 Lujan, R.A. and Street, R.A. (2012). Flexible X-ray Detector Array Fabricated with Oxide Thin-Film Transistors. *IEEE Electron. Device Lett.* 33: 688–690.

14 Marrs, M., Bawolek, E., Smith, J.T., Raupp, G.B., and Morton, D. (2013). Flexible amorphous silicon PIN diode x-ray detectors. *SPIE Def. Secur. Sens.* 87300C.

15 Baeg, K.-J., Binda, M., Natali, D., Caironi, M., and Noh, -Y.-Y. (2013). Organic light detectors: Photodiodes and phototransistors. *Adv. Mater.* 25: 4267–4295.

16 Halls, J.J.M., Walsh, C.A., Greenham, N.C., Marseglia, E.A., Friend, R.H., Moratti, S.C., and Holmes, A.B. (1995). Efficient photodiodes from interpenetrating polymer networks. *Nature* 376: 498.

17 Yu, G., Gao, J., Hummelen, J.C., Wudl, F., and Heeger, A.J. (1995). Polymer photovoltaic cells: Enhanced efficiencies via a network of internal donor-acceptor heterojunctions. *Science* 270: 1789.

18 Ng, T.N., Wong, W.S., Chabinyc, M.L., Sambandan, S., and Street, R.A. (2008). Flexible image sensor array with bulk heterojunction organic photodiode. *Appl. Phys. Lett.* 92: 213303.

19 Kielar, M., Dhez, O., Pecastaings, G., Curutchet, A., and Hirsch, L. (2016). Long-term stable organic photodetectors with ultra low dark currents for high detectivity applications. *Sci. Rep.* 6: 39201. doi: 10.1038/srep39201.

20 Weisfield, R.L., Hartney, M.A., Street, R.A., and Apte, R.B. (1998). New amorphous-silicon image sensor for X-ray diagnostic medical imaging applications. *SPIE Proc. Med. Imag. Phys. Med. Imag.* 3336. doi: 10.1117/12.317044.

21 Takahashi, T., Yu, Z., Chen, K., Kiriya, D., Wang, C., Takei, K., Shiraki, H., Chen, T., Ma, B., and Javey, A. (2013). Carbon nanotube active-matrix backplanes for mechanically flexible visible light and X-ray imagers. *Nano Lett.* 13: 5425–5430.

22 Gelinck, G.H., Kumar, A., Moet, D., Van Der Steen, J.-L., Shafique, U., Malinowski, P.E., Myny, K., Rand, B.P., Simon, M., Ruetten, W., Douglas, A., Jorritsma, J., Heremans, P., and Andriessen, R. (2013). X-ray imager using solution-processed organic transistor arrays and bulk heterojunction photodiodes on thin, flexible plastic substrate. *Org. Electron.* 14: 2602–2609.

23 Gelinck, G.H., Van Breemen, A.J.J.M., Shanmugam, S., Langen, A., Gilot, J., Groen, P., Andriessen, R., Simon, M., Ruetten, W., Douglas, A.U., Raaijmakers, R., Malinowski, P.E., and Myny, K. (2016). X-ray detector on plastic with high sensitivity using low cost, solution-processed organic photodiodes. *IEEE Trans. Electron. Dev.* 63: 197–204.

24 Moy, J.P. (1999). Large area x-ray detectors based on amorphous silicon technology. *Thin Solid Films* 337: 213–221.

25 Sazonov, A., Striakhilev, D., Lee, C.-H., and Nathan, A. (2005). Low temperature materials and thin film transistors for flexible electronics. *Proc. IEEE* 93: 1420–1428.

26 ASU (2015). Press release: Largest flexible x-ray detector manufactured with thin film transistors. https://phys.org/news/2015-12-largest-flexible-x-ray-detector-thin.html#jCp. Accessed Jan 13th, 2022.

27 Smith, J.T., Couture, A.J., Stowell, J.R., and Allee, D.R. (2014). Optically Seamless Flexible Electronic Tiles for Ultra Large-Area Digital X-ray Imaging. *IEEE Trans. Components. Packag. Manuf. Techn.* 4: 1109.

28 Dong, H., Fu, X., Liu, J., Wang, Z., and Hu, W. (2013). Key points for high-mobility organic field-effect transistors. *Adv. Mater. Weinheim* 25: 6158–6183. doi: 10.1002/adma.201302514.

29 Gao, X. and Zhao, Z. (2015). High mobility organic semiconductors for field-effect transistors. *Sci. China Chem.* 58: 947–968.

30 Sirringhaus, H. (2014). Organic Field-Effect Transistors: The Path Beyond Amorphous Silicon. *Adv. Mat.* 26: 1319–1325.

31 Someya, T., Kato, Y., Iba, S., Noguchi, Y., Sekitani, T., Kawaguchi, H., and Sakurai, T. (2005). A large-area, flexible, and lightweight sheet image scanner integrated with organic field-effect transistors and organic photodiodes. *IEEE Trans. Electron Dev.* 52: 2502.

32 Nausieda, I., Ryu, K., Kymissis, I., Akinwande, A.I., Bulovic, V., and Sodini, C.G. (2007). An organic active-matrix imager. *IEEE Int. Solid-State Circuit. Conf.* 446: 72.

33 Huitema, H.E.A., Gelinck, G.H., Van Veenendaal, E., Touwslager, F.J., and Van Lieshout, P.J.G. (2005). Rollable active-matrix displays with organic electronics. In: *Flexible Flat Panel Book*, 1st e (ed. G.P. Crawford), 245–262. Wiley, Hoboken USA.

34 Zhao, C. and Kanicki, J. (2014). Amorphous In–Ga–Zn–O thin-film transistor active pixel sensor x-ray imager for digital breast tomosynthesis. *Med. Phys.* 41: 091902–4.

35 Li, F.M., Unnikrishnan, S., Van De Weijer, P., Van Assche, F., Shen, J., Ellis, T., Manders, W., Akkerman, H., Bouten, P., and Van Mol, T. (2013). Flexible barrier technology for enabling rollable AMOLED displays and upscaling flexible OLED lighting. *SID Sympos. Digest.* 44: 199–202.

36 Büchele, P., Richter, M., Tedde, S.F., Matt, G.J., Ankah, G.N., Fischer, R., Biele, M., Metzger, W., Lilliu, S., Bikondoa, O., Macdonald, J.E., Brabec, C.J., Kraus, T., Lemmer, U., and Schmidt, O. (2015). X-ray imaging with scintillator-sensitized hybrid organic photodetectors. *Nat. Photon.* 9: 843–848.

37 Yakunin, S., Sytnyk, M., Kriegner, D., Shrestha, S., Richter, M., Matt, G.J., Azimi, H., Brabec, C.J., Stangl, J., Kovalenko, M.V., and Heiss, W. (2015). Detection of x-ray photons by solution-processed lead halide perovskites. *Nat. Photon.* 9: 444.

38 Wei, H., Fang, Y., Mulligan, P., Chuirazzi, W., Fang, H.H., Wang, C., Ecker, B.R., Gao, Y., Loi, M.A., Cao, L., and Huang, J. (2016). Sensitive x-ray detectors made of methylammonium lead tribromide perovskite single crystals. *Nat. Photon.* 10: 333.

39 Kim, Y.C., Kim, K.H., Son, D.-H., Jeong, D.-N., Seo, J.J.Y., Choi, Y.S., Han, I.T., Lee, S.Y., and Park, N.N.G. (2017). Printable organometallic perovskite enables large-area, low-dose x-ray imaging. *Nature* 550: 87.

40 Shrestha, S., Fischer, R., Matt, G.J., Feldner, P., Michel, T., Osvet, A., Levchuk, I., Merle, B., Golkar, S., Chen, H., Tedde, S.F., Schmidt, O., Hock, R., Rührig, M., Göken, M., Heiss, W., Anton, G., and Brabec, C.J. (2017). High-performance direct conversion x-ray detectors based on sintered hybrid lead triiodide perovskite wafers. *Nat. Photon.* 11: 436–440.

41 Marrs, M.A. and Raupp, G.B. Substrate and passivation techniques for flexible amorphous silicon-based X-ray detectors sensors. *Sensors* 16: 1162. doi: 10.3390/s16081162.

15

Interacting with Flexible Displays

Darran R. Cairns[1] and Anthony S. Weiss[2]

[1] *West Virginia University, Statler College of Engineering and University of Missouri – Kansas City, School of Science and Engineering*
[2] *University of Missouri – Kansas City in the School of Science and Engineering, USA*

15.1   Introduction

Interacting with devices through displays has become more and more common in the twenty-first century. Such functionality is no longer limited to a narrow range of applications, display types, or sizes. However, as flexible displays become more commercially widespread there are some specific design criteria that may influence the development of devices for determining touch interactions in flexible devices. Much of the early work on flexible devices for touch input has focused on modifying technologies that are widely used for non-flexible applications to implement them in flexible devices. As such there has been a significant amount of work on developing materials and processes compatible with flexible substrates. A large proportion of these have been for electronic sensing modalities.

In this chapter we will review some widely used technologies for sensing touch in non-flexible applications; introduce some novel uses of touch-sensing devices for flexible devices; describe some work on developing materials and processes for touch-sensing devices that are compatible with flexible displays; and discuss some questions that may guide future development of touch-input interactions in flexible applications.

15.2   Touch Technologies in Non-Flexible Displays

A very useful review of touch technology primarily for rigid surfaces has been compiled by Geoff Walker [1]. It is a useful source for understanding touch technologies and has more details on touch-screen controllers than we have included in this chapter. We have tried to highlight some of the lessons that can be learned that may allow for the design of sensors for use with flexible devices.

15.2.1   Resistive Touch Sensors

While resistive touch sensors are perhaps not used as widely today as they once were, they are illustrative of some of the challenges and design criteria applicable to electrical sensing more broadly and specifically to flexible devices. The general principle of operation is the detection of touch location at the point where a switch is closed.

Flexible Flat Panel Displays, Second Edition. Edited by Darran R. Cairns, Dirk J. Broer, and Gregory P. Crawford.
© 2023 John Wiley & Sons Ltd. Published 2023 by John Wiley & Sons Ltd.

15.2.2   4-Wire Resistive

Touch input with 4-wire resistive was used widely in portable devices, kiosks, and machine inter-faces in the 1990s and 2000s in part because the controller chips were relatively simple and widely available. A 4-wire sensor is constructed from two substrates with a transparent conductive coat-ing on each. The conductive coatings on each of the substrates are separated by insulating spacers as shown in Figure 15.1. On each of the conductive layers a pair of conductive bus bars are printed parallel with each other. When a finger touches the topmost substrate it deforms in the region of touch and contacts the opposite substrate, thus making a connection between the con-ducting layers. The location of touch is then determined by using each pair of bus bars to create a voltage divider—this is shown schematically in Figure 15.2.

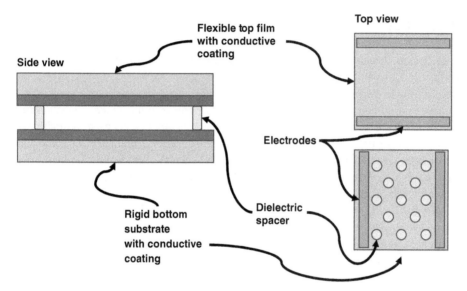

Figure 15.1   Schematic diagram showing the construction of 4-wire resistive sensors.

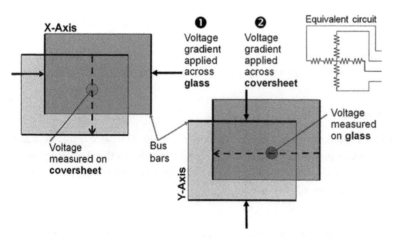

Figure 15.2   Schematic diagram showing operation of 4-wire resistive sensors from [1].

It is useful if the conductive coating is uniform and has a measurable resistance. For many applications a sheet resistance of 10 to 1000 ohms/square is suitable. Applying a voltage difference between the bus bars leads to a voltage gradient between the bus bars which can be measured and used to give one coordinate of position. By aligning the bus bars on the two substrates to be orthogonal then one substrate can be used to sense the horizontal and one the vertical position of a touch.

Inhomogeneity in the conductive coating could lead to variability in the resistance across the screen and make the voltage gradient non-uniform. Inhomogeneity can occur during manufacturing but can also develop over time due to damage to the conductive layers. The susceptibility to damage can depend on the manufacture of the conductive layer and the materials.

While the controllers are relatively simple the construction of 4-wire resistive sensors includes multiple components with multiple layers. For example, the top component of the sensor must be able to be deformed and therefore in many cases polymer films are used. Typically, these films have a lower transmission than glass, which can reduce the brightness and contrast of the underlying display. In addition, widely used transparent conductive coatings such as indium tin oxide have high refractive indices which leads to significant reflections at their interface with air such as the space between the top and bottom conductive coatings. To mitigate for this many resistive touch screens have an antiglare coating applied to the outermost surface. This antiglare coating is often designed so as to provide additional scratch resistance due to the relative softness of the polymer film. If the gap between the top and bottom layers is relatively narrow, then Newton rings can occur and to mitigate for this it is common to roughen the surface underneath the top conductive coating. Finally, the top conductive coating must also have circuitry, i.e. the bus bars, and the processing temperature for conductive inks is constrained by the maximum processing temperature the plastic film may be subjected to. This can result in relatively high resistivity of the conductive traces, which can reduce accuracy of the touch measurement. A schematic of an example top flexible layer is shown in Figure 15.3.

15.2.3 5-Wire Resistive

For applications with high levels of repeated touching and stylus use there is a likelihood of damage to the conductive coatings significant enough that errors in touch location and even dead spots can occur. The need for more durable resistive sensors has led to the use of 5-wire resistive. One major reason for this is that the resistance of the top conductive layer is not used to measure

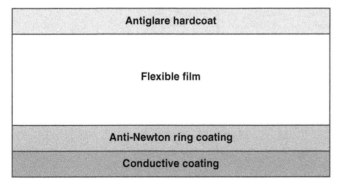

Figure 15.3 Schematic diagram showing top flexible layer from a 4-wire resistive sensor.

position but rather as a probe to measure voltage when voltages are applied sequentially in orthogonal directions on the lower substrate. One added complication with this approach is the need to account for non-linearity that would occur in the voltage gradient if the bus bars were continuous and so modified electrode designs are often used. These modified electrode designs can also include the absence of conductive coating in some regions. The operation and an example linearization pattern are shown in Figure 15.4.

While resistive touchscreens are not as widely used today and may not be the technology of choice for flexible displays, the top flexible layer highlights some of the design criteria that need to be taken into account and the factors that need to be considered for devices that use flexible touch input and particularly for sensors that use some form of electrical measurement.

1) Optics: including transmission, reflection, glare, contrast, and interference.
2) Electrical properties: including resistivity and capacitance.
3) Mechanical properties: including hardness, stiffness, degradation, and reliability.

15.2.4 Capacitive Sensing

While projected capacitive has grown significantly in popularity for touchscreens in recent years it has been around commercially in different forms since the 1990s. The term *projected capacitive* has changed a little in its meaning over the years and there can be conflation of the term with mutual-capacitive or even more specifically with grid-based mutual-capacitive sensors for some people working in the field of displays. However, at its most general, projected capacitive refers to capacitive sensors that project an electric field and monitor how a person interacts with this projected field. Surface capacitive, which itself has really only been used as a term since the rapid rise of projected capacitive, refers to capacitive sensors that have an electric field across the entire surface of the sensor.

We will describe surface-capacitive technologies and projected-capacitive sensors in subsequent sections, but we could equally have used a classification of self-capacitance versus mutual-capacitance or single-touch versus multi-touch capacitive sensors. These sets of terms are often conflated, which unfortunately can lead to some confusion over sensing, such as whether self-capacitance is limited to single touch (it is not) or that mutual-capacitive sensors are

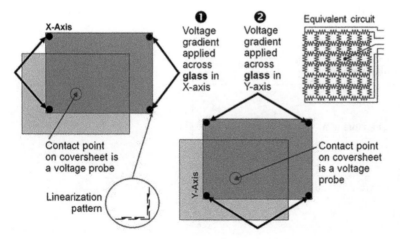

Figure 15.4 Schematic diagram showing operation of 5-wire resistive sensors from [1].

always capable of detecting multiple touches (they are not). To complicate matters further, in many cases the exact same sensor design can be used using both self- and mutual capacitive and for single or multi-touch.

One useful way to distinguish how a capacitive sensor operates is by considering if it measures changes in capacitance of a single electrode to ground (self-capacitance) changes in the capacitance between two electrodes (mutual capacitance). In May 2004, Apple filed what became US Patent 7,663,607 which describes a multi-point touch screen [2]. They describe multi-point touch screens that measure self-capacitance and multi-point touch screens that measure mutual capacitance. In the self-capacitive versions, the capacitance of a grid of individual electrodes are measured sequentially and any changes in the capacitance of an individual electrode used to determine the location of a touch. The electrodes can be small enough that multiple pads are covered by a single finger and image-processing techniques can be used to interpolate the central location of the touch. In mutual-capacitive versions it is the change in capacitance between two electrode that is used. We will discuss this in more detail later.

15.2.5 Surface Capacitive

Surface capacitive has been widely used in kiosks and gaming applications because of improved durability and optics compared with resistive technologies. The technology typically operates by applying an alternating current signal to the corners of a conductive coating on a sheet of glass or plastic to create an electric field across the surface. To make the electric field more uniform it is common to modify the conductivity along the edges by printing an electrode pattern—sometimes called a linearization pattern. It is also possible to design linearization patterns that add conductive regions through printing and remove conductive areas through deletion.

When a user touches the surface current is drawn from the corners proportionally to the distance from the touch to each corner and the four measured currents can be used to determine the location of touch. The uniformity of the coating is therefore important and alternatives to indium tin oxide are commonly used. Indium tin oxide is indium oxide doped with small amounts of tin whereas it was more typical to use tin oxide doped with fluorine or antimony to produce a uniform coating with a relatively high sheet resistance (of the order of a 1000 ohms/square). To protect the conductive coating from damage a protective coating is often applied such as a sol-gel silica coating. To reduce reflections from the sensor particulate can be included in the coating to provide a roughened antiglare surface or the glass or plastic substrate can itself be roughened. Some variation in the conductivity can be accommodated by storing correction coefficients determined during testing as part of the manufacturing process.

15.2.6 Projected Capacitive

Projected-capacitive sensors have been around since at least the mid-1990s for niche applications. In particular projected capacitive was used for applications where a relatively thick rugged protective layer such as toughened glass could be applied over the sensor and an electric field projected above the surface to determine a user's interaction. One particular example of this still in use today is the Zytronics range of projected capacitive sensors that have used thin wire mesh sandwiched between sheets of glass to produce a highly durable product. While projected-capacitive sensors have been used since the 1990s it was not until the release of the iPhone that projected capacitive moved beyond these niche applications.

With the release of the iPhone Apple introduced a touch interface that enabled users to interact with mobile devices in ways they had not before. Actions like "swipe left" and "pinch-to-zoom" have become ubiquitous and raised users' expectations on what touch screens need; for example, today they should:

1) Not degrade the optics of the underlying display in ways noticeable to the consumer.
2) Allow for complex gesture control of the device including the use of multiple fingers, varying pressures, and gestures to control interaction.
3) The touch interaction should not degrade during the time a consumer owns the device.
4) Have no bezels and allow complete freedom in the design of the touch area including curved surfaces.

While projected-capacitive sensors can utilize self-capacitance, mutual capacitance, or both, the most widely used has been mutual capacitive. As discussed, in mutual-capacitive measurements it is the change of capacitance between a pair of electrodes that changes in response to a "touching" event. Therefore, for any sensor design there must be a layout whereby there are pairs of electrodes between which a reference capacitance can be established. This can be done by arranging electrodes in either a single layer or two separate layers, with perhaps the most important distinction being whether the electrodes of the sensor that form the plates of a capacitor are primarily parallel to the substrate surface or primarily orthogonal to the substrate. For electrodes in a single layer these have been used for a number of years as buttons with each pair of electrodes connected individually to a controller.

For projected-capacitive sensors with one set of capacitor plates on a second layer this separation can be through an insulating film or substrate as shown in Figure 15.5 or it can be through an insulating coating at the crossing points where conductive traces cross as shown in Figure 15.6.

For the majority of projected-capacitive applications, the conductive traces are placed behind a rigid glass substrate. If projected capacitive is to be used widely in flexible applications then new protection schemes will be crucial.

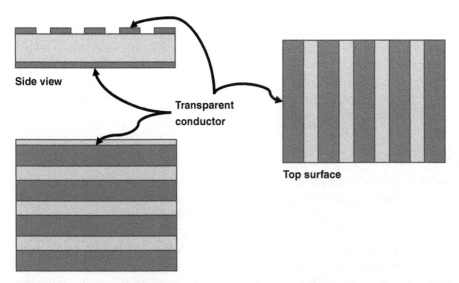

Side view

Transparent conductor

Top surface

Figure 15.5 Schematic diagram showing separated rows and columns for projected-capacitive sensor with dielectric at crossing points.

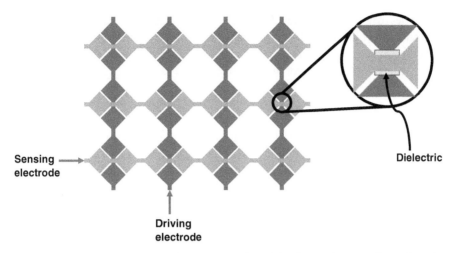

Figure 15.6 Schematic diagram showing operation of the diamond pattern for projected-capacitive sensor with dielectric at crossing points.

15.2.7 Infrared Sensing

Early infrared sensing used simple schemes for determining an interaction based on the blocking of beams. This can be done by attaching orthogonal rows of emitters with corresponding rows of detectors on the opposite edges as shown in Figure 15.7. When the light traveling between an individual emitter/detector pair is blocked this is registered as one coordinate of a touch location. Therefore, by scanning through each pair of emitters and detectors a touch location can be determined. Ambient light can interfere with the operation of infrared sensors and it is common to place the detectors and emitters behind a bezel to limit the impact of ambient light. It is also common to use scanning systems that limit the potential for light from neighboring emitters to interfere with detection by a sensor.

Traditional infrared sensing has been primarily used for applications which require relatively simple interactions due to the inability of correlating more than one simultaneous touch event. Since each x-coordinate is calculated separately from the y-coordinates it is not possible to differentiate which x correlates to which y. It is possible to detect multiple touches with more sophisticated signal processing while using similar emitter/sensor layouts by designing each sensor to illuminate multiple detectors and using more complex scanning systems and more extensive signal processing.

One of the major challenges with using infrared sensors in flexible applications is maintaining alignment between pairs of emitters and detectors when the device is flexed.

15.2.8 Surface Acoustic Wave

Surface acoustic wave has also been used widely in kiosks and gaming applications because of improved durability and optics compared with resistive technologies. The technology typically operates by emitting a pulsed sound wave along two orthogonal edges of the screen and reflecting a portion of the sound energy across the screen at intervals determined

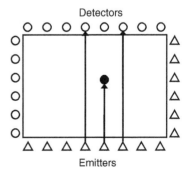

Figure 15.7 Schematic diagram showing operation of a traditional infrared sensor.

by the spacing of a series of reflectors positioned along the edges. Touch is then calculated by determining if any of these sound pulses have been attenuated and correlating the timing of an attenuated pulse to the reflector from which the pulse was reflected across the screen. Doing this for detectors on two orthogonal edges allows for the location to be found.

15.2.9 Bending Wave Technologies

Bending wave technologies use the mechanical sound waves generated when an object hits a rigid substrate to determine touch location. When an object strikes a rigid surface sound waves of varying frequencies are generated and propagate out from the touch location. Due to dispersion, whereby the speed of sound depends on frequency, the further the sound propagates to a sensor the broader the detected sound signal is. In addition, reflected signals from the display edges interfere with the propagating signal resulting in a complex detected signal. One way to deal with this complexity is to use a robot to touch the screen at many locations and to store the expected signal at each location in a lookup table often stored in nonvolatile RAM (NOVRAM). Alternatively, filtering and signal processing can be utilized to analyze the wave fronts in real time. Both approaches have been utilized commercially for rigid sensors. With respect to flexible applications there are at least two significant challenges. The first is the need for an audible tap to generate detectable sound waves and the second is the importance of clamping the sensor edges, which increases the sound produced by a tap.

15.3 Touch Technologies in Flexible Displays

In this section we will highlight a few examples of touch interaction on flexible displays. Many of these applications have focused on using printable conductive coatings in place of vacuum-deposited transparent conductive layers to create flexible projected-capacitive touch sensors. Some examples of printable conductive coatings are silver nanowires, carbon nanotubes, and conductive polymers. The ability to print and pattern are very useful and allow for a reduction in the number of layers used in a touch-integrated display. Liu et al. [3] used printability to good effect by printing directly onto a circular polarizer used with an organic light-emitting diode (OLED) display.

The theme of using silver nanowires (see Figure 15.8) together with active-matrix OLED displays is continued by Chen et al. [4] who highlight two particularly interesting things. Firstly, colorless polyimide layers can be used to improve coating properties of the silver nanowire inks; secondly, sensors can be readily made by laminating films together with optically clear adhesive as shown in Figure 15.9. The use of widely used polymer film processing can be attractive but the need for laser ablation to pattern the printed conductive layers is not always seen as ideal.

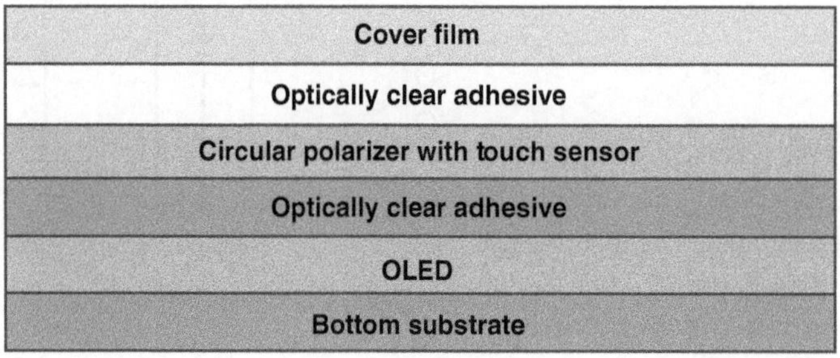

| Cover film |
| Optically clear adhesive |
| Circular polarizer with touch sensor |
| Optically clear adhesive |
| OLED |
| Bottom substrate |

Figure 15.8 Schematic diagram showing integrated silver nanowire touch as described in [3].

Polyimide is often used as a coating or a substrate in the manufacture of flexible displays and touch screens but it can have severe limitation on the thin-film transistor (TFT) performance when compared with those fabricated on glass. Ke et al. [5] have attempted to solve this problem using a transfer process as shown in Figure 15.10.

Silver nanowires are a popular material in developing touch sensors in flexible devices due to their ability to be printed but there are other conductive coatings. One interesting material is carbon nanobud which consists of a mixture of carbon nanotubes and fullerenes that can be deposited in the gas phase to produce transparent films at low sheet resistance with low haze.

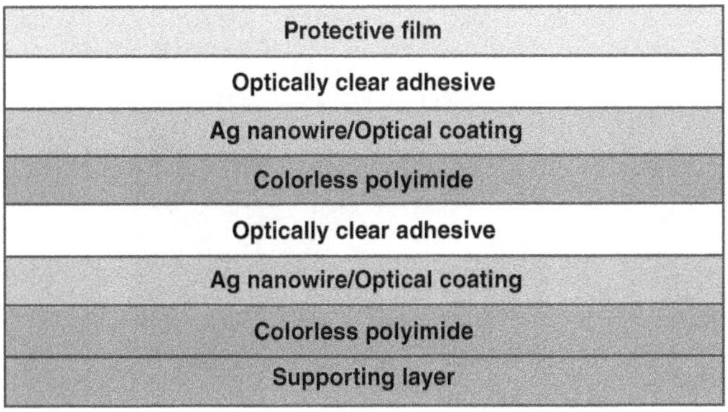

Protective film
Optically clear adhesive
Ag nanowire/Optical coating
Colorless polyimide
Optically clear adhesive
Ag nanowire/Optical coating
Colorless polyimide
Supporting layer

Figure 15.9 Schematic diagram showing a laminated silver nanowire touch sensor as described in [4].

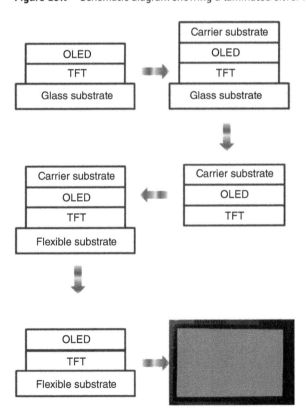

Figure 15.10 Process flow of OLED thin-film transferred to flexible substrate and the final light on image as shown in [5].

While printable transparent conductors have improved dramatically over the last few years, haze can still be an issue. Anisimov et al. [6] have been able to produce touch systems with ultra-low reflectance using nanobuds and a glass overlay as shown in Figure 15.11.

So far we have only discussed projected-capacitive sensors but there are alternatives for flexible applications. Be Dieu et al. [7] have constructed a hybrid piezo-capacitive sensor by incorporating flexible pVDF-TrFE between two electrode layers. In this way the two modes of sensing can be used to determine if the touching object is conducting or not. The sensor construction is shown in Figure 15.12.

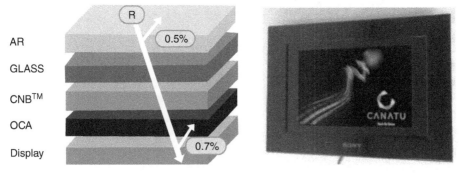

AR
GLASS
CNB™
OCA
Display

R
0.5%
0.7%

Figure 15.11 Ultra-low reflectance G1 touch system with 1.2% specular reflection as shown in [6].

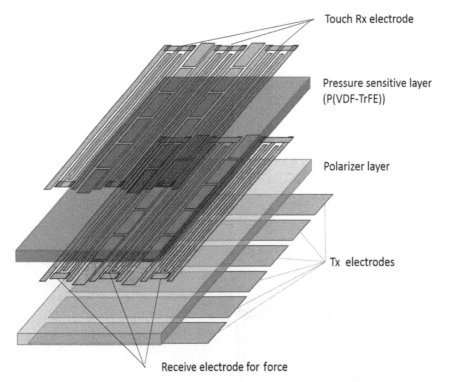

Touch Rx electrode

Pressure sensitive layer
(P(VDF-TrFE))

Polarizer layer

Tx electrodes

Receive electrode for force

Figure 15.12 Hybrid-sensing scheme to differentiate conductive from non-conducting touching objects as shown in [7].

An interesting device utilizing flexible components has been reported by Schneider et al. [8]. It is a bistable polymer-cholesteric liquid crystal dispersion sandwiched between two conductive polymer (PEDOT) coated films. The device can be written on with a stylus and then erased by pushing a button (see Figure 15.13).

For something completely different we can turn to a pair of interactive trousers as shown in Figure 15.14. Heller et al. [9] have created touch-sensing textiles by printing on fabric layers to produce capacitive "FabriTouch" pads.

A number of these examples may seem exploratory but flexible touch screens are now a commercial reality. Samsung have released both foldable phones and phones with curved edges to the display, and it is this curved shape that has led to widespread use of flexible touch sensors which are protected by the glass overlay. Some interesting details on curved sensors is reported by Ikeda et al. [10] with a nice example of a prototype phone shown in Figure 15.15.

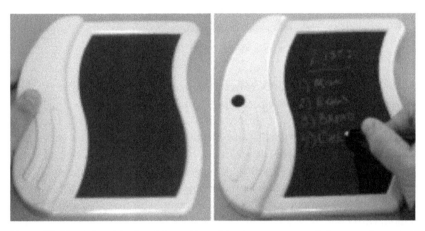

Figure 15.13 A 6- x 4-inch writing tablet with a laser-cut curved shape as shown in [8].

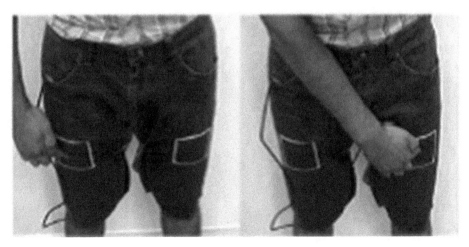

Figure 15.14 Two FabriTouch pads integrated into a pair of trousers. Users tried both parallel (left) and crossed touch gestures (right) as shown in [9].

Figure 15.15 Curved OLED display with touch screen as shown in [10].

We have highlighted a few examples of flexible touch technologies and it is apparent that one particular area of research is in the conductive coating, including how to pattern the conductive layer. Choi et al. [11] have developed a process to print a mesh conductive coating directly without using any laser removal processes. To accomplish this they have used reverse-offset printing of silver nanoparticle ink with a microfabricated cliché to allow for the printing of fine lines of about 10 micrometers. An example of a printed touch sensor using this method is shown in Figure 15.16.

As we have discussed, silver is not the only potential conducting material for flexible devices and carbon nanobuds are also receiving much attention. However, patterning remains to be an important consideration; with laser etching a useful technology. An example of a laser-patterned carbon nanobud substrate from [12] is shown in Figure 15.17.

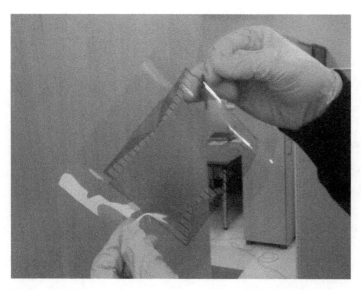

Figure 15.16 5-inch single-layer touch-screen sensor module printed on a flexible polyimide film as shown in [11].

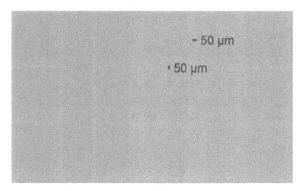

Figure 15.17 A micrograph of a laser patterned Carbon NanoBud® deposit for a touch sensor product as shown in [12].

15.4 Summary

While flexible touch sensors are incorporated into curved applications today there are myriad opportunities for growth. To enable this to happen there is a significant number of materials and manufacturing work needed to develop conductive coatings that can be easily patterned and have the required optical and electrical properties. In addition there are important considerations in protecting functional layers in a flexible package and in overcoming temperature limitations during processing.

References

1 Walker, G. (2012). A review of technologies for sensing contact location on the surface of a display. *Journal of the Society for Information Display* 20 (8): 413–440.

2 Hotelling, S., Strickon, J.A., and Huppi, B.Q. (2010). *U.S. Patent No. 7,663,607*. Washington, DC: U.S. Patent and Trademark Office.

3 Liu, Z., Mao, X., Wang, M.H., Gu, X., Shi, S.M., Zhou, W.F., Shan-chen, K., Guang-cai, Y., and Wang, D.W. (2017, May). 63-1: A stack of bendable touch sensor with silver nanowire for flexible AMOLED display panel. *SID Symposium Digest of Technical Papers* 48 (1): 927–929.

4 Chen, J., Qiao, G., Li, W., and Zhu, S. (2018). 40.1: *Invited paper*: Recent development on flexible touch sensor for flexible AMOLED display. *SID Symposium Digest of Technical Papers* 49: 426–427. doi: 10.1002/sdtp.12744.

5 Ke, T.Y., Kang, T., Lee, C.T., Chen, C.Y., Su, W.J., Wang, W.T., Huang, Z., Wang, J-C., Hsu, S., Wang, C-L., Lai, Y-H., Wang, W., Liu, C-H., and Lin, K.Y. (2020). Flexible OLED display with 620° C LTPS TFT and touch sensor manufactured by weak bonding method. *Journal of the Society for Information Display* 28 (5): 392–400.

6 Anisimov, A.S., Brown, D.P., Mikladal, B.F., Súilleabháin, L.Ó., Parikh, K., Soininen, E., Sonninen, Dewei, T.,Varjos, I., and Vuohelainen, R. (2014, June). 16.3: Printed touch sensors using carbon NanoBud® material. *Sid Symposium Digest of Technical Papers* 45 (1): 200–203.

7 de Dieu, B., Mugiraneza, J., Maruyama, T., Yamamoto, T., and Sugita, Y. (2019, June). 44-3: 3D piezo-capacitive touch with capability to distinguish conductive and non-conductive touch objects for on-screen organic user interface in LCD and foldable OLED display application. *SID Symposium Digest of Technical Papers* 50 (1): 608–611.

8 Schneider, T., Magyar, G., Barua, S., Ernst, T., Miller, N., Franklin, S., Montbach, E., Davis, D., Khan, A., and Doane, J.W. (2008, May). P-171: A flexible touch-sensitive writing tablet. *SID Symposium Digest of Technical Papers* 39 (1): 1840–1842. Oxford, UK: Blackwell Publishing Ltd.

9 Heller, F., Ivanov, S., Wacharamanotham, C., and Borchers, J. (2014, September). FabriTouch: Exploring flexible touch input on textiles. In: *Proceedings of the 2014 ACM International Symposium on Wearable Computers*, 59–62.

10 Ikeda, T., Nakamura, D., Ikeda, M., Iwaki, Y., Ikeda, H., Watanabe, K., Miyake, H., Hirakata, Y., Yamazaki, S., Kurosaki, D., Ohno, M., Bower, C., Cotton, D., Matthews, A., Andrew, P., Gheorghiu, C., and Bergquist, J. (2014, June). 11.1: A 4-mm radius curved display with touch screen. *SID Symposium Digest of Technical Papers* 45 (1): 118–121.

11 Choi, Y.M., Kim, K.Y., Lee, E., and Lee, T.M. (2014, June). 16.2: Reverse-offset printed single-layer metal-mesh touch screen panel. *SID Symposium Digest of Technical Papers* 45 (1): 197–199.

12 Mikladal, B.F., Anisimov, A.S., Brown, D.P., Haajanen, J., Soininen, E.L., Varjos, I., and Vuohelainen, R. (2013, June). 57.5 L: Late-news paper: Flexible transparent conductors and touch sensors for high contrast displays. *SID Symposium Digest of Technical Papers* 44 (1): 795–798. Oxford, UK: Blackwell Publishing Ltd.

16

Mechanical Durability of Inorganic Films on Flexible Substrates

Yves Leterrier

Laboratory for Processing of Advanced Composites (LPAC), Ecole Polytechnique Fédérale de Lausanne (EPFL), CH-1015 Lausanne, Switzerland

16.1 Introduction

Flexible electronics such as flexible displays are multilayer structures based on substrates foils engineered with a diversity of thin-film architectures [1, 2]. Flexibility goes together with thinness: present devices (liquid crystals, electrophoretic displays, organic light-emitting diodes) are intrinsically thin, and corresponding displays are thus flexible when using thin substrates. Nevertheless, the extent to which, and the number of times flexible display structures can be safely bent are essential, but highly challenging design features [3]. An outstanding issue related to mechanical integrity is the lack of understanding of the key factors that control damage processes such as film cracking and interfacial delamination. The reason is the considerable property contrast between substrate and coating materials and associated complex distortion and damage problems. Polymer materials used as substrates are much less rigid and much more sensitive to changes of temperature and relative humidity than inorganic materials such as oxides used as e.g. passivation, diffusion barrier, and transparent conducting layers. In contrast polymers are usually much more robust than inorganic materials. In addition, the small dimensions of the film material with thickness often in the sub-micron range add a great deal of complications for experimental analyses of mechanical properties. Mechanical models and test methods relevant for thin films on polymers have emerged in the last 20 years [4]. Progress in the field of thin-film mechanics is very active, stimulated by the development of novel designs, improved fabrication processes and new materials with increased mechanical stability. In spite of such knowledge, design and process engineers working on the implementation of flexible electronics may still lack confidence due to a general lack of understanding, or lack of input data for reliable modeling tools.

The aim of this chapter is to introduce basic mechanical concepts and provide key ingredients for rational design of flexible display structures, with focus on critical strain for damage and associated critical radius of curvature. Section 16.2 introduces important materials for flexible displays and highlights their property contrast. Section 16.3 is devoted to the analysis of stresses and strains and related critical radius of multilayer structures, with attention paid to process-induced internal strains. Section 16.4 details the main test methods and the analysis of failure mechanisms under tensile and compressive loading, and are illustrated with case studies from the literature. Section 16.5 is devoted to durability influences, including the effect of temperature, fatigue, and corrosion. Section 16.6 is a short review of the development of robust, "unbreakable" films and layer materials relevant to flexible displays. Section 16.7 closes this chapter.

Flexible Flat Panel Displays, Second Edition. Edited by Darran R. Cairns, Dirk J. Broer, and Gregory P. Crawford.
© 2023 John Wiley & Sons Ltd. Published 2023 by John Wiley & Sons Ltd.

16.2 Flexible Display Materials

16.2.1 Property Contrast between Coating and Substrate Materials

Table 16.1 compiles representative flexible display substrates and functional layers, their process methods, and representative thickness. Substrates are generally polymer foils, with thermomechanical properties in striking contrast with those of functional display layers (essentially brittle oxides and nitrides) as shown in Figure 16.1. Steel and glass foils are interesting alternatives to polymers, being thermally stable and perfect diffusion barriers. However, steel is not transparent and both steel and glass yields/fractures at very low strains. As will be discussed later, a key property regarding the mechanical stability of flexible display structures is the critical strain, at which mechanical damage occurs, be it in the form of cohesive failure of brittle films, or adhesive failure at a film/substrate interface.

16.2.2 Determination of Mechanical Properties of Inorganic Coatings

The determination of critical conditions (curvature, stretch, temperature) for mechanical failure of a display requires knowledge of several features and properties of individual layer materials, namely their Young's modulus, Poisson's ratio, coefficients of thermal and hygroscopic expan-

Table 16.1 Substrates foils and coating layers used in flexible displays.

Function	Material	Process method	Typical thickness
Substrate	Elastomers (PDMS)	Solution cast	1 mm
[5]	Polymers (PI, PET, PEN, PES ...)	Solution cast, extrusion – stretched	10–200 μm
	Paper	Water slurry + lamination	100 μm
	Steel	Cold rolling	100 μm
	Glass	Float and fusion	30–600 μm
Passivation/ diffusion barrier	SiN_x	Chemical vapor deposition	50–1000 nm
	SiO_x	Chemical vapor deposition	10–500 nm
[6, 7]	Al_2O_3, ZrO_2	Atomic layer deposition	10–300 nm
	Parylene C	Vapor deposition	0.2–100 μm
	Polymer nanocomposite	Solution cast	1–10 μm
Transparent	Sn-doped In oxide, ITO	Vapor deposition	100–200 nm
electrode [8, 9]	Al, Ga or In-doped zinc oxide (AZO, GZO, IZO)	Vapor deposition/Solution cast + annealed	0.1–5 μm
	Graphene and carbon nanotube polymer composites	Compounding	1–10 μm
	Conducting polymers (e.g. PEDOT-PSS[1])	Solution cast	10–100 nm
TFT [10, 11]	Si (amorphous, microcrystalline, polycrystalline)	Chemical vapor deposition	100 nm

(1) poly(3,4-ethylenedioxythiophene):poly(styrene sulfonate)

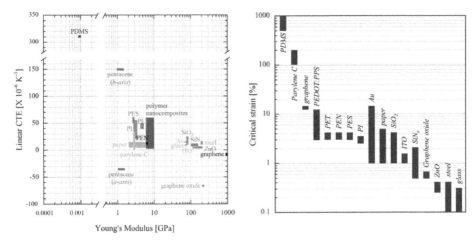

Figure 16.1 Young's modulus (log-scale) versus coefficient of thermal expansion (left) and critical strain for tensile failure (fracture strain or yield strain, right) of relevant flexible display materials. Data ranges reflect the diversity of process-induced microstructures in these various materials.

sion (CTE and CHE), thickness, and critical strain. Methods to determine the latter in the case of tensile and compressive loading situations will be detailed in Section 16.4.

Nanoindentation techniques are well established to measure the elastic modulus of films on substrates [12] including flexible polymer substrates [13]. Nanoindentation is also used to determine the fracture [14] and adhesion [15] properties of coatings on flexible substrates. In a nanoindentation experiment a specimen is indented with a sharp tip while measuring the indentation load and displacement during both loading and unloading [16–18]. Doerner and Nix have proposed the following empirical dependence of the effective modulus of the sample, E, on the normalized indentation depth h_{ind}/h_f, using an adjustable parameter ξ [19]:

$$\frac{1}{E} = \frac{w}{E_f} + \frac{1-w}{E_s} \quad where \quad w = 1 - \exp\left\{ -\xi \frac{h_f}{h_s} \right\} \tag{16.1}$$

In case of high elastic contrast (stiff film on soft substrate) the indentation depth is recommended to be much less than 10% of the film thickness, possible 1% to minimize substrate influence [12]. The accuracy of the method is compromised by the presence of "third-body interactions" such as indenter-film friction and the occurrence of so-called "piling up". These issues and data-processing models were reviewed in detail by [20] and [21]. Figure 16.2 compares experimental data for films deposited on stiff substrates with Equation 16.1. It is evident that as indentation depth increases the effective modulus progressively converges toward that of the substrate. The application of Equation 16.1 leads to accurate determination of the Young's modulus of films with sub-micron thickness.

Alternative methods to determine the Young's modulus of thin films include tensile tests using thin substrates and the classic laminate plate theory [22], buckling analyses [23, 24], acoustic and heterodyne atomic force microscopy [25], surface acoustic wave methods [26, 27], and scanning local acceleration microscopy [28].

Conventional mechanical and dilatometry test methods can be used to measure the Poisson's ratio, CTE, and CHE of film samples with thicknesses above a few microns (the Poisson's ratio

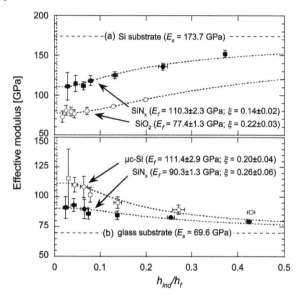

Figure 16.2 Effective modulus versus normalized indentation depth for selected films on Si (a) and glass (b) substrates. Symbols are experimental data and dashed lines represent Equation (16.1). Best-fit values for film modulus E_f and indentation parameter ξ are indicated.

has a small influence on calculated properties and approximate values are usually sufficient). However, these measurements are highly challenging for films with thickness in the sub-micron range. Approaches to determine the CTE and CHE are described in Section 16.3.1.

16.3 Stress and Strain Analyses

The strain in a flexible display structure being curved is the combination of internal strains and bending strains as depicted in Figure 16.3 and detailed in the following. Excessive strain levels may relax through cracking and delamination. Internal strains are primarily controlled by the fabrication process, and by changes of e.g. temperature during service, whereas bending strains result from the applied curvature.

16.3.1 Intrinsic, Thermal, and Hygroscopic Stresses and Strains

Internal stresses (often referred to as residual stresses) and strains in vapor-deposited and solution-processed films and coatings generally include intrinsic, thermal, and hygroscopic contributions whose process dynamics are sketched in Figure 16.4 [29–33]. Intrinsic stresses are associated with process-induced disorder (tensile or compressive in inorganic films) and a number of shrinkage mechanisms (generally tensile in organic films). Thermal stresses develop upon cooldown from process temperature or upon temperature variations during service because material constituents have mismatched thermal expansion coefficients. Thermal stresses are generally compressive in inorganic films and maybe tensile or compressive in organic films, when using polymers as substrates. Hygroscopic stresses also develop upon exposure to ambient humidity, due to a mismatch in hygroscopic expansion between material constituents. Hygroscopic stresses

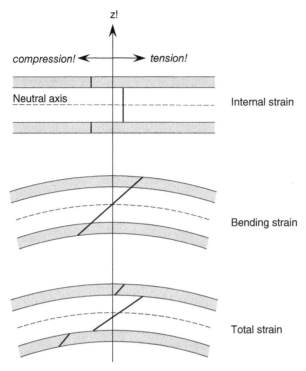

Figure 16.3 Sketches of internal, bending, and total strain profiles in a symmetric three-layer structure (substrate coated on both sides with identical coating). The total strain is the sum of the internal and bending strains.

are generally tensile in inorganic films and can be tensile or compressive in organic films. Additional stresses may develop during post-deposition processes due to further dimensional changes of the polymer substrate, such as upon unloading from roll-to-roll manufacturing [34].

The internal strain $\varepsilon_i = \varepsilon_i^{in} + \varepsilon_i^{th} + \varepsilon_i^{hy}$ includes intrinsic, thermal, and hygroscopic contributions, with:

$$\varepsilon_i^{th} = \Delta\alpha\,\Delta T = (\alpha_s - \alpha_f)(T - T_0) \tag{16.2a}$$

$$\varepsilon_i^{hy} = \Delta\beta\,\Delta RH = (\beta_s - \beta_f)(RH - RH_0) \tag{16.2b}$$

where α_s and α_f are the CTE of substrate and film, T is the actual temperature, and T_0 is the process temperature (which would correspond to a stress-free temperature in the case where the intrinsic strains would be zero), β_s and β_f are the CHE of substrate and film, RH is the actual relative humidity, and RH_0 the relative humidity of the process (zero in the case of vacuum deposition).

The two main methods to measure the internal strain in films on substrates (from which the internal stress can be calculated providing that the elastic constants of the film are known) are X-ray diffraction and bilayer curvature measurements. X-ray diffraction is limited to crystalline materials and provides accurate determination of internal strain from the change of lattice spacing, and associated change of scattering angle. The $\sin^2\psi$ method, where ψ is the declination

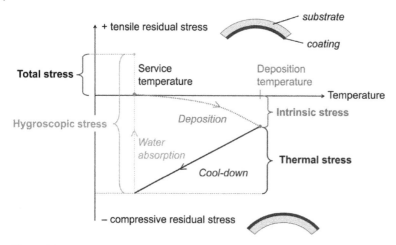

Figure 16.4 Residual stress dynamics during vacuum deposition of films on polymer substrates. The intrinsic stress (compressive in the figure) can be either tensile or compressive. The tensile stress is usually compressive. The hygroscopic stress is usually tensile. The total stress (tensile in the figure) can be either tensile or compressive. Reproduced with permission from [35].

angle between the normal to the film surface and the scattering vector, moreover enables measurement of the elasticity coefficients of the film for isotropic materials [36, 37].

The bilayer curvature measurements rely on the analysis of the equilibrium curvature of a film/substrate system resulting from the presence of internal strains. The film/substrate system can adopt the shape of a spherical cap (isotropic strain and stiff substrate), a saddle shape, or a roll (compliant substrate) depending on the geometry of the system, the thickness and elastic properties of the constituents, and on the degree of in-plane anisotropy of the internal strain. For samples in the form of narrow strips (i.e. with main curvature along the length of the strip) the in-plane film strain can be calculated using the classic Stoney equation (actually its in-plane biaxial stress version) [38]:

$$\varepsilon_i = -\frac{\left(1-\nu_f\right)E_s h_s^2}{6\left(1-\nu_s\right)E_f h_f}\left(\frac{1}{R_2}-\frac{1}{R_1}\right) \tag{16.3}$$

where h_i, ν_i, and E_i represent the thickness, Poisson's ratio, and Young's modulus of layer i (subscripts f and s for film and substrate, respectively), and R_2 and R_1 are the radii of curvature of the coated substrate and of the plain substrate, respectively. The "–"sign is a convention (compressive strains are negative and tensile strains are positive). In case of a large elastic contrast between film and substrate the following correction to Stoney's equation should be used [39]:

$$\varepsilon_i = -\frac{\left(1-\nu_f\right)\left(1+\eta(4\chi-1)\right)h_s}{6\left(1-\nu_s\right)\eta\chi}\left(\frac{1}{R_2}-\frac{1}{R_1}\right) \tag{16.4}$$

where $\eta = h_f/h_s$ and $\chi = E_f/E_s$. This equation is valid for small deflections (compared with the sample dimensions). Refined models are detailed in [32] and the case of large deflections is treated in [40].

Figure 16.5 regroups internal stress data for transparent conductive oxides (TCOs) [41] and diffusion barrier [35] films, which highlights the influence of process conditions (Figure 16.5a

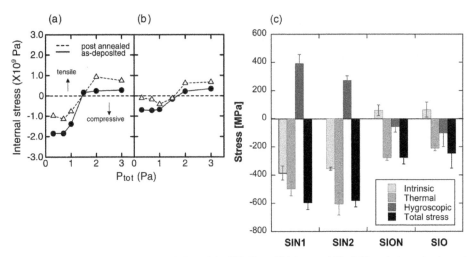

Figure 16.5 Internal stress in (a) ITO and (b) IZO films (thickness: 180–240 nm) deposited on glass substrates by r.f. magnetron sputtering without substrate heating under various total gas pressures (P_{tot}) before and after post-annealing in air at 250°C for 1 hour (reproduced with permission from [41]) and internal stress components (c) in 400-nm thick silicon nitride (SIN 1, SIN2), silicon oxinitride (SION) and silicon oxide (SIO) coatings on a 125 µm PI substrate at 22°C and 50% RH (positive stresses are tensile and negative stresses are compressive; reproduced with permission from [35]).

and b) and composition (Figure 16.5c) on the magnitude and nature (tensile or compressive) of the stress. These data are particularly useful to reduce internal stresses and even produce stress-free layers. The individual components of the in-plane film strain such as shown in Figure 16.5c are identified using the protocol detailed in [33]. The intrinsic strain is obtained in a first step, from the analysis of the film/substrate curvature measured under the conditions prevalent at the end of the process cycle (e.g. vacuum, temperature). The temperature-dependent thermal strain and humidity-dependent hygroscopic strain are obtained from the change of curvature of the film/substrate subjected to iso-hygric temperature jumps and isothermal relative humidity jumps, respectively. This approach combined with modeling tools enabled derivations of the CTE and CHE of thin-film materials, using known CTE and CHE of substrate materials (see also [42]).

16.3.2 Strain Analysis of Multilayer Films under Bending

The strain (stress) in multilayer film structures under bending is a linear superposition of the internal process-induced strain (stress) previously detailed and externally applied bending strain (stress), as sketched in Figure 16.3:

$$\varepsilon = \varepsilon_b + \varepsilon_i \tag{16.5}$$

The bending strain ε_b is proportional to the distance z from the neutral axis with maximum tensile strain on the top (convex) surface and maximum compressive strain on the bottom (concave) surface:

$$\varepsilon_b = a_0 + a_1 z \tag{16.6}$$

where a_0 and a_1 are constants. The neutral axis is the line within the multilayer structure where the strain does not change upon pure bending. The position z_{NA} of the neutral axis in a multilayer (taking the free surface of the first layer as the origin $z = 0$) is given by:

$$z_{NA} = \frac{\sum_{i=1}^{N} \bar{E}_i h_i \bar{z}_i}{\sum_{i=1}^{N} \bar{E}_i h_i} \tag{16.7}$$

where $\bar{E}_i = E_i / (1 - \nu_i^2)$ is the plane strain modulus of layer i (E_i and ν_i are its Young's modulus and Poisson's ratio, respectively), $\bar{z}_i$ is the position of the mid-plane of layer i, h_i is the thickness of layer i, and N is the number of layers.

16.3.3 Critical Radius of Curvature

The critical radius of curvature, R_{crit}, at which device failure occurs is among the key design parameters for flexible electronics. For the elastic case (dissipative processes such as substrate yielding are not considered) and in the case of pure bending R_{crit} is inversely proportional to the critical strain ε_{crit} (related to either cohesive failure of a layer, or interfacial failure) [43]:

$$R_{crit} = \left(\frac{h_f + h_s}{2\varepsilon_{crit}} \right) \left[\frac{1 + 2\eta + \chi\eta^2}{(1 + \eta)(1 + \chi\eta)} \right] \tag{16.8}$$

which simplifies in the case $h_f \ll h_s$ ($\eta \ll 1$) and $E_f < \sim 10\, E_s$ ($\chi < \sim 10$):

$$R_{crit} \approx \frac{h_s}{2\varepsilon_{crit}} \tag{16.9}$$

Figure 16.6 plots the normalized strain in the film versus film/substrate thickness ratio, calculated using Equation 16.8. Two kinds of substrates are compared: steel (Young's modulus ratio

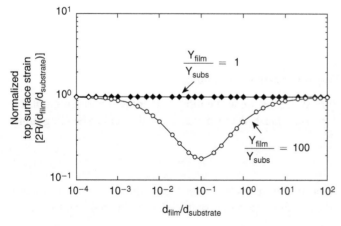

Figure 16.6 Normalized strain in the film as a function of film/substrate thickness ratio $d_{film}/d_{substrate}$ for two substrates: steel (ratio of Young's modulus $Y_{film}/Y_{subs} = 1$) and plastic ($Y_{film}/Y_{subs} = 100$). Reproduced with permission from [43].

$Y_{film}/Y_{substrate}$ equal to 1) and polymer (Young's modulus ratio equal to 100). For given radius of curvature (R in the figure) and film/substrate thickness ratio ($d_{film}/d_{substrate}$), the compliant substrate can reduce the strain by as much as a factor of 5 owing to the shift of the neutral axis toward the film.

Notice that the analytical models introduced in this chapter are accurate enough to determine the critical radius of curvature for thin flexible devices, for which the critical strain is < 1–2 % and R_{crit} is much greater than the total device thickness h. These models, however, are unable to capture mechanical nonlinearities associated with the geometry of actual devices (edge effects, thickness changes, etc.) and also with large displacements ($R_2 < 10\ h$, see e.g. [40]).

It is important to point out that R_{crit} is not a material property as it depends on substrate thickness. The key property is, again, the critical strain, which is controlled by the toughness of the brittle films, or by the toughness of the film/substrate interface as developed in Section 16.4.

16.4 Failure Mechanics of Brittle Films

As sketched in Figure 16.3 the film located on the top, convex side of the bend multilayer experiences tensile strains and the film located on the bottom, concave side of the bend multilayer experiences compressive strains. The magnitude of these strains depends on the applied curvature and on the internal strains. Failure occurs as soon as the strain in either film reaches the critical value ε_{crit} (tensile failure for the film on the top with formation of channeling cracks, and compressive failure for the film on the bottom with buckling and delamination). A clear and unambiguous analysis of the failure mechanisms of thin films requires identification of the *locus of failure*, i.e. whether the failure is cohesive (within a layer) or adhesive (at an interface between adjacent layers) [44]. The reader is also referred to the comprehensive fracture mechanics treatment in multilayers by [4].

16.4.1 Damage Phenomenology under Tensile and Compressive Loading

The failure process of brittle films on substrates under tensile loading reveals three damage stages depicted in Figure 16.7 [45].

Stage I: crack onset and random cracking (Figure 16.7a and d). Cracks initiate in the film at defect sites and start propagating perpendicular to the loading direction at a critical strain, ε_{crit} (also termed crack onset strain, COS). The interaction between cracks is negligible and the generation of new cracks is governed by the statistical distribution of defects within the film.

Stage II: mid-point cracking (Figure 16.7b and e). The crack density (CD) increases whereas the generation of new cracks diminishes. Transverse buckling with localized interfacial delamination is observed across fragments due to Poisson's ratio effects (visible in Figure 16.7c and f).

Stage III: delamination and saturation (Figure 16.7c and f). No further cracks are generated in this stage and CD reaches a saturation value, CD_{sat}, related to the so-called critical stress transfer length and interfacial shear strength [46]. Extensive transverse buckling is evident and delamination becomes the dominant failure mechanism.

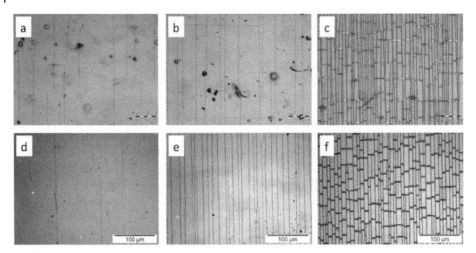

Figure 16.7 Optical micrographs of the fragmentation process of brittle films on polymer substrates under tensile loading (SiN$_x$ on polyimide at (a): 1.2%, (b): 1.5%, and (c): 14.7% strain; ITO on PET at (d): 1.1%, (e): 2.2%, and (f): 11.6% strain). The scale bar in micrographs (a), (b), and (c) represents 20 µm.

Under compressive loading the dominant failure mode for films on substrates is buckling and associated adhesive failure at the film/substrate interface, with examples shown in Figure 16.8. Several morphologies can be observed such as the well-known "telephone cord" pattern [47] depending on the film thickness, in-plane stress state (uniaxial or biaxial), interfacial toughness, and film toughness. Detailed buckling analysis described next enables one to identify the source of problems and thereby to optimize the processing of the individual layers. For instance, the adhesion quality, and influences of processing steps, can be quantified, using the size of the buckle.

16.4.2 Experimental Methods

Two main methods are available to analyze the failure of brittle films, namely fragmentation in situ in a microscope (and related electro-fragmentation) for the tensile loading case and electro-bending, for both tensile and compressive loading cases. In a fragmentation test, a coated substrate is loaded under uniaxial tension, and the damage state in the film due to interfacial stress

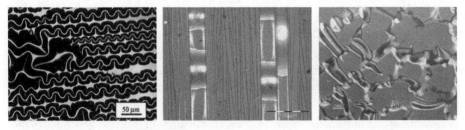

Figure 16.8 Buckling morphologies in films on substrates: (a) telephone cord buckling delamination of a compressed W film (reproduced with permission from [48]), (b) buckling in 800-nm-thick SiN$_x$ fragments on polyimide under tensile stress due to lateral Poisson's contraction, and (c) buckling with and without cracks in a 400-nm-thick SiN$_x$ film on polyimide (courtesy of Philips Research).

transfer from the substrate is analyzed as a function of strain. In-situ tests in an optical or scanning electron microscope are usually employed for the detection of cracks (e.g. Figure 16.6). The method is free of third-body interactions and is used to quantify the cohesive properties (which control critical strain) and the adhesive properties (which control delamination) of films on substrates [32, 49]. The fragmentation test is limited to high elongation substrates (i.e. with a strain to failure several times higher than that of the film). Electro-fragmentation tests carried out in situ (using special clamps to measure electrical resistance of the sample) enable correlating a macroscopic damage state variable (e.g. electrical resistance) to the actual damage at the microscopic scale [35]. Extension of such tests to dielectric films is possible with a conductive probe layer [50].

Bending test methods reproduce the loading state of flexed devices under quasi-static or cyclic loading conditions, using template cylinders of known radius [51,52], or through loading between two parallel plates [53]. This loading geometry is adequate to analyze the behavior of tensile loaded and compressed layers (i.e. layers located on the convex and concave sides of the bent multilayer, respectively). Bending tests are fast and are thus useful for statistical analyses of critical strain and for rapid screening of a series of materials. However, they are usually limited to conductive films, using electrical measurements as a signature of damage, since a direct observation of the damage state in situ in a microscope is hardly feasible. Again, extension of bending tests to dielectric films is possible with a conductive probe layer [54]. In the parallel plate geometry the sample curvature is not constant, with lowest radius of curvature and largest strain $\varepsilon_{\max} = 1.1985 h / (L - h)$ in the middle of the bend, where h is the sample thickness and L is the distance between the two plates [55].

16.4.3 Fracture Mechanics Analysis

The cohesive properties of the film (critical strain, toughness, Weibull modulus) are derived from the early stages of tensile failure (initiation stage I) [46, 56–60]. The film toughness G_{coh} can be calculated assuming that it is equal to the energy release rate at critical strain [61–63]. For a semi-infinite substrate one has:

$$G_{coh} = \frac{\pi}{2} h_f \bar{E}_f \, \varepsilon_{crit}^2 \, g\left(\alpha_D, \beta_D\right) \tag{16.10}$$

where h_f and $\bar{E}_f = E_f / \left(1 - \nu_f^2\right)$ are the thickness and plane strain modulus of the film (E_f and ν_f are the Young's modulus and Poisson's ratio of the film) and $g\left(\alpha_D, \beta_D\right)$ is a function of the Dundurs parameters α_D and β_D [64], which describe the elastic mismatch of the film/substrate system. In the case of plane strain problems:

$$\alpha_D = \frac{\bar{E}_f - \bar{E}_s}{\bar{E}_f + \bar{E}_s} \quad \text{and} \quad \beta_D = \frac{\mu_f \left(1 - 2\nu_s\right) - \mu_s \left(1 - 2\nu_f\right)}{2\mu_f \left(1 - \nu_s\right) + 2\mu_s \left(1 - \nu_f\right)} \tag{16.11}$$

where $\bar{E}_s = E_s / \left(1 - \nu_s^2\right)$ is the plane strain modulus of the substrate (E_s and ν_s are the Young's modulus and Poisson's ratio of the substrate), and $\mu = E_f / \left(2 + 2\nu_f\right)$ and $\mu_s = E_s / \left(2 + 2\nu_s\right)$ are the shear moduli of the film and substrate, respectively. For films with the same properties as their substrate, $\alpha_D = \beta_D = 0$. A stiff film on a soft substrate results in $\alpha \to 1$, whereas a soft film on a stiff substrate results in $\alpha_D \to -1$. The function g is primarily dependent on parameter α_D, which is therefore more representative of film/substrate elastic contrast than parameter β_D. For most film/substrate combinations $0 < \beta_D < \alpha_D/4$.

A practical consequence of Equation 16.10 is that the critical strain for film failure scales with the inverse of square root of film thickness and elastic contrast (providing that the elastic modulus and toughness of the film are independent of film thickness). It was, for example, found that the critical strain for silica films on a steel substrate was a factor of almost five times higher than on a polymer substrate [65]. This huge difference in critical strain was largely due to the difference in elastic contrast, and also to different internal strains. Reducing film thickness or film/substrate elastic contrast both lead to an increase in critical strain, hence increasing the admissible curvature of the multilayer device. For substrates of finite thickness, Equation 16.10 remains accurate when the substrate to film stiffness ratio $\bar{E}_s h_s = \bar{E}_f h_f > 10$, otherwise the approximation devised by [65] may be used.

The interfacial toughness is obtained from an energy release rate analysis for steady-state tunneling delamination and buckling, again assuming that it is equal to the energy release rate at critical (buckling) strain:

$$G_{adh} = \frac{\bar{E}_f \varepsilon^2 h_f}{2} \left(1 - \frac{\varepsilon_{crit}}{\varepsilon}\right)^2 f\left(\Psi; \alpha_D; \beta_D\right) \tag{16.12}$$

where ε is the applied compressive strain and $f\left(\Psi; \alpha_D; \beta_D\right)$ is a function of mode-mixity ψ and Dundurs parameters. Notice that Equation 16.12 is valid for buckling without film cracking. The case where cracking occurs is treated in [66]. Explicit relations between the adhesion energy and the buckle morphology were developed in [67]. An alternative based on the analysis of edge delamination in a fragmentation experiment is treated in [68].

16.4.4 Role of Internal Stresses

In the presence of internal stresses, the measured critical strain is a linear combination of an intrinsic failure strain, ε_{crit}^*, and the internal strain, ε_i:

$$\varepsilon_{crit} = \varepsilon_{crit}^* - \varepsilon_i \tag{16.13}$$

The analysis of internal strains thus enables determination of the intrinsic failure strain of the film. An alternative electromechanical method was proposed to directly obtain these two strains simultaneously [69]. The knowledge of the various contributions to the actual critical strain of a film on a substrate can be useful to optimize both film composition (and associated intrinsic strain) and processing conditions (and associated internal strain). Large internal tensile strains may lead to premature cracking if the film is loaded in tension. On the contrary large internal compressive strains combined with insufficient adhesion are often the cause of buckling failure [47, 70]. Further details are given in [71], where the authors emphasize the importance of increasing interfacial adhesion and decreasing film/substrate elastic contrast to compensate for the presence of internal compressive strains and prevent premature occurrence of buckling failure.

16.4.5 Influence of Film Thickness on Critical Strain

Figure 16.9 regroups fragmentation data for silicon nitride and silicon oxide films with different thickness [72] and shows the relevance of Equation 16.10 to predict the influence of film thickness on critical strain. It is evident that the fragmentation technique is very sensitive: thicker films crack at a lower strain, and their crack density at saturation is lower. Notice that this behavior is quite different from

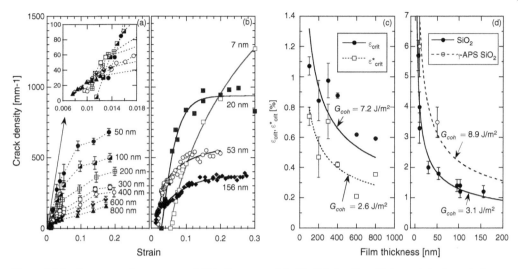

Figure 16.9 Fragmentation process of SiN$_x$ films on polyimide ((a) the film thickness is indicated; the inset shows the early fragmentation stage) and SiO$_2$ films on PET ((b) the film thickness is indicated), and critical strain versus film thickness for the SiN$_x$ films (c) and SiO$_2$ and silylated SiO$_2$ films (d). The lines in (c) and (d) represent the fracture mechanics scaling (Equation 16.10) with best fit values of G_{coh} indicated.

that of thin metal films due to absence of yielding [73]. The toughness G_{coh} of the two materials is indicated in Figure 16.9c and d. In the case of the SiN$_x$ films with significant compressive stresses [72], the toughness was found to be equal to 7.2 J/m^2 (using ε_{crit} data) or 2.6 J/m^2 (using ε^*_{crit} data, i.e. correcting for the presence of internal stresses). This marked difference resulting from the presence of compressive stresses in SiN$_x$ films was confirmed using nanoindentation [74]. In the case of the SiO$_2$ films the toughness was found to be equal to 3.1 J/m^2, a typical value for fused silica [75]. Modification of the OH-terminated SiO$_2$ surface with organo-silane molecules (γ-APS) enabled a threefold increase the toughness of the film, resulting in significant increase of critical strain [76–78].

16.5 Durability Influences

16.5.1 Influence of Temperature

Increasing the temperature impacts the critical strain of films due to the combined action of substrate softening and expansion behavior on heating [79, 80]. The former effect increases Dundurs parameter α_D and elastic contrast function g, leading to a decrease of ε_{crit} (Equation 16.10). The latter generates tensile thermal strains in the film (Equation 16.2a, the CTE of polymer substrates is generally higher than that of inorganic films), also contributing to a decrease of ε_{crit}. Figure 16.10 shows the considerable influence of temperature on critical strain of an oxide film on a PET substrate. The temperature dependence of ε^*_{crit} and ε_i is also shown. Upon heating from room temperature to 140°C ε_{crit} decreased from 0.5% to 0.2%. Such 0.3% decrease was controlled by the 0.1% decrease of ε^*_{crit} due to substrate softening and the 0.2% increase of tensile strain ε_i. Further increasing the temperature to 180°C led to an unexpected increase of ε_{crit} back to 0.5% at 180°C. This was essentially due to shrinkage of the polymer substrate (evaporation of residual water and relaxation of process-induced molecular orientation) and corresponding buildup of compressive strain in the film.

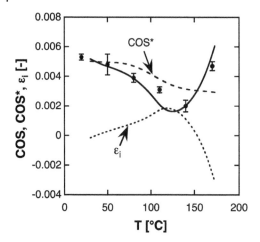

Figure 16.10 Predicted ε_{crit} (COS, solid line) versus measured ε_{crit} (dots) as a function of temperature for 209-nm-thick silicon oxide films on a PET substrate. Internal strain ε_i and ε_{crit}^* (COS*) predictions are also shown. Reproduced with permission from [79].

Additional temperature influences relate to the viscoelastic nature of polymer substrates, which may manifest itself through relaxation and creep phenomena, i.e. problems of dimensional stability. Not much effort has been made on this issue, for which numerical simulation tools could be powerful to design devices with improved stability.

16.5.2 Fatigue

The critical strain for film failure determined under quasi-static loading [49, 53] might not be representative of the actual loading present during operational life. In fact, evidence of slow crack growth was reported for indium tin oxide (ITO) films under fatigue loading at strain levels below the critical strain [81, 82]. High cycle fatigue of thin films on polymer substrates has been studied in detail for metallic films (especially Cu and Al) [83–87] and transparent conductive oxide films [52, 88]. Figure 16.11 shows the peculiar damage morphology of tensile fatigue loaded ITO films on polyester substrates, where tensile cracks and edge delamination are evident. The extensive

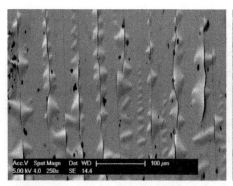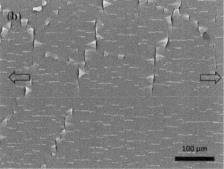

Figure 16.11 Electron micrographs of damage state in ITO films on PET (tensile fatigue loaded to 100'000 cycles at 0.58% strain amplitude, left) and PEN (immersed in 0.1 M acid and tensile tested to 20% strain, right. Reproduced with permission from [90].

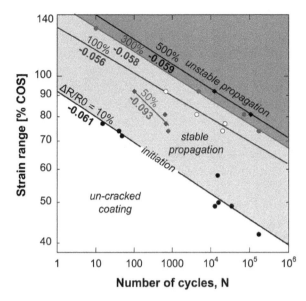

Figure 16.12 Fatigue damage map of a TCO coating tested under cyclic loading to predefined strain levels (normalized with respect to the critical strain, COS). Reproduced with permission from [35]. The lines are power law fits to the experimental data for selected resistance increase $\Delta R/R_0$ levels (indicated in percentages in the figure together with the fatigue strength exponents).

edge delamination is the sign of damage accumulation. It was reported that fatigue under compressive loading is more critical than under tensile loading, which may be a result of compressive residual stresses [89].

Figure 16.12 shows a fatigue endurance map of a TCO coating on a PET substrate [35], where iso-resistance levels are shown versus number of cycles and maximum strain. Initiation of tensile cracks corresponded to a change of electrical resistance $\Delta R/R_0 = 10\%$, where R and R_0 are the resistances of the strained and unstrained samples, respectively. Stable propagation occurred until $\Delta R/R_0 \sim 300\%$ after which catastrophic failure took place. The same power-law scaling with a fatigue strength exponent equal to -0.06 was found between maximum strain and critical number of cycles for both initiation and propagation of tensile cracks. The present scaling follows the modified Basquin law for non-zero mean stress [91], which enables predicting damage events in such coatings upon fatigue loading to any strain levels. For instance, at maximum strain equal to 70% of the critical strain, cracks will initiate after ca. 70 cycles. They will progressively grow upon further cycling until catastrophic failure, which will occur after ca. 800 000 cycles. The existence of a threshold strain was not investigated. It would be $< 40\%$ of the critical strain and would correspond to an endurance limit well beyond 10^5 cycles for this TCO coating.

16.5.3 Corrosion

The presence of corrosive environments (moisture, acids, bases) has been reported to negatively impact the mechanical integrity of inorganic films on polymers. Moisture plays a significant role in film delamination due to interfacial diffusion and reduction of interfacial toughness, combined with buildup of hygroscopic strains in the polymer substrate [92]. Oxide films may also corrode when exposed to moisture [93], acids, and bases, the corrosion state being controlled by the pH as described in the form of *Pourbaix diagrams* [94], eventually decreasing the critical strain for tensile failure and increasing the rate of crack growth [90]. Corroding environments and tensile mechanical stresses exacerbate each other to initiate and propagate cracks, a phenomenon known as *stress corrosion cracking* (SCC). SCC is controlled by microscopic processes occurring at crack tips, characterized by high concentrations of corroding medium. Acrylic acid such as that found in pressure-sensitive adhesives causes cracks to initiate at strains as low as a quarter of those observed for films with no corrosion [95]. As shown in Figure 16.13, the combination of fatigue and corrosion severely reduces the critical strain for tensile failure of ITO films and speeds up the degradation of the electrical conductivity under cyclic loading [96].

In real life multiple degradation factors (temperature, moisture, mechanical, and electrical stresses) often act simultaneously and in synergy. Only limited data are available to understand these coupled phenomena and it would thus be useful to increase knowledge on the resulting lifetime of display devices. Inspiring approaches to this end are presented in the Duracosys conference series, primarily devoted to structural fiber reinforced polymer composites, indeed with similar time-dependent degradation features to the present flexible composite devices ([97] and following years).

Prevention of SCC follow three main principles: (i) decrease the tensile stress (internal and applied) through process and multilayer optimization, (ii) move the pH of the corroding medium towards 7, and (iii) improve the composition of the (oxide) film or use passivation layers. As pointed out in Section 16.4.5, controlled corrosion of a silica gas barrier film combined with crosslinking of an aminosilane passivation layer enabled tremendous improvement of both the critical strain for tensile failure and the barrier performance of the silica film [76, 77]. The key was the healing of crack initiation defects in the oxide resulting from the dissolution of superficial layers [98, 99].

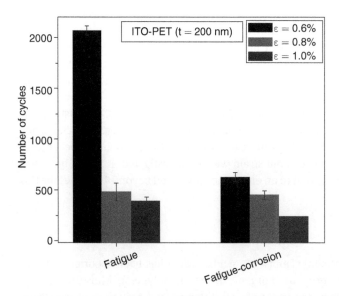

Figure 16.13 Critical number of cycles for fatigue and fatigue corrosion with acrylic acid of 200-nm-thick ITO-coated PET films. Reproduced with permission from [96].

16.6 Toward Robust Layers

Much effort is being paid to develop unbreakable flexible devices, either by adequate designs to prevent premature failure of brittle device components or by the substitution of fragile layers by robust ones. Design approaches are implemented to reduce film strain, such as neutral axis designs (Equation 16.7). The idea is to place the brittle film at the neutral axis, using an additional layer on the opposite side to the substrate, with appropriate thickness and Young's modulus [43]. In this case negligible strain will build up in the film upon bending the multilayer assembly. The critical radius of curvature will no longer be controlled by the critical strain of the film, but rather by delamination problems. Other designs relevant for brittle films are in the form of wavy patterns, which are able to accommodate tensile strains considerably larger than the intrinsic critical strain of the film [100, 101].

Robust layers include films with improved fracture toughness, controlled compressive internal stresses, as well as reduced thickness (see Section 16.4.5). Examples are multilayer structures, which combine several of the aforementioned factors. In the case of ITO films, the additions of Ti [102] and Ag [103] interlayers were reported to improve the crack and delamination resistances of the films, owing to an increase of the crystallinity of the ITO. Such efforts may benefit from numerical analyses of the influence of interlayer toughness and thickness on the critical strain of the film [104]. Planarization "hard coats" reduce the film/substrate elastic contrast and enable significant increase of the critical strain of ITO films [105]. Research is also extremely active to substitute fragile layers by intrinsically compliant layers. Graphene, with a strain to failure of 13% [106] provides an unbreakable alternative to ITO [107]. Further examples include semitransparent Cu nanowire meshes [108] and Ag grids embedded in PEDOT-PSS [109]. Last but not least, self-healing materials with the ability of autonomous damage recovery emerge as new challenging perspectives for the development of robust functional nanosystems [110,111]. Self-healing encapsulation of light-emitting diodes [112], self-healing gate dielectrics in flexible graphene field-effect transistors [113], and self-healing Ag interconnects [114, 115] are recent examples of such developments relevant for flexible display technologies.

16.7 Final Remarks

The allowable radius of curvature of flexible displays is limited by mechanical failures, such as cohesive cracking of functional films and interfacial delamination between adjacent device layers. Cohesive failure occurs predominantly under tensile loading, with critical failure strain usually close to 1%. Buckling delamination results from compressive loading and insufficient adhesion, and may also initiate around 1% strain. In both tensile and compressive modes, the corresponding critical radius of curvature is of the order of few cm, for display thickness of few 100 μm.

Adequate designs should be based on knowledge of critical strains, internal stress state, and a number of material properties. A variety of methods have been developed to this end, such as the electro-fragmentation and electro-fatigue tests in situ in a microscope. These two methods reproduce the thermomechanical loads present during processing and service life, hence they enable identifying and modeling the critical conditions for failure. In fact, critical strains were reported to decrease with time under e.g. fatigue and SCC conditions, which ultimately limits the flexibility of the display. Possible degradation processes are multiple and often act in synergy, and further research on the long-term endurance of flexible display multilayers is advisable. Together with continuous improvements in mechanical models, numerical simulations and test methods, novel designs, improved fabrication processes, and emergence of new materials with increased mechanical robustness, improved knowledge on degradation processes will be key for the manufacturing of truly reliable flexible displays.

Acknowledgments

The author is indebted to Piet Bouten for nanoindentation tests and fruitful discussions. He would like to thank Gil Rochat, Léonard Médico, Fabio Demarco, Gregory Tornare, Pierre Dumont, Damien Gilliéron, Albert Pinyol, and Judith Waller for experimental support.

Nomenclature

a_0, a_1	Constants in strain profile (Equation 16.6).
CD, CD_{sat}	Crack density, crack density at saturation
CHE	Coefficient of hygroscopic expansion
CTE	Coefficient of thermal expansion
E, E_f, E_s	Young's modulus, of film, of substrate
$\bar{E}_f, \bar{E}_s$	Plane strain moduli of film, of substrate
$g(\alpha D; \beta D)$	Normalized energy release rate
G_{coh}, G_{adh}	Film toughness, interfacial toughness
h, h_f, h_s	Thicknesses of multilayer, of film, of substrate
R, R_0	Electrical resistance, of unstrained film
R_1, R_2	Radii of curvature of substrate, of coated substrate
R_{crit}	Critical radius of curvature
RH	Relative humidity
SCC	Stress corrosion cracking
T	Temperature
TCO	Transparent conducting oxide
w	Indentation depth function (Equation 16.1)
α_D, β_D	Dundurs parameters
α_s, α_f	CTE of substrate, of film
β_s, β_f	CHE of substrate, of film
χ	Film-to-substrate Young's modulus ratio
ε	Strain
ε_{crit}, COS	Critical failure strain, crack onset strain
ε^*_{crit}	Intrinsic crack onset strain
ε_i	Film internal strain
$\varepsilon_i^{in}, \varepsilon_i^{th}, \varepsilon_i^{hy}$	Intrinsic, thermal, hygroscopic contributions to the internal strain

ν, ν_f, ν_s	Poisson's ratio of multilayer, of film, of substrate
η	Film-to-substrate thickness ratio
μ_f, μ_s	Shear modulus of film, of substrate
Ψ	Mode-mixity
ξ	Adjustable factor in nanoindentation tests

References

1 Crawford, G.P. (2005). *Flexible Flat Panel Displays*. Chichester: John Wiley & Sons.

2 Nathan, A., Ahnood, A., Cole, M.T., Lee, S., Suzuki, Y., Hiralal, P., Bonaccorso, F. et al. (2012). Flexible electronics: The next ubiquitous platform. *Proc. IEEE* 100: 1486–1517.

3 Leterrier, Y., Pinyol, A., Dumont, P., Gillieron, D., Mewani, V., Månson, J.-A.E., Andersons, J., Bouten, P., Timmermans, P. et al. (2008). Invited paper: Models and experiments of mechanical integrity for flexible displays. *SID Sympos. Digest Tech. Papers* 39: 310–313.

4 Hutchinson, J.W. and Suo, Z. (1992). *Adv. Appl. Mech.* 29: 63–191.

5 MacDonald, W.A., Looney, M.K., MacKerron, D., Eveson, R., and Rakos, K. (2008). Designing and manufacturing substrates for flexible electronics. *Plast. Rubber Compos.* 37: 41.

6 Park, J.S., Chae, H., Chung, H.K., and Lee, S.I. (2011). Thin film encapsulation for flexible AM-OLED: A review. *Semicond. Sci. Technol.* 26.

7 Gokhale, A.A. and Lee, I. (2014). Recent advances in the fabrication of nanostructured barrier films. *J. Nanosci. Nanotechnol.* 14: 2157.

8 Szyszka, B., Dewald, W., Gurram, S.K., Pflug, A., Schulz, C., Siemers, M., Sittinger, V. and Ulrich, S. (2012). Recent developments in the field of transparent conductive oxide films for spectral selective coatings, electronics and photovoltaics. *Curr. Appl. Phys.* 12: S2–S11.

9 Wang, P.-C., Dewald, W., Gurram, S.K., Pflug, A., Schulz, C., Siemers, M., Sittinger, V. and Ulrich, S. (2013). Transparent electrodes based on conducting polymers for display applications. *Displays* 34: 301.

10 Fortunato, G., Pecora, A., and Maiolo, L. (2012). Polysilicon thin-film transistors on polymer substrates. *Mater. Sci. Semicond. Process.* 15: 627.

11 Fortunato, E., Barquinha, P., and Martins, R. (2012). Oxide semiconductor thin-film transistors: A review of recent advances. *Adv. Mater.* 24: 2945.

12 Bull, S.J. (2005). Nanoindentation of coatings. *J. Phys. D: Appl. Phys.* 38: R393.

13 Sun, Y.J., Padbury, R.P., Akyildiz, H.I., Goertz, M.P., Palmer, J.A., and Jur, J.S. (2013). Influence of Subsurface hybrid material growth on the mechanical properties of atomic layer deposited thin films on polymers. *Chem. Vapor Depos.* 19: 134.

14 Chang, R.-C., Tsai, F.-T., and Tu, C.-H. (2013). A direct method to measure the fracture toughness of indium tin oxide thin films on flexible polymer substrates. *Thin Solid Films* 540: 118.

15 Kassavetis, S., Logothetidis, S., and Zyganitidis, I. (2012). Nanomechanical testing of the barrier thin film adhesion to a flexible polymer substrate. *J. Adhes. Sci. Technol.* 26: 2393.

16 Chalker, P.R., Bull, S.J., and Rickerby, D.S. (1991). A review of the methods for the evaluation of coating-substrate adhesion. *Mater. Sci. Eng.* A140: 583.

17 Pharr, G.M. and Oliver, W.C. (1992). Measurement of thin film mechanical properties using nanoindentation. *MRS Bull.* July: 28–33.

18 Oliver, W.C. and Pharr, G.M. (2004). Measurement of hardness and elastic modulus by instrumented indentation: Advances in understanding and refinements to methodology. *J. Mater. Res.* 19: 3.

19 Doerner, M.F. and Nix, W.D. (1986). A method for interpreting the data from depth-sensing indentation instruments. *J. Mater. Res.* 1: 601.

20 Bhushan, B. and Li, X.D. (2003). Nanomechanical characterisation of solid surfaces and thin films. *Int. Mater. Rev.* 48: 125.

21 Fischer-Cripps, A.C. (2006). Review of analysis and interpretation of nanoindentation test data. *Surf. Coat. Technol.* 200: 4153.

22 Reddy, J.R. (2004). *Mechanica of Laminated Composite Plate and Shells: Theory and Analysis.* Boca Raton: CRC Press.

23 Hahm, S.W., Hwang, H.S., Kim, D., and Khang, D.Y. (2009). Buckling-based measurements of mechanical moduli of thin films. *Electron. Mater. Lett.* 5: 157.

24 Tahk, D., Lee, H.H., and Khang, D.Y. (2009). Elastic moduli of organic electronic materials by the buckling method. *Macromolecules* 42: 7079.

25 Cuberes, M.T., Assender, H.E., Briggs, G.A.D., and Kolosov, O.V. (2000). Heterodyne force microscopy of PMMA/rubber nanocomposites: Nanomapping of viscoelastic response at ultrasonic frequencies. *J. Phys. D-Appl. Phys.* 33: 2347.

26 Wittkowski, T., Jorzick, J., Seitz, H., Schroder, B., Jung, K., and Hillebrands, B. (2001). Elastic properties of indium tin oxide films. *Thin Solid Films* 398: 465.

27 Lefeuvre, O., Kolosov, O.V., Every, A.G., Briggs, G.A.D., and Tsukahara, Y. (2000). Elastic measurements of layered nanocomposite materials by brillouin spectroscopy. *Ultrasonics* 38: 459.

28 Rochat, G., Leterrier, Y., Plummer, C.J.G., Månson, J.-A.E., Szoszkiewicz, R., Kulik, A.J., and Fayet, P. (2004). Effect of substrate crystalline morphology on adhesion of PECVD thin SiOx coatings on polyamide. *J. Appl. Phys.* 95: 5429.

29 Ohring, M. (1992). *The Materials Science of Thin Films.* New-York: Academic Press.

30 Tamulevicius, S. (1998). Stress and strain in the vacuum-deposited thin films. *Vacuum* 51: 127.

31 Spaepen, F. (2000). Interfaces and stresses in thin films. *Acta Mater.* 48: 31.

32 Leterrier, Y. (2003). Durability of nanosized gas barrier coatings on polymers. *Prog. Mater. Sci.* 48: 1.

33 Dumont, P., Tornare, G., Leterrier, Y., and Månson, J.-A.E. (2007). Intrinsic, thermal and hygroscopic residual stresses in thin gas-barrier films on polymer substrates. *Thin Solid Films* 515: 7437.

34 Leterrier, Y., Wyser, Y., and Månson, J.-A.E. (2001). Internal stresses and adhesion of thin silicon oxide coatings on poly(ethylene terephthalate). *J. Adhes. Sci. Technol.* 15: 841.

35 Leterrier, Y., Mottet, A., Bouquet, N., Gillieron, D., Dumont, P., Pinyol, A., Lalande, L., Waller, J.H., and Månson J.-A.E. (2010). Mechanical integrity of thin inorganic coatings on polymer substrates under quasi-static, thermal and fatigue loadings. *Thin Solid Films* 519: 1729.

36 Noyan, I.C., Huang, T.C., and York, B.R. (1995). Residual-stress strain analysis in thin-films by x-ray-diffraction. *Crit. Rev. Solid State Mater. Sci.* 20: 125.

37 Withers, P.J. and Bhadeshia, H. (2001). Overview – Residual stress part 1 – Measurement techniques. *Mater. Sci. Technol.* 17: 355.

38 Stoney, G.G. (1909). The tension of metallic films deposited by electrolysis. *Proc. Roy. Soc. London* a82.

39 Röll, K. (1976). Analysis of stress and strain distribution in thin films and substrates. *J. Appl. Phys.* 47: 3224.

40 Masters, C.B. and Salamon, N.J. (1993). Geometrically nonlinear stress deflection relations for thin-film substrate systems. *Int. J. Eng. Sci.* 31: 915.

41 Sasabayashi, T., Ito, N., Nishimura, E., Kon, M., Song, P.K., Utsumi, K., Kaijo, A., and Shigesato, Y. (2003). Comparative study on structure and internal stress in tin-doped indium oxide and indium-zinc oxide films deposited by r.f. magnetron sputtering. *Thin Solid Films* 445: 219.

42 Fang, W.L. and Lo, C.Y. (2000). On the thermal expansion coefficients of thin films. *Sens. Actuat. Phys.* 84: 310.

43 Suo, Z., Ma, E.Y., Gleskova, H., and Wagner, S. (1999). Mechanics of rollable and foldable film-on-foil electronics. *Appl. Phys. Lett.* 74: 1177.

44 Pitton, Y., Hamm, S.D., Lang, F.-R., Leterrier, Y., Mathieu, H.J., and Månson, J.-A.E. (1995), An adhesion study of SiOx/PET films: A comparison between scratch and fragmentation tests, 1st ICAST, Amsterdam, Holland, October 16–20.

45 Wheeler, D.R. and Osaki, H. (1990). Intrinsic bond strength of metal-films on polymer substrates - a new method of measurement. *ACS Sympos. Ser.* 440: 500.

46 Leterrier, Y., Boogh, L., Andersons, J., and Månson, J.-A.E. (1997). Adhesion of silicon oxide layers on poly(ethylene terephthalate). I: Effect of substrate properties on coating's fragmentation kinetics. *J. Polym. Sci. B: Polym. Phys* 35: 1449.

47 Moon, M.W., Jensen, H.M., Hutchinson, J.W., Oh, K.H., and Evans, A.G. (2002). The characterization of telephone cord buckling of compressed thin films on substrates. *J. Mech. Phys. Solids* 50: 2355.

48 Volinsky, A.A. and Waters, P. (2013). Delaminated film buckling microchannels. In: *Mechanical Self-Assembly 2013: Science and Applications* (Ed. X. Chen), 153–170. New York: Springer.

49 Plojoux, J., Leterrier, Y., Månson, J.-A.E., and Templier, F. (2007). Mechanical integrity analysis of multilayer insulator coatings on flexible steel substrates. *Thin Solid Films* 515: 6890.

50 Pinyol, A., Meylan, B., Gilliéron, D., Mewani, V., Leterrier, Y., and Månson, J.-A.E. (2009). Electro-fragmentation analysis of dielectric thin films on flexible polymer substrates. *Thin Solid Films* 507: 2007.

51 Grego, S., Lewis, J., Vick, E., and Temple, D. (2005). Development and evaluation of bend-testing techniques for flexible-display applications. *J. Soc. Inf. Disp.* 13: 575.

52 Cairns, D.R. and Crawford, G.P. (2005). Electromechanical properties of transparent conducting substrates for flexible electronic displays. *Proc. IEEE* 93: 1451.

53 Abdallah, A.A., Kozodaev, D., Bouten, P.C.P., den Toonder, J.M.J., Schubert, U.S., and de With, G. (2006). Buckle morphology of compressed inorganic thin layers on a polymer substrate. *Thin Solid Films* 503: 167.

54 Guan, Q., Laven, J., Bouten, P.C.P., and De With, G. (2013). Subcritical crack growth in SiNx thin-film barriers studied by electro-mechanical two-point bending. *J. Appl. Phys.* 113.

55 Matthewson, M.J., Kurkjian, C.R., and Gulati, S.T. (1986). Strength measurement of optical fibers by bending. *J. Am. Ceram. Soc.* 69: 815.

56 Leterrier, Y., Andersons, J., Pitton, Y., and Månson, J.-A.E. (1997). Adhesion of silicon oxide layers on poly(ethylene terephthalate). II: Effect of coating thickness on adhesive and cohesive strengths. *J. Polym. Sci. B: Polym. Phys* 35: 1463.

57 Hui, C.Y., Shia, D., and Berglund, L.A. (1999). Estimation of interfacial shear strength: An application of a new statistical theory for single fiber composite test. *Compos. Sci. Technol.* 59: 2037.

58 Nairn, J.A. (2000). Matrix microcracking in composites. In: *Comprehensive Composite Materials, Vol. 2. Polymer-Matrix Composites* (ed. A. Kelly and C. Zweben), 403–432. Oxford: Elsevier.

59 Kim, S.R. and Nairn, J.A. (2000). Fracture mechanics analysis of coating/substrate systems Part I: Analysis of tensile and bending experiments. *Eng. Fract. Mech.* 65: 573.

60 Ochiai, S., Iwamoto, S., Nakamura, T., and Okuda, H. (2007). Crack spacing distribution in coating layer of galvannealed steel under applied tensile strain. *ISIJ Int.* 47: 458.

61 Beuth, J.L. (1992). Cracking of thin bonded films in residual tension. *Int. J. Solids Struct.* 29: 1657.

62 Ambrico, J.M. and Begley, M.R. (2002). The role of initial flaw size, elastic compliance and plasticity in channel cracking of thin films. *Thin Solid Films* 419: 144.

63 Andersons, J. et al. (2008). Evaluation of toughness by finite fracture mechanics from crack onset strain of brittle coatings on polymers. *Theor. Appl. Fract. Mech.* 49: 151.

64 Dundurs, J. (1969). Edge-bonded dissimilar orthogonal elastic wedges. *J. Appl. Mech.* 36: 650.

65 Leterrier, Y., Pinyol, A., Gillieron, D., Månson, J.-A.E., Timmermans, P.H.M., Bouten, P.C.P., and Templier, F. (2010). Mechanical failure analysis of thin film transistor devices on steel and polyimide substrates for flexible display applications. *Eng. Fract. Mech.* 77: 660.

66 Cotterell, B. and Chen, Z. (2000). Buckling and cracking of thin films on compliant substrates under compression. *Int. J. Fract.* 104: 169.

67 Cordill, M.J., Fischer, F.D., Rammerstorfer, F.G., and Dehm, G. (2010). Adhesion energies of Cr thin films on polyimide determined from buckling: Experiment and model. *Acta Materialia* 58: 5520.

68 Tarasovs, S., Andersons, J., and Leterrier, Y. (2010). Estimation of interfacial fracture toughness based on progressive edge delamination of a thin transparent coating on a polymer substrate. *Acta Materialia* 58: 2948.

69 Vellinga, W.P., De Hosson, J.T.M., and Bouten, P.C.P. (2011). Direct measurement of intrinsic critical strain and internal strain in barrier films. *J. Appl. Phys.* 110.

70 Evans, A.G., Drory, M.D., and Hu, M.S. (1988). The cracking and decohesion of thin-films. *J. Mater. Res.* 3: 1043.

71 Bouten, P.C.P., Leterrier, Y., and Slikkerveer, P.J. (2005). *Flexible Flat Panel Displays* (Ed. G.P. Crawford). New-York: Wiley.

72 Andersons, J., Leterrier, Y., Tornare, G., Dumont, P., and Månson, J.-A.E. (2007). Evaluation of interfacial stress transfer efficiency by coating fragmentation test. *Mech. Mater.* 39: 834.

73 Lu, N., Suo, Z., and Vlassak, J.J. (2010). The effect of film thickness on the failure strain of polymer-supported metal films. *Acta Materialia* 58: 1679.

74 King, S., Chu, R., Xu, J., and Huening, J. (2009). Impact of film stress on nanoidentation fracture toughness measurements for PECVD SiNx:H films. *ECS Trans.* 19: 455.

75 Freiman, S.W. (1980). Fracture mechanics of glass. In : *Glass Science and Technology, Vol 5, Elasticity and Strength of Glasses* (ed. D.R. Uhlmann and N.J. Kreidl). New York: Academic Press.

76 Bouchet, J., Rochat, G., Leterrier, Y., Månson, J.-A.E., and Fayet, P. (2006). The role of the amino-organosilane/SiOx interphase in the barrier and mechanical performance of nanocomposites. *Surf. Coat. Technol.* 200: 4305.

77 Singh, B., Bouchet, J., Leterrier, Y., Månson, J.-A.E., Rochat, G., and Fayet, P. (2007). Durability of aminosilane-silica hybrid gas-barrier coatings on polymers. *Surf. Coat. Technol.* 202: 208.

78 Leterrier, Y., Singh, B., Bouchet, J., Månson, J.-A.E., Rochat, G., and Fayet, P. (2009). Supertough UV-curable silane/silica gas barrier coatings on polymers. *Surf. Coat. Technol.* 203: 3398.

79 Waller, J.H., Lalande, L., Leterrier, Y., and Månson, J.-A.E. (2011). Modelling the effect of temperature on crack onset strain of brittle coatings on polymer substrates. *Thin Solid Films* 519: 4249.

80 Hu, K., Cao, Z.H., Wang, L., She, Q.W., and Meng, X.K. (2013). The anomalous temperature effect on the ductility of nanocrystalline Cu films adhered to flexible substrates. *Chin. Phys. Lett.* 30.

81 Pinyol, A., Meylan, B., Gilliéron, D., Mottet, A., Mewani, V., Leterrier, Y., and Månson, J.-A.E. (2008). Electro-fragmentation analysis of dielectric thin films on flexible polymer substrates. In: *Large-Area Processing and Patterning for Active Optical and Electronic Devices (Mater. Res. Soc. Symp. Proc. Volume 1030E, Warrendale, PA, 2008), 1030-G03-12* (ed. -S.C.-S.V. Buloviæ, I. Kymissis, J. Rogers, M. Shtein, and T. Someya).

82 Oh, J.S., Cho, Y.R., Cheon, K.E., Karim, M.A., and Jung, S.J. (2007). Failure mechanism of patterned ITO electrodes on flexible substrate under static and dynamic mechanical stresses. *Sol. State Phenom.* 124-126: 411.

83 Martynenko, E., Zhou, W., Chudnovsky, A., Li, R.S., and Poglitsch, L. (2002). High cycle fatigue resistance and reliability assessment of flexible printed circuitry. *J. Electron. Packag.* 124: 254.

84 Eberl, C., Spolenak, R., Kraft, O., Kubat, F., Ruile, W., and Arzt, E. (2006). Damage analysis in Al thin films fatigued at ultrahigh frequencies. *J. Appl. Phys.* 99: 113501.

85 Sun, X.J., Wang, C.C., Zhang, J., Liu, G., Zhang, G.J., Ding, X.D., Zhang, G.P., and Sun, J. (2008). Thickness dependent fatigue life at microcrack nucleation for metal thin films on flexible substrates. *J. Phys. D-Appl. Phys.* 41: 6.

86 Zhang, G.P., Sun, K.H., Zhang, B., Gong, J., Sun, C., and Wang, Z.G. (2008). Tensile and fatigue strength of ultrathin copper films. *Mater. Sci. Eng. A* 483-484: 387.

87 Eve, S., Huber, N., Last, A., and Kraft, O. (2009). Fatigue behavior of thin Au and Al films on polycarbonate and polymethylmethacrylate for micro-optical components. *Thin Solid Films* 517: 2702.

88 Lewis, J., Greco, S., Chalamala, B., Vick, E., and Temple, D. (2004). Highly flexible transparent electrodes for organic light-emitting diode-based displays. *Appl. Phys. Lett.* 85: 3450.

89 Potoczny, G.A., Bejitual, T.S., Abell, J.S., Sierros, K.A., Cairns, D.R., and Kukureka, S.N., (2013). Flexibility and electrical stability of polyester-based device electrodes under monotonic and cyclic buckling conditions. *Thin Solid Films* 528: 205.

90 Bejitual, T.S., Compton, D., Sierros, K.A., Cairns, D.R., and Kukureka, S.N. (2013). Electromechanical reliability of flexible transparent electrodes during and after exposure to acrylic acid. *Thin Solid Films* 528: 229.

91 Morrow, J. (1968). Fatigue properties of metals. In: *Fatigue Design Handbook, Section 3.2*, Vol. AE-4 (ed. J.A. Graham). Warrendale, PA: Society of Automotive Engineers.

92 Abdallah, A.A., Bouten, P.C.P., Den Toonder, J.M.J., and De With, G. (2008). The effect of moisture on buckle delamination of thin inorganic layers on a polymer substrate. *Thin Solid Films* 516: 1063.

93 Dhakal, T.P., Hamasha, M.M., Nandur, A.S., Vanhart, D., Vasekar, P., Lu, S., Sharma, A., and Westgate, C.R. (2012). Moisture-induced surface corrosion in AZO thin films formed by atomic layer deposition. *IEEE Trans. Dev. Mater. Reliabil.* 12: 347.

94 Pourbaix, M. (1974). *Atlas of Electrochemical Equilibria in Aqueous Solutions*, 2nd English ed. Houston: National Association of Corrosion Engineers.

95 Sierros, K.A., Morris, N.J., Ramji, K., and Cairns, D.R. (2009). Stress–corrosion cracking of indium tin oxide coated polyethylene terephthalate for flexible optoelectronic devices. *Thin Solid Films* 517: 2590.

96 Bejitual, T.S., Morris, N.J., Cronin, S.D., Cairns, D.R., and Sierros, K.A. (2013). Mechano-chemical degradation of flexible electrodes for optoelectronic device applications. *Thin Solid Films* 549: 251.

97 Cardon, A.H., Fukuda, H., Reifsnider, K., and Verchery, G. (eds.) (2000). *Recent Developments in Durability Analysis of Composite Systems*. Rotterdam: A.A. Balkema.

98 Bouchet, J., Pax, G.M., Leterrier, Y., Michaud, V., and Månson, J.-A.E. (2006). Formation of aminosilane-oxide interphases. *Compos. Interf.* 13: 573.

99 Singh, B., Bouchet, J., Rochat, G., Leterrier, Y., Månson, J.-A.E., and Fayet, P. (2007). Ultra-thin hybrid organic/inorganic gas barrier coatings on polymers. *Surf. Coat. Technol.* 201: 7107.

100 Rogers, J.A., Someya, T., and Huang, Y.G. (2010). Materials and mechanics for stretchable electronics. *Science* 327: 1603.

101 Park, K., Lee, D.K., Kim, B.S., Jeon, H., Lee, J.E., Whang, D., Lee, H.J., Kim, Y.J., and Ahn, J.H. (2010). Stretchable, transparent zinc oxide thin film transistors. *Adv. Funct. Mater.* 20: 3577.

102 Kim, E.-H., Yang, C.-W., and Park, J.-W. (2012). Designing interlayers to improve the mechanical reliability of transparent conductive oxide coatings on flexible substrates. *J. Appl. Phys.* 111.

103 Yang, C.W. and Park, J.W. (2010). The cohesive crack and buckle delamination resistances of indium tin oxide (ITO) films on polymeric substrates with ductile metal interlayers. *Surf. Coat. Technol.* 204: 2761.

104 Miller, D.C., Foster, R.R., Zhang, Y., Jen, S.-H., Bertrand, J.A., Lu, Z., Seghete, D. et al. (2009). The mechanical robustness of atomic-layer- and molecular-layer-deposited coatings on polymer substrates. *J. Appl. Phys.* 105.

105 Leterrier, Y., Medico, L., Demarco, F., Månson, J.-A.E., Betz, U., Escola, M.F., Olsson, M.K., and Atamny, F. (2004). Mechanical integrity of transparent conductive oxide films for flexible polymer-based displays. *Thin Solid Films* 460: 156.

106 Zhang, Y.Y. and Gu, Y.T. (2013). Mechanical properties of graphene: Effects of layer number, temperature and isotope. *Comput. Mater. Sci.* 71: 197.

107 De Arco, L.G., Zhang, Y., Schlenker, C.W., Ryu, K., Thompson, M.E., and Zhou, C.W. (2010). Continuous, highly flexible, and transparent graphene films by chemical vapor deposition for organic photovoltaics. *Acs Nano* 4: 2865.

108 Kang, M.G., Park, H.J., Ahn, S.H., and Guo, L.J. (2010). Transparent Cu nanowire mesh electrode on flexible substrates fabricated by transfer printing and its application in organic solar cells. *Solar Energy Mater. Solar Cells* 94: 1179.

109 Li, Y., Mao, L., Gao, Y., Zhang, P., Li, C., Ma, C., Tu, Y., Cui, Z., and Chen, L. (2013). ITO-free photovoltaic cell utilizing a high-resolution silver grid current collecting layer. *Solar Energy Mater. Solar Cells* 113: 85.

110 Amendola, V. and Meneghetti, M. (2009). Self-healing at the nanoscale. *Nanoscale* 1: 74.

111 Mauldin, T.C. and Kessler, M.R. (2010). Self-healing polymers and composites. *Int. Mater. Rev.* 55: 317.

112 Lafont, U., Van Zeijl, H., and Van Der Zwaag, S. (2012). Increasing the reliability of solid state lighting systems via self-healing approaches: A review. *Microelectron. Reliabil.* 52: 71.

113 Lu, C.C., Lin, Y.C., Yeh, C.H., Huang, J.C., and Chiu, P.W. (2012). High mobility flexible graphene field-effect transistors with self-healing gate dielectrics. *Acs Nano* 6: 4469.

114 Baliga, S.R., Ren, M., and Kozicki, M.N. (2011). Self-healing interconnects for flexible electronics applications. *Thin Solid Films* 519: 2339.

115 Odom, S.A., Chayanupatkul, S., Blaiszik, B.J., Zhao, O., Jackson, A.C., Braun, P.V., Sottos, N.R., White, S.R., and Moore, J.S. (2012). A self-healing conductive ink. *Adv. Mater.* 24: 2578.

17

Roll-to-roll Production Challenges for Large-area Printed Electronics

Dr. Grzegorz Andrzej Potoczny

President, Miami University, USA

17.1 Introduction

For new emerging printed electronic technologies like flexible displays (flexible organic light-emitting diodes – FOLEDs), lighting panels, flexible organic photovoltaics (OPVs), sensors, and radio frequency identification tags (RFID), roll-to-roll (R2R) manufacturing possibility became an ultimate goal. Its advantages are described in most of the published literature under-taking the subject of printed electronics on polymer foil [1–3]. In comparison to batch sheet-to-sheet (S2S) processes, R2R process offers costs reduction due to high production throughput and low material wastage, easy control, and reduced process steps as many materials can be deposited and dried simultaneously [1, 2]. The ideal production process can be limited to three basic steps: deposition, patterning, and encapsulation of devices. These steps can be integrated to a single continuous production line [2] as Figure 17.1 represents. In addition, the size of the sub-strate scales up is one dimension that is based on increase of web width rather than width and length [4]. Furthermore, rolled-up web significantly prevents contaminations entering the interior surface of devices, which helps to achieve higher yields and cost reduction for cleanroom requirements [4, 5]. For the photovoltaics industry, R2R process of thin modules allows significant reduction in energy payback time and increase packaging density for transportation in comparison to standard silicone-based panels manufactured by S2S processes.

However, large-area R2R production for flexible, printed electronics is challenging and all obstacles need to be overcome to fully realize its potential. Printed electronic technologies are complex and multi-disciplinary expertise, including chemistry, physics, materials, electronic, and engineering sciences, is required [6]. The performance, efficiency, and lifetime of produced devices depends on properties of materials used in the stack, which need to meet specific and advanced requirements such as optical transparency, electrical conductivity, structural stability, layer-to-layer adhesion, and compatibility etc. [2]. These properties are determined and con-trolled by chemistry and surface and interface nanostructure. Hence, their knowledge is of high importance as they affect production steps [2]. Their processing requirements can add processing steps and it may be necessary to utilize different manufacturing techniques. This way, they con-trol the degree of integrity of manufacturing techniques in the R2R production line.

In this chapter we have selected four categories – (i) infrastructure, (ii) equipment, (iii) mate-rials, and (iv) processing – where major production challenges, presented in Figure 17.2, are discussed in detail.

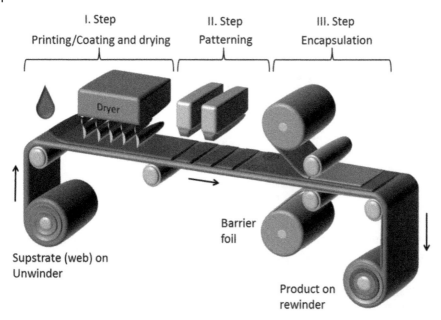

Figure 17.1 Schemat of ideal R2R production process that can be limited to three integrated basic process steps: I. printing/coating, II. patterning, and III. encapsulation. In this process, substrate roll – usually PET foil (commonly called web) or conductive film (e.g. ITO) on PET – is placed on the unwinder. The substrate is connected to a leader web (e.g. PET) that already passes through the machine and connects with roll on rewinder. Then the substrate web is unrolled and passes through machine supported by the rolls where printing/coating processes are conducted followed by drying, patterning, and lamination steps. At the end of the machine the web is rolled-up on the rewinder as a product. After the R2R process the product can go for trimming to demanded size, packaging, and shipping.

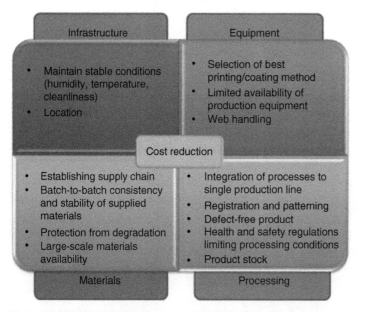

Figure 17.2 Major R2R large-area production challenges considering infrastructure, equipment, materials, and processing.

17.2   Infrastructure

For large-scale production of flexible printed electronics, a proper location and building size should be carefully planned. Plant located in vicinity of transit or some heavy industry, where big machines are under operation might be subjected to external vibrations. The vibrations may cause printing/coating defects in a form of layer nonuniformity – chatter defect as Figure 17.3(a) shows. Hence, it impacts product quality and performance. The thickness uniformity needs to be controlled within ±2 – 5% [5]. It is especially important for OLED and OPV technologies where very thin functional layers are deposited. Thickness variations can cause irregular illumination intensities and power conversion efficiency levels in OLED and OPV applications, respectively [7]. If a production facility is exposed to vibrations the additional costs may apply for special floor preparation that is able to absorb any external or internal vibrations.

Maintaining stable conditions of humidity, temperature, and high cleanliness for large production halls can be difficult and expensive. The contamination is brought on the production line mostly by personnel and therefore larger contamination can be expected for large production where more people are necessary to conduct operations. Since the flexible printed electronics is relatively new technology, equipment and processes might still be under development, hence infrastructure with enough space to allow the later addition of new equipment should be considered. Bigger or more equipment might also be necessary as the market continues to grow.

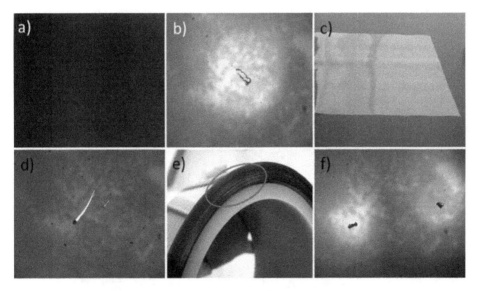

Figure 17.3   Substrate and coating defects related to R2R production process of printed electronics: (a) chatter due to machine vibrations, (b) pinholes of conductive substrate causing local de-wetting and layer thickness nonuniformity, (c) slitting dust entrapped in the web, (d) polymer substrate deformation (wrinkling), (e) impurities entrapped in coating layer causing local de-wetting and thickness nonuniformity, and (f) scratches due to machine and operator handling.

17.3 Equipment

Large-scale R2R production requires decisions on which equipment, process conditions, and size and step processes are the most suitable for product requirements at low costs to achieve production success. Criteria like flexibility (versatile production options), quality (e.g. process control), reliability (e.g. yield), productivity (e.g. throughput), and operations (e.g. equipment maintenance or material handling) should be applied for production machine or other equipment considerations [7]. Moreover, printed electronic applications have different requirements. For example, OPVs and OLEDs require uniform thin layers, thin film transistors (TFTs) needs high-resolution line width and uniformity, sensors and microelectromechanical systems (MEMS) require more complex patterns, and so on [3]. It can also be found that within the same technology different production methods can be applied. The decision on what method to choose can be driven not only by the abovementioned criteria, costs, or application requirements but also by internal knowledge on specific methods gained or developed within a company. Continuous production process can be realized by integrating different coating and printing techniques like gravure printing, gravure-offset printing, flexographic printing, screen printing, inkjet printing, or slot-die coating to R2R machine. Nonetheless, slot-die coating, screen printing, and inkjet printing gained industrial attraction due to their simplicity, high resolution, and easy control [1]. Slot-die coating and screen printing were found to be the most utilized techniques for large-area OLEDs and OPVs production. Figure 17.4 represents operational principles of these techniques. Working principles of other printing/coating techniques are described here [1]. Furthermore, printing/coating equipment can be used as closed or open system, where deposition processes can be conducted in inert or ambient atmosphere, respectively. For OLED production, closed systems, where the process can be conducted under vacuum or inert gas, can be chosen to provide high-quality products with

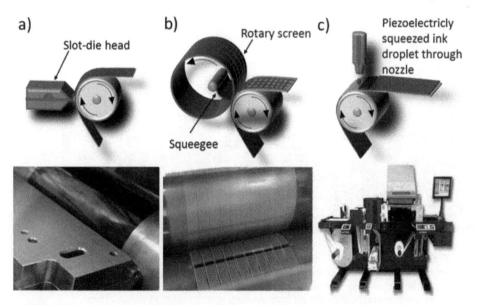

Figure 17.4 Operational principles of printing/coating techniques: (a) slot-die coating,(b) rotary screen printing, and (c) inkjet printing. Images (a) and (b) on second row printed with permission from ARMOR solar power films GmbH.

minimum defects caused by impurities or degradation by presence of oxygen and moisture. However, these systems add complexity and costs to production process. In contrast, being open to ambient condition systems significantly reduces complexity and costs but printed/coated materials are exposed to particles and ambient conditions that change based on weather and seasons. This may cause coating and product quality issues. For OPV applications, slot-die coating in controlled ambient atmosphere was found to be a convenient deposition technique for R2R production as it provides high layer uniformity and ease of equipment maintenance and handling.

Machine web handling should be gentle to avoid device damage during production. In addition, it is difficult to find off-the-shelf R2R machines that can be applied straight away for production of flexible electronic devices. Organic electronic manufacturers rely on equipment producers, whereas equipment producers wait for information from device manufacturers to define the required function of equipment [8]. Machine design may also differ within the same technology for different companies due to differences in materials and process steps. When customizing equipment with equipment manufacturers one may also face limitations imposed by health and safety regulations that take into account e.g. risks of fire and explosion due to solvent evaporation at drying stages. In general, the purchased equipment will require additional modifications to suit the specific process requirements, which might be timely and costly. It should be noted that modification of new equipment can invalidate the manufacturer's guarantee. It is important to establish, before purchasing equipment, where the process limitations are by testing materials and processes on a laboratory or pilot line scale.

17.4 Materials

In general, materials for printed electronics can be classified as conductors, semiconductors, and insulators. Their dispersions in solutions should be chemically and physically stable to avoid agglomeration and allow printing/coating on polymer foil. These materials should also be designed to have specific range of viscosity, surface tension, environmental stability, solvents compatibility with the underling layers in multilayer devices, well-defined structure at nano scale to provide proper morphology (grain size, roughness, etc.), adhesion, and mechanical integrity [1, 3]. It should be noted that solvent changes in formulations are not only driven by improvement of properties or cost reduction but mostly by law. In addition, directives on limitation of volatile organic compounds in coatings can differ by country and location. Ink manufacturers are obligated to continuously move toward formulations based on green solvents that are more environmentally friendly. However, there might be trade-off between solvent used, processability, and performance. A proper ink formulations are of heigh importance especially for inkjet printing technology where not properly selected solvents, surfactants and tuned viscosities can cause nozzle clogging and inkjet heads degradation. None compatible solvents with heads' components will significantly reduce their lifetime. As a result, new heads need to be installed even a few times per year leading to heigh maintenance costs. The availability of materials suitable for R2R production processing can be limited due to materials being under development stage. Therefore, they may require further scale-up to provide large quantities. It can be found that material suppliers might not be ready to deliver large volumes and material received from scaling-up process may not guarantee batch-to-batch reproducibility. Lead time for ordered materials can also be extended. This complicates production in the case when a market is not well established and manufactured product is based on projects that differ in size with often tight deadlines. The costs for new and even mature materials on the market can be high. For instance, mature precious metal pastes (e.g. silver based) commonly used as electrodes or electrical contacts, due to their high conductivity, are costly and their price may vary depending on silver stock commodity pricing. Another costly electrode is transparent indium tin oxide (ITO) that is vacuum deposited on polymer foil and used as conductive substrate for OLEDs, photovoltaics, thin-film transistors (TFT), etc. [9, 10]. Furthermore, flexible electronic devices require encapsulation for protection. OLEDs, OPVs and perovskite solar cells (PSC) have the highest requirements. They require encapsulating materials with high barrier properties to protect from moisture and oxygen that causes corrosion of functional layers in the devices [2]. The barrier requirements for oxygen transmission rate (OTR) and water vapor transmission rate (WVTR), regarding applications, are presented in Figure 17.5. Barriers and conductive substrates need to be of high quality exhibiting the lowest number of spikes and pinhole defects as possible per area. It is well known that the number of defects increases with area and maintaining consistency for high product quality might be challenging. Figure 17.3(b) shows coating defect, local de-wetting, and thickness nonuniformity due to pinholes in the conductive substrate. Mechanical properties of materials should also be considered. In addition, photoactive materials that convert sunlight into electricity used in OPVs can reach up to 60% of the total product costs due to low materials variety and limited availability in large scale on the market. It is challenging to find low-cost materials to reduce total production costs. On the other hand, for some materials one may find it problematic to purchase a low quantity as minimum orders can be quite high. This applies especially for start-up companies entering the market. Being "forced" to get large volumes of materials may unnecessarily stretch the budget and increase the risk of stored materials expiring. In general, production costs can be reduced when production scale increases. However, it is dictated by market size and cost reduction might still be listed as one of the highest priorities, among other challenges.

Establishing a robust material supply chain is challenging as for some materials it is difficult to find more than one supplier. Furthermore, there might be a situation where the supplier is moving out of the business due to low sales. It should also be noted that materials from different suppliers may differ significantly and could require changes in material preparation for production or/and process conditions. Even if there are suppliers with materials that match required specifications, they can be expensive and do not meet the target price set by device maker. Hence, due to lack of choice, price of a product will naturally increase.

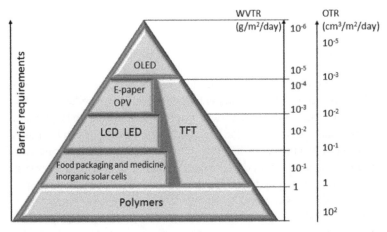

Figure 17.5 Pyramid diagram showing barrier requirements in terms of WVTR and OTR for printed electronic applications.

17.5 Processing

To realize R2R production, the same performance and processes matured on laboratory scale need to be transferred to larger production scale. This requires integration of equipment to a single line and finding a match between equipment, materials, and processes. Tuning the processes is not trivial task due to existing boundary conditions like materials, solvents, layer-to-layer registration, high speed, equipment, drying temperature, etc. [1]. In reality, ideal R2R production process presented in Figure 17.1 can be difficult to achieve as subsequent production steps often require additional sub-processes that need to be optimized. These processes can differ from each other, for instance, by throughput, and in general can be realized separately out of a single production line [2]. As an example, taking 1 m2 area of foil substrate, laser patterning processes can take tens of minutes depending on device design complexity and printed electronics technology. Whereas other processes like coating or printing can be completed below 1 minute. Besides R2R processing, the separated processes can also be conducted via roll-to-sheet (R2S) or S2S where different substrate speeds can be applied [7]. Figure 17.6 represents the basic principles of these processes. In addition, the final product on the roll is not necessarily desired and further sheeting or trimming steps are usually required [7]. In reality therefore, manufacturing processes are conducted not only on the primary R2R machine but also on other secondary machines, hence causing throughput reduction.

Another R2R production challenge is patterning and registration control for multiple layers on a moving web. Passing a web through a machine that does not have physical web guides complicates registration control. Complex equipment needs to be added to the production line to compensate unwanted lateral movement [7]. Inkjet printing technique significantly reduces patterning complexity due to a drop-on-demand process where the ink can be printed on the intended location

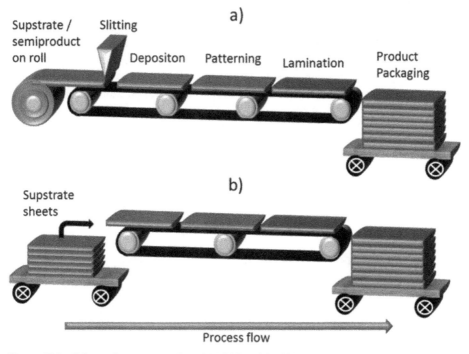

Figure 17.6 Schematic representation of (a) R2S and (b) S2S process.

without any physical mask [3]. However, polymer substrates may undergo deformation due to elevated temperatures [4] at drying stages together with applied web tension. Polymer substrates possess relatively low glass transition temperature (Tg) and soften at elevated temperatures above Tg, which alters physical and mechanical properties [8]. As a result they elongate in machine direction (MG) and shrink in transverse direction (TD). Figure 17.3(c) shows the wavy shape of polyethylene terephthalate (PET) foil due to shrinkage in TD that passed through a R2R machine under elevated temperatures above 120°C. For instance, the dimensional stability of 0.1% for polyimide results in 100 μm of misalignment for every 10-cm width of the web, which is close to display pixel size [4]. Figure 17.7 represents printing misalignment in multiple processes due to substrate deformation. The heat-stabilization process of polymer substrates minimizes dimensional changes at elevated temperatures [11], however, it does not eliminate the issue completely. Cracking of brittle materials like ITO may also occur, which will reduce conductivity [12, 13]. Figure 17.8 shows cracking of ITO film on PET foil and increase of electrical resistance upon applied mechanical stress. Mismatch in coefficient of thermal expansion (CTE) between substrate and subsequently printed/coated layers on top during thermal cycling may also cause cracking of these layers [1, 8]. Therefore, it is highly desirable to engineer polymer substrates with minimum CTE and minimize processing temperatures to reduce stresses [5]. Both misregistration and cracking are classified as functional defects, which means that they can lead to low- or non-functional devices. The choice of polymer substrate may differ for different technologies due to different processing and application requirements. In general, polymer substrates should possess high dimensional stability, high resistance to solvents, low CTE, low OTR, and WVTR, and exhibit low surface roughness. In addition, displays and photovoltaic applications require high transparency in the visible region (400–800 nm wavelength) [1].

Registration accuracy of drain-source electrodes to the gate electrode is of high importance for TFTs fabrication that are used for flexible displays, sensors, tags, etc. At the current state, printing processes of organic TFTs cannot reproduce the high-quality structures that are achieved with vacuum and photolithographic processes [14]. To be more precise, short channels < 10 μm with 100 kHz operating frequencies cannot be manufactured via R2R processes as the current limit for registration accuracy is ± 20 μm. Morphological and structural imperfections like non-uniform thickness, surface roughness, edge waviness, or non-uniform contact resistance between electrode and semiconductor of printed structures, besides web deformation, significantly reduces process yield. For example, process yield is < 70% for gravure R2R printed organic TFT as integration density is < 100 TFTs [14]. Low mobility of printed semiconductors and high resistance of materials increase operating voltages – typically above 20 V [6]. Therefore, there is a doubt that fully printed TFTs can be integrated to high operational devices like smartphones or

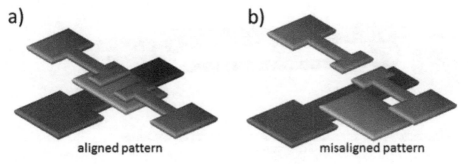

a)

b)

aligned pattern   misaligned pattern

Figure 17.7 Representation of printing misalignment in multiple process due to substrate deformation on R2R process: (a) correctly aligned pattern and (b) misaligned pattern.

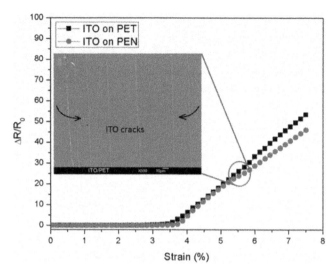

Figure 17.8 Increase of electrical resistance of ITO film on PET foil due to ITO cracking upon applied strain.

tablet PCs that require high integration density. Low integration level of printed TFTs per area limits applications to niche markets, for example, disposable devices including integrated wireless sensors [14].

Theoretically, R2R processes allows high-speed production. This is realized in a situation when all processes can be conducted at high throughput. In other words, the slowest process dictates entire production throughput. In practice, achieving high production speeds might be limited not only by deposition technique or drying hardware but also by type of material and its requirements for drying. Processing speeds for different printing/coating techniques are reported here [1]. Allowed processing drying temperatures for some solvent types can also be regulated by law. For most applications, deposited materials require just being dried; therefore high production speed can be achieved. However, for other technologies like OPV applications, drying of photoactive material is more complicated and brings another dimension as solvents need to be evaporated under specific control to allow the right morphology to be formed. This is necessary to guarantee maximum electrical functionality for the product being used. It can be found that a wrongly selected drying system or too short drying zone length may not enable achievement of high production speeds. This applies especially in situations where deposited materials require gradual and long drying to achieve the proper functionality. Therefore, the dream of high throughput that the R2R process offers can be blurred here. Long drying process steps limit process-window parameters and may require finding new material candidates or making machine modifications (e.g. new web path and/or new drying equipment to allow longer or faster heating). Selection of a drying system should be considered carefully as it is, in most cases, the most expensive and possibly the most complex part of the whole R2R machine [5].

Production of defect-free product that guarantee high yield is not trivial. Non-consistency for batch-to-batch reproducibility of ink or paste formulations provided by vendors can lead to processability issues like local de-wetting spots, retracting, non-continuous lines for rotary screen printing, thinning or streaks for slot-die coating due to higher or lower viscosity, respectively, etc. In these conditions, it is difficult to establish reliable standard operating procedures. Furthermore, conductive substrates or barrier films can have spikes or pinholes (see Figure 17.3(b)). This will affect quality and functionality of the next layers printed/coated on top. Other defects that can be present on R2R processing are scratches on the coated web due to machine handling. They can be generated when the functional side of the web is in direct contact with the rolls or during winding or unwinding of the material from the roll. Figure 17.3(d) shows scratches on coating caused by machine or human handling. A proper web path that avoids direct contact or use of contact-free rolls should be taken into account at production machine planning stage. An additional challenge that contributes, and in many cases is the root cause of scratches, de-wettings, or mura defects, is to keep environments where there could be a lot of working personnel and movements clean. Moreover, conductive substrates and barrier foils must go through slitting process by vendor or converting companies where the master roll is cut to required width and length before it is sent to the production plant. This process can create slitting dust. Its presence in a winded roll, as Figure 17.3(e) shows, can contaminate processing equipment causing scratches and local coating de-wettings or mura defects, as Figures 17.3(d) and (f) show respectively, that reduce quality and functionality of the final product. These defects lead to low yield, hence increased production costs. In-line quality control is complex as it is difficult to select and reject non-functioning parts from the web. In addition, fixing defective products is not possible [7]. Regular quality and cleanliness control is necessary for incoming goods. Detailed processing challenges for specific printing/coating techniques are described here [1].

Frequent changeover, due to small production runs or different product, may increase overall production time. Equipment cleaning and new machine settings preparation might be time consuming and as a side-effect changeover time can increase. For example, cleaning a rotary screen after printing is difficult and time consuming. Moreover, extra care needs to be taken to avoid screen damage by twisting or applying too much force.

Relatively short vitality of organic printed electronics limits the possibility to create product stocks in large quantities [7]. It is rather favorable to manufacture products on demand. This requires good materials planning and logistics.

17.6 Summary

Complexity of large volume R2R manufacturing of organic electronics has been presented and discussed. It is difficult to rate the challenges presented in Figure 17.2 based on their level of importance to be overcome as they may differ according to technology, coating/printing equipment, and applications. However, it can be agreed that cost reduction may remain the biggest challenge regardless of technology. For instance, in order to compete with existing technology, flexible flat panel display production cost needs to be reduced four times [8]. R2R production challenges presented and discussed in this chapter may look discouraging or to have high investment risks at first glance when thinking to move from laboratory to production stage. In addition, as J. Sheats states [5]: "If the market is not there, the best technology in the world will not prosper." Nevertheless, the market forecast for printed electronics for the near future looks very promising, as Figure 17.9 presents, and increase in printed electronic manufactures is observed. If the organic electronic products are well conceptualized and designed, "killer applications" can be made and the market broadened [5]. Moreover, existing producers deploy production scale-ups as market demand is increasing. As an example, Figure 17.10 shows urban installation of R2R-produced free-shape and large-area OPVs fabricated by ARMOR solar power films GmbH. In addition, new candidates for conductive

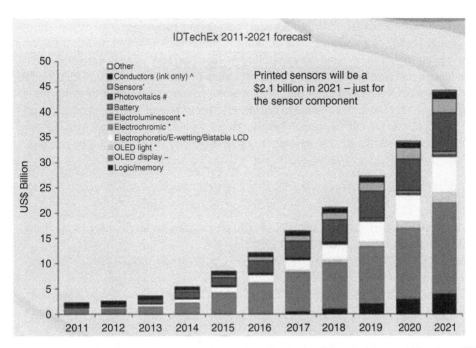

Figure 17.9 Market forecast for printed electronic technologies. Printed with permission from IDTechEx.

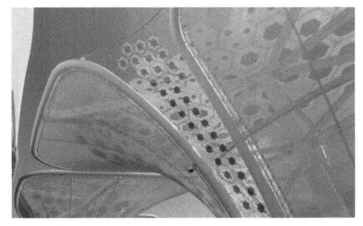

Figure 17.10 Urban installation of fully patterned and R2R-produced OPV modules. Printed with permission from ARMOR solar power films GmbH.

substrates based on metal grid or silver nanowire have started to replace ITO-based electrodes allowing for cost reduction and improved process flexibility. In terms of fully printed organic TFTs, it was expected that registration accuracy of ± 5 μm will be achieved by 2020 and create new IT market for low-cost radiofrequency (RF) tags, RF-sensor tags, and smart packaging [14]. Despite the fact that there are high resolution printing techniques (e.g. gravure or nano-imprint) allowing to achieve high density, further research is still necessary to demonstrate process integration of TFT devices and circuits. In addition, initial research on 3D integration technology can be considered as one of the future techniques that allows to overcome printing resolution limits, but this technology needs to mature before integration to large volume production processes [15]. It should be noted that solutions to some technical equipment or processing issues very often can be adopted from different technology platforms. Furthermore, R2R equipment suppliers strive to improve equipment flexibility, simplicity, and web handling to meet product processing requirements. Testing equipment and processes are also offered by variety of institutions or innovation centers that help to obtain useful data to support the decision-making process on adequate equipment.

There is no doubt that the future belongs to organic electronics where fast R2R production process is the backbone. We should also remember that technologies we are commonly using today were driven by challenges faced in the past according to the saying: *no problems = no progress.*

References

1 Khan, S., Lorenzelli, L. et al. (2015). Technologies for printing sensors and electronics over large flexible substrates: A review. *IEEE Sens. J.* 15 (6): 3164–3185.

2 Logothetidis, S. (2008). Flexible organic electronic devices: Materials, process and applications. *Mater. Sci. Eng. B* 152 (1–3): 96–104.

3 Yin, Z., Huang, Y. et al. (2010). Inkjet printing for flexible electronics: Materials, processes and equipments. *Chin. Sci. Bull.* 55 (30): 3383–3407.

4 Kim, H.-J., Almanza-Workman, M., et al. (2009). Roll-to-roll manufacturing of electronics on flexible substrates using self-aligned imprint lithography (SAIL). *J. Soc. Infor. Display.* 17 (11): 963.

5 James, R.S. (2002). Roll-to-roll manufacturing of thin film electronics. *SPIE's 27th Annual International Symposium on Microlithography*, SPIE.

6 Chang, J., Ge, T., et al. (2012). Challenges of printed electronics on flexible substrates. *2012 IEEE 55th International Midwest Symposium on Circuits and Systems (MWSCAS)*. IEEE. 582–585.

7 Willmann, J.R., Stocker, D. et al. (2014). Characteristics and evaluation criteria of substrate-based manufacturing. Is roll-to-roll the best solution for printed electronics? *Org. Electron.* 15 (7): 1631–1640.

8 Schwartz, E. (2006). Roll to roll processing for flexible electronics. Warrendale, PA: Cornell University.

9 Han, H., Adams, D. et al. (2005). Characterization of the physical and electrical properties of Indium tin oxide on polyethylene napthalate. *J. Appl. Phys.* 98 (8): 083705–8.

10 Potoczny, G. (2012). *Electro-Mechanical Behaviour of Indium Tin Oxide Coated Polymer Substrates for Flexible Electronics*. School of Metallurgy and Materials. Birmingham: University of Birmingham. Doctor of Philosophy.

11 MacDonald, W.A., Looney, M.K. et al. (2008). Designing and manufacturing substrates for flexible electronics. *Plast. Rubber Compos.* 37 (2-4): 41–45.

12 Cairns, D.R., Sachsman, S.M. et al. (1999). Mechanical behavior of indium oxide thin films on polymer substrates. *Mater Res. Soc. Symp. Proc. Lib.* 594: 401–406.

13 Potoczny, G.A., Bejitual, T.S. et al. (2013). Flexibility and electrical stability of polyester-based device electrodes under monotonic and cyclic buckling conditions. *Thin Solid Films* 528: 205–212.

14 Noh, J., Jung, M. et al. (2015). Key issues with printed flexible thin film transistors and their application in disposable RF sensors. *Proc. IEEE* 103 (4): 554–566.

15 Bonnassieux, Y., Brabec, C. J. et al. (2021). The 2021 flexible and printed electronics roadmap. *Flex. Print. Electron.* 6 (2): 023001.

18

Direct Ink Writing of Touch Sensors and Displays

Current Developments and Future Perspectives

Konstantinos A. Sierros[1] and Darran R. Cairns[2]

[1] West Virginia University, Department of Mechanical & Aerospace Engineering, WV
[2] University of Missouri – Kansas City, School of Science and Engineering, Kansas City, MO

18.1 Introduction

The recent advent of additive manufacturing methods opens up new possibilities for flexible optoelectronics manufacture. Among these methods direct ink writing (DIW), which was patented as a deposition/patterning approach at Sandia Labs in 1996 [1], has gained increased interest for utilization in display and touchscreen manufacturing. DIW is a solution-based robotic patterning, of a viscoelastic "ink" through a nozzle, on digitally predefined substrate locations [2]. It allows for scalable two- and three-dimensional (2D/3D) materials assembly, and even vertical component integration, with minimal waste, thus reducing or eliminating the need for additional tooling. With a resolution of 1–200 μm it is comparable to other printing processes [3, 4].

Critical functional components such as metal flexible electrodes can be patterned using DIW in 2D/3D with in-situ laser annealing for optimizing microstructural properties and electrical conductivity on the fly [5]. Although inkjet printing of similar structures has been successfully demonstrated in the past [6], even utilizing roll-to-roll manufacturing for solar cells [7], there are still limitations associated with this approach. An important limitation associated with inkjet printing, often overlooked, is the need for low viscosity jetting solutions [8], which constrain ink design and synthesis. On the other hand, DIW allows for the utilization of a wide range of ink viscosities through the extrusion of continuous and uniform features. Consequently, this leads to novel and versatile materials as well as ink synthesis and development, in particular for low-temperature substrate utilization and post-processing, which is particularly applicable to flexible optoelectronics [9]. In addition, feature uniformity (e.g. line edges) is important in display and touchscreen applications and it can be restricted by the drop-by-drop ink-coalescing nature of the inkjet process [10]. It is therefore important to explore new printing processes and understand their main characteristics and attributes.

In this chapter the goal is two-fold. First, to discuss some notable, recent, examples of ink formulation and post-processing approaches as well as DIW of display and touch sensor components/devices. Second, to present current opportunities while pointing at future directions for further utilizing this versatile direct-write assembly approach.

Flexible Flat Panel Displays, Second Edition. Edited by Darran R. Cairns, Dirk J. Broer, and Gregory P. Crawford.
© 2023 John Wiley & Sons Ltd. Published 2023 by John Wiley & Sons Ltd.

18.2 DIW and Ink Development

DIW is based on room-temperature controlled extrusion of an ink through an orifice using externally applied pressure [2]. Figure 18.1 shows a schematic of the DIW process along with an in-house-built printer of our research group [11].

Upon exit from the nozzle the ink experiences relatively fast solvent evaporation on the underlying substrate, which is usually mounted on an XYZ moving stage. Ink rheological properties are therefore crucial for successful printing and printed feature fidelity. DIW inks are usually multicomponent and often include solvents, binders, and functional nanoparticles but they can also be nanoparticle free [12]. In addition, it is possible to have more than one ink available for patterning in a multi-nozzle system configuration. Whether particle based or particle free these ink systems should exhibit shear thinning behavior [13] for successful printing. Ink rheological data are frequently fitted using well-established models such as the Herschel-Bulkley [14] for yield-stress non-Newtonian fluids. There are three main parameters that can be determined from this model including consistency k, flow index n, and yield stress τ_0. The latter is of high importance since it determines the threshold stress beyond which the ink starts flowing under applied pressure (and thus retains its shape upon deposition) while the flow index governs whether the ink is shear thinning or thickening.

Recently, there is great emphasis placed on developing models that "quantify" printability by linking rheology and printing parameters. Most of these efforts are currently focusing on the printing of highly dense ceramic architectures [13]. Along with ink rheology, DIW parameters are also important for printed feature fidelity. Among them, applied pressure, printing speed, nozzle length and diameter, and nozzle-to-substrate distance are currently studied with respect to printing quality for a range of ink viscosities [15]. Depending on ink formulation and viscoelastic properties (i.e. storage and loss moduli) self-supporting, relatively thin structures can be

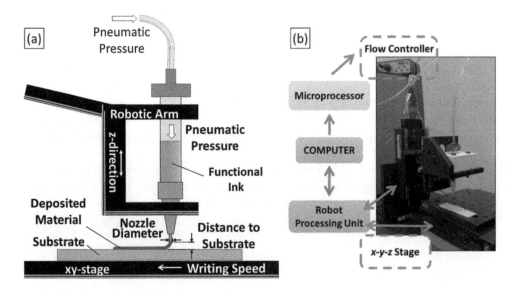

Figure 18.1 (a) DIW apparatus schematic. (b) Lab-built robotic printer image and configuration (arrows in green indicate communications).

explored [16]. Furthermore, ink-substrate interactions – in particular the first printed layer – play an important role along with ink-solvent concentration [12]. Such interactions have an impact on drying processes, which can lead to undesired thermomechanical responses such as cracking and delamination [17]. The latter may be the result of – usually rapid – solvent evaporation upon deposition, leading to multidirectional anisotropic shrinkage [18]. Typically, inks of high nanoparticle loading (> 20% w.t.) [19] are required for forming dense structures but increase in concentrations may lead to optical properties degradation (e.g. transparent to opaque transition in the visible range). Therefore, it is important to consider the effect that lowering ink viscosity has on the layer-by-layer printing. However, the inherent versatility in ink design and synthesis can alleviate such optical and mechanical challenges. For example, patterning of conductive grids of silver (Ag) nanoparticles (5–50 nm in size)-based inks has been reported [2] using DIW, as shown in Figure 18.2.

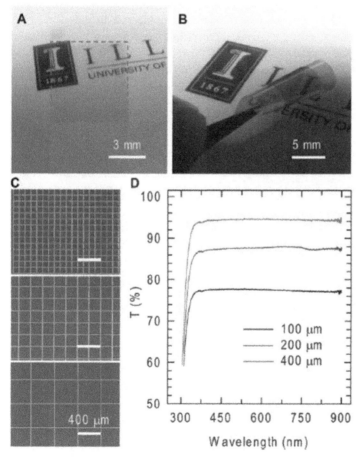

Figure 18.2 Optical images of the transparent conductive grids with center-to-center line spacing of 200 mm, printed on (A) glass and (B) polyimide substrates. (C) SEM micrographs of the conductive grids, printed with center-to-center spacing of 100 mm (top), 200 mm (middle), and 400 mm, respectively. (D) Transmittance of the conductive grids of varying center-to-center spacing patterned on glass substrates. Reprinted with permission from [3]. Copyright 2011, American Chemical Society.

The grids can be used in touchscreen applications and flexible displays since they can be deposited on glass or plastic substrates with adequate optical transmittance (> 94%) due to their center-to-center spacing of 400 μm. The ink synthesis approach is also important to discuss since it starts with growing the Ag nanoparticles using a metal salt precursor ($AgNO_3$), a capping agent (poyacrylic acid, PAA), and diethanolamine (DEA) for reducing purposes. The resulting ink is of high solid concentration (> 70% w.t.) whereas it exhibits electrical resistivity equal to 3.64×10^{-5} Ω cm.

We modified [20] this synthesis method by using ethanolamine as the reducing agent of the Ag ions to nucleate nanoparticles. Here the lower viscosity of ethanolamine, as compared to diethanolamine, allows for relatively easy extraction of the grown Ag nanoparticles and its lower pH aids toward smaller amounts of utilized metal salt precursor. This modified approach allows for high nanoparticle yield (~ 61.6%). After printing Ag lines on polyethylene naphthalate (PEN) substrates using DIW, we conducted cyclic mechanical characterization of the printed samples in tension (Figure 18.3). The cyclic tensile strain was kept at 2%, which is reported to be within the indium-tin oxide's (ITO) failure initiation threshold [21]. In general, the Ag patterns do not experience failure initiation up to ~ 10% strain whereas the cumulative increase in electrical resistance due to cyclic tensile stress remains under the failure threshold. This increase was attributed to the particle binders' viscoelastic effects.

Solution-based, metal precursor reduction approaches yield Ag inks with appropriate rheological properties and allow for patterning of flexible 2D/3D geometries and self-supported struc-

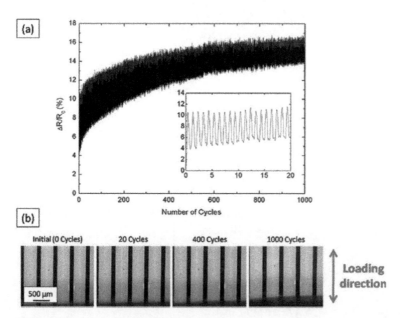

Figure 18.3 (a) Normalized change in electrical resistance of Ag line patterns cyclically tested up to 2% strain for 1000 cycles. Inset graph shows the first 20 cycles. (b) Optical images depicting the micro-patterned surface at different loading cycles. Reprinted with permission from [12]. Copyright 2015, Elsevier.

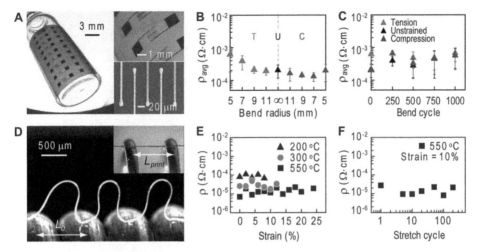

Figure 18.4 (A) Optical and SEM images of silver microelectrodes patterned on a polyimide substrate with a bend radius of 14 mm. (B) Electrical resistivity of the silver microelectrodes as a function of bend radius under tension (T), unstrained (U), and compression (C). (C) Electrical resistivity of the silver microelectrodes as a function of bend cycle at a bend radius of ± 5 mm. (D) Optical image of stretchable silver arches printed onto a spring. (E) Electrical resistivity of the stretchable silver microelectrode arches as a function of strain and annealing temperature. (F) Electrical resistivity of the stretchable silver microelectrode arches as a function of strain cycle. Error bars indicate the SD measured from 10 electrodes. Reprinted with permission from [22]. Copyright 2009, AAAS.

tures (Figure 18.4) [22]. The metal structures can be printed on pre-stretched flexible substrates (such as polyimide-PI) for enabling Ag mechanical flexibility while maintaining appropriate electrical conductivities. Again, control of nanoparticle size, shape, and concentration (> 70% w.t.) in the ink is critical for successful DIW.

Furthermore, there are different approaches reported in the past about the post-processing of patterned metal electrodes on flexible substrates. Here we are not discussing conventional thermal treatments since they have been extensively investigated. Among novel methods, chemical sintering [23] is an interesting approach. Specifically, Ag nanowires were directly written on plasma-treated polycarbonate substrates and then chemically sintered utilizing NaCl, KCl, and $CaCl_2$ solutions (0.8–1.2 M) at room temperature. After sintering, the nanowires formed junctions whereas the non-sintered counterparts were not observed to exhibit such behavior. Figure 18.5 depicts the junctions formation and the resulting electrical resistance of the chemically treated flexible electrodes.

Another interesting, and scalable, approach to post-treat particle-free inks on low-temperature plastic substrates is photonic sintering, which typically takes advantage of a laser or flash lamp to provide energy to the metal ink for solvent evaporation and precursor decomposition and/or reduction [24]. In addition, plasma treatment for sintering the metal electrodes has also been reported. In particular, it is reported that plasma (160 W for 8 minutes) leads to acceptable electrical resistivities (7.3 ± 0.2 μΩ) in printed Cu patterns [25]. Hong et al. [26] reported on the localized laser sintering of Ag metallic meshes on plastics. The resulting directly written grid lines (25 cm² area) exhibited ~ 130 nm height, ~ 11 μm width, 85%

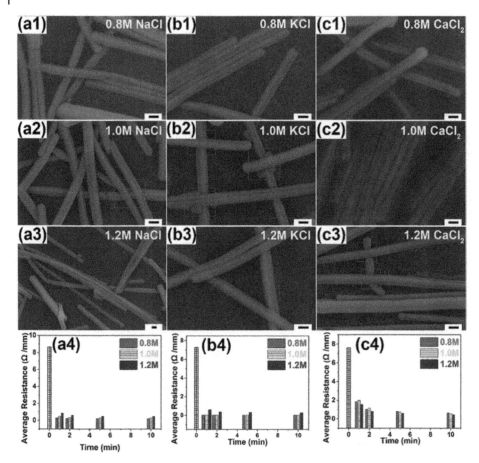

Figure 18.5 (a1–3), (b1–3), (c1–3), SEM images of Ag NW electrodes sintered by different electrolyte solution with different concentration, the scale bars are 100 nm. (a4)–(c4) The dependence of the average resistance of sintered electrodes on time. Reprinted with permission from [23]. Copyright 2017, IOP Publishing Ltd.

optical transmittance, and < 30 Ω/sq. sheet resistance, respectively. Figure 18.6 depicts the printed conductive grid as well as related mechanical characterization results (i.e. bending and scotch tape testing).

Furthermore, electrical sintering of metallic structures has also been studied in the past. In this method, a voltage is applied in situ on the printed nanoparticle-based ink structure resulting in significant conductivity increase that is two orders of magnitude higher than for thermally treated counterparts for the case of Ag inks on paper substrates. The process is rapid with the major transition occurring in 2 μs [27]. Finally, microwave sintering is another rapid method with high potential since polymer substrates are not affected by microwave radiation whereas metal nanoparticles (e.g. Ag) are highly absorbent to microwave radiation due to their high dielectric loss factor [28].

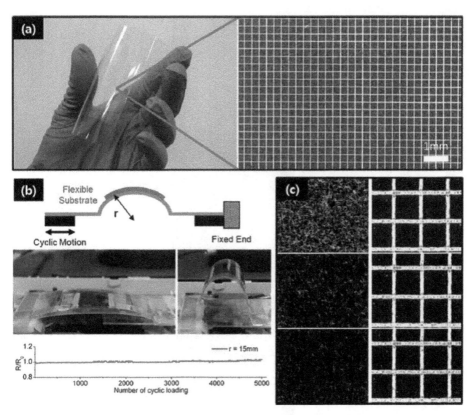

Figure 18.6 (a) Transparent conductor fabricated on a flexible (PEN) substrate (metallic grid on a 5 cm5 cm region). (b) Test setup for cyclic bending, photograph of minimum and maximum bending deformation of the flexible substrate, and resistance change according to the number of bending cycles. (c) Optical image of a transparent conductor made of Ag NW mesh (left column) and Ag NP sintering (right column) as prepared (first row), after first adhesive tape test (second row), and after second adhesive tape test (third row). Reprinted with permission from [26]. Copyright 2013, American Chemical Society.

18.3 Applications of DIW for Displays and Touch Sensors

Currently, there are a few notable examples of DIW of optoelectronics on the device level. This approach is still in its infancy as related to displays and touchscreens. Although some of the following examples are not intended to directly utilize flexible substrates, they nonetheless indicate the possible future direction and high potential of the DIW method.

Kong et al. [29] demonstrated the DIW of quantum dot light-emitting diodes (QD-LEDs). In this exemplar the LEDs are embedded and interconnected in a 3D fashion (Figure 18.7). The latter demonstrates the versatility of the DIW approach toward vertical integration of optoelectronic devices. All components of the device were 3D printed (with potential for fabrication on curved surfaces) and the QD-LED active materials exhibited tunability and pure-color emission.

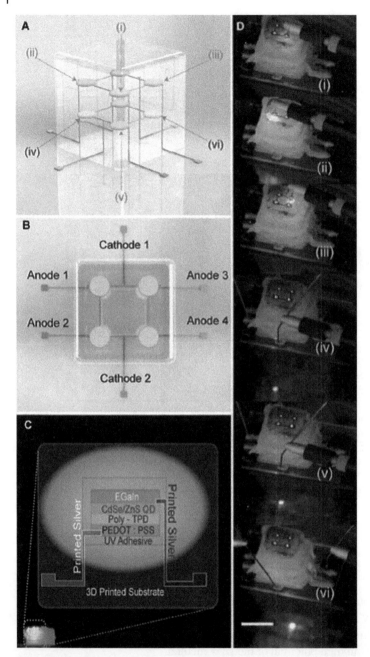

Figure 18.7 3D printed 2 × 2 × 2 multidimensional array of embedded QD-LEDs. (A) Layout of the multicolor 3D QD-LED array design. (B) Schematic showing the QD-LED interconnect assignments, in which the 3D array can be individually controlled via six external contact points. (C) Design schematic of the unit cell of each 3D printed QD-LED. (D) Electroluminescence from three top-layer and three bottom-layer QD-LEDs in the 3D matrix. The six locations (i)–(vi) correspond to the locations indicated by the arrows in panel A. Scale bar is 1 cm. Reprinted with permission from [29]. Copyright 2014, American Chemical Society.

Recently, Liu et al. [30] demonstrated stretchable electroluminescent devices based on DIW of coaxial fiber structures (Figure 18.8). This approach requires detailed nozzle design for enabling multicore-shell DIW. However, it allows the fabrication of customizable 3D macrostructures with light-emitting properties. A variety of inks can be coaxially DIW including luminescent centers, electrodes, and encapsulation one-dimensional (1D) structures. The printing head as well as the coaxial structure are schematically shown in Figure 18.9.

In addition to DIW of displays discussed earlier, there is some work reported on DIW of touch-sensor devices. One example is the report of Li et al. [31] for the DIW fabrication of capacitive touchpads on paper (Figure 18.10). Capitalizing on Ag ink developments and photonic sintering (see section 18.2) they achieved robust (electrical stability was retained for 5000 rolling cycles) and disposable touch devices.

In another more recent work, stable Ag nanowire inks were studied during the development of transparent capacitive touchpads [32]. Figure 18.11 shows a 2 × 2 transparent touchpad. The insulating layer of the device comprised of polydimethylsiloxane. The ink's printability was established by tuning its rheology and controlling nanowire dimensions. Optical transparency in the visible range and electrical properties were adjusted by optimizing the number of directly written layers.

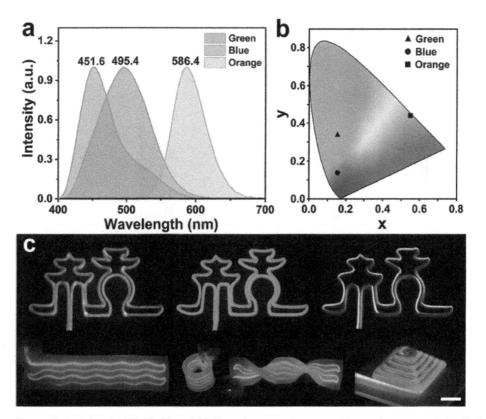

Figure 18.8 Printed ACEL 1D, 2D and 3D fibres incorporating green, blue and orange-emitting ZnS microparticles. (a) Normalized EL spectra of ACEL devices incorporating ZnS:Cu(G), ZnS:Cu(B) and ZnS:Cu,Mn microparticles emitting green, blue, and orange colors and (b) the corresponding CIE1931 diagram. (c) Photographs of the printed ACEL devices with different structures and colors obtained from blue, green and orange light-emitting phosphors: Chinese characters under bending and twisting, and the 3D pyramid. The scale bar is 10 mm. Reprinted with permission from [30]. Copyright 2020, The Royal Society of Chemistry.

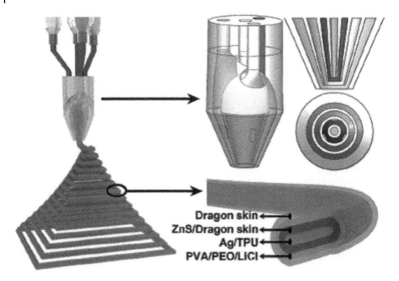

Figure 18.9 Schematic illustration of the multicore-shell printheads and printed 1D stretchable ACEL devices. Reprinted with permission from [30]. Copyright 2020, The Royal Society of Chemistry.

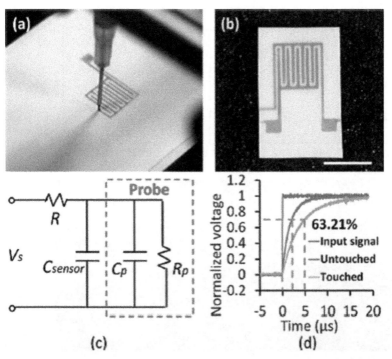

Figure 18.10 Images of (a) a touchpad during silver nanowire ink writing and (b) a completed touchpad on paper substrate. (c) Schematic circuit for the detection of touchpad capacitance with a signal generator and an oscilloscope (labeled only as probe in the image). (d) When applying a square wave across the RC circuit, the observed response with an increasing voltage between two ends of the touchpad. The orange dashed lines in d mark the time constant equivalent to the product of the resistance and capacitance of the circuit. Reprinted with permission from [30]. Copyright 2014, The Royal Society of Chemistry.

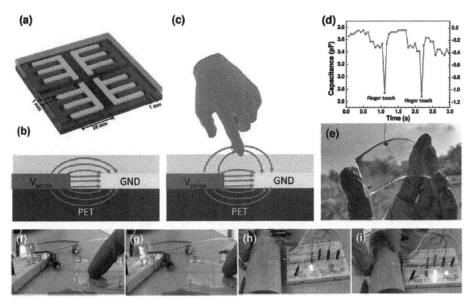

Figure 18.11 The capacitive touch panel implemented using AgNW ink. (a)Schematic of the 2 × 2 touchpad system. The variation in the electric field in the absence and presence of the finger is shown schematically in (b) and (c) respectively. (d) There is a local change in the field, which manifests as a capacitive change and this is plotted as a function of time. (e) The transparency of the touch panel is demonstrated in natural light. (f) and (g) shows the response of the panel with touch in the flat condition and (h) shows the functioning of the touchpad in the curved condition (40 mm diameter). (i) Multitouch response is also possible even in the curved condition. Reprinted with permission from [32]. Copyright 2019, IOP Publishing Ltd.

18.4 Future Challenges and Opportunities

Although the current chapter is not an exhaustive review of additive manufacturing methods of flexible displays and touchscreens, it provides a perspective for low-cost fabrication of flexible displays and touchscreens by capitalizing on DIW fabrication. The advantages of DIW are numerous and adequately discussed. However, an important attribute that needs to be highlighted is the ability to print inks of a wide range of viscosities, which by itself is an enabler for vertical device integration. The latter could be particularly important in the display industry where expensive, multi-step lithography processes are currently used to fabricate multilayer structures. A challenge arises here and is related to ink synthesis, formulation, and characterization. Multi-material shear-thinning inks need to be further explored by combining nanoparticles of different types. Such inks will further enable multiple functionalities (e.g. electrical conductivity and light emission) in a single printed layer. Such approach may require voxelated DIW for controlling spatially the embedded functionalities in the layer. A following challenge is to increase DIW resolution in order to achieve voxel-by-voxel printed architectures. Small nozzle sizes coupled with appropriate ink rheology and application of external stimuli (e.g. voltage-although this qualifies as e-jet) may be the most efficient approach. For the near future it is anticipated that printing technologies will act in a complementary fashion specifically on the device level.

Furthermore, DIW equipment design and development is needed for more efficient combination of patterning and post-processing options. This will enable truly one-step printing of final components and/or devices. Nozzle design here is important for allowing multi-material printing. In addition, the library of available printable inks needs to be further expanded. Lessons learned from traditional display-printing methods (e.g. flexographic, screen printing) need to be applied in terms of ink drying and post-processing as well as to issues related to ink-substrate interactions and substrate surface treatments. Another topic that currently offers a wide range of research opportunities is the quantification of printability. There is a fundamental question here on defining printability. This is challenging to directly answer due to the complex interrelations between ink properties, DIW parameters and post-processing, and resulting microstructural, optical, electrical, and mechanical properties. Most current approaches are focusing on dimensionless modeling of the process and specific ink groups, mostly ceramic. Therefore, the opportunity to explore novel ink systems with multiple functionalities and their printing behavior is arising. The latter may be more effective if combined with machine-learning approaches.

DIW of flexible displays and touchscreens is in its infancy but the lithography-free alternative has already exhibited potential for further exploration and utilization. The versatility in ink design coupled with the low cost for developing DIW equipment renders this approach attractive for 3D integration, which is of paramount importance in the industry. Further developments may focus on ink-development studies, DIW equipment design and development, and quantification of printability.

References

1 Cesarano III, J. and Calvert, P.D. (1997). Freeforming objects with low-binder slurry. US Patent 6027326.
2 Lewis, J.A. (2006). Direct ink writing of 3D functional materials. *Adv. Funct. Mater.* 16: 2193–2204.
3 Ahn, B.Y., Lorang, D.J., and Lewis, J.A. (2011). Transparent conductive grids via direct writing of silver nanoparticle inks. *Nanoscale* 3: 2700–2702.
4 Lewis, J.A. and Gratson, G.M. (2004). Direct writing in three dimensions. *Mater. Today* 7: 32–39.
5 Skylar-Scott, M.A., Gunasekaran, S., and Lewis, J.A. (2016). Laser-assisted direct ink writing of planar and 3D metal architectures. *PNAS* 113: 6137–6142.
6 Kinner, L., Nau, S., Popovic, K., Sax, S., Burgues-Ceballos, I., Hermerschmidt, F., Lange, A., Boeffel, C., Choulis, S.A., and List-Kratochvil, E.J.W. (2017). Inkjet-printed embedded Ag-PEDOT:PSS electrodes with improved light out coupling effects for highly efficient ITO-free blue polymer light emitting diodes. *Appl. Phys. Lett.* 110: 101107.
7 Angmo, D., Larsen-Olsen, T.T., Jorgensen, M., Sondergaard, R.R., and Krebs, F.C. (2013). Roll-to-roll inkjet printing and photonic sintering of electrodes for ITO free polymer solar cell modules and facile product integration. *Adv. Energy Mater.* 3: 172–175.
8 Krainer, S., Smit, C., and Hirn, U. (2019). The effects of viscosity and surface tension on ink-jet printed picoliter dots. *RSC Adv* 54: 31708–31719.
9 Scheideler, W. and Subramanian, V. (2019). Printed flexible and transparent electronics: Enhancing low-temperature processed metal oxides with 0D and 1D nanomaterials. *Nanotechnology* 30: 272001.
10 Nayak, L., Mohanty, S., Nayak, S.K., and Ramadoss, A. (2019). A review on inkjet printing of nanoparticle inks for flexible electronics. *J. Mater. Chem.* 7: 8771–8795.

11 Torres Arango, M.A. and Sierros, K.A. (2017). New paradigms on materials synthesis and additive manufacturing of flexible electronics for energy applications. *TechConnect Briefs* 178–181.

12 Torres Arango, M.A., Abidakun, O.A., Korakakis, D., and Sierros, K.A. (2017). Tuning the crystalline microstructure of Al-doped ZnO using direct ink writing. *Flex. Print. Electron.* 2: 035006.

13 M'Barki, A., Bocquet, L., and Stevenson, A. (2017). Linking rheology and printability for dense and strong ceramics by direct ink writing. *Sci. Rep.* 7: 6017.

14 Herschel, W.H. and Bulkley, R. (1926). Konsistenzmessungen von Gummi-Benzollösungen. *Kolloid Zeitschrift* 39: 291–300.

15 Cordonier, G.J. and Sierros, K.A. (2020). Unconventional application of direct ink writing: Surface force-driven patterning of low viscosity inks. *ACS Appl. Mater. Interfaces* 12 (13): 15875–15884.

16 Rocha, V.G., Saiz, E., Tirichenko, I.S., and Garcia-Tunon, E. (2020). Direct ink writing advances in multi-material structures for a sustainable future. *J. Mater. Chem. A* 8: 15646–15657.

17 Mguyen, D.T., Meyers, C., Yee, T.D., Dudukovic, N.A., Destino, J.F., Zhu, C., Duoss, E.B., Baumann, T.F., Suratwala, T., Smay, J.E., and Dylla-Spears, R. (2017). 3D-printed transparent glass. *Adv. Mater.* 29: 1701181.

18 Liu, D.-M. (1997). Influence on porosity and pore size on the compressive strength of porous hydroxyapatite ceramic. *Ceram. Int.* 23: 135–139.

19 Camposeo, A., Persano, L., Farsari, M., and Pisignano, D. (2019). Additive manufacturing: Applications and directions in photonics and optoelectronics. *Adv. Optical Mater.* 7: 1800419.

20 Torres Arango, M.A., Cokeley, A.M., Beard, J.J., and Sierros, K.A. (2015). Direct writing and electro-mechanical characterization of Ag micro-patterns on polymer substrates for flexible electronics. *Thin Solid Films* 596: 167–173.

21 Cairns, D.R., Witte II, R.P., Sparacin, D.K., Sachsman, S.M., Paine, D.C., and Crawford, G.P. (2000). Strain-dependent electrical resistance of tin-doped indium oxide on polymer substrates. *Appl. Phys. Lett.* 76: 1425.

22 Ahn, B.Y., Duoss, E.B., Motala, M.J., Guo, X., Park, S.-I., Xiong, Y., Yoon, J., Nuzzo, R.G., Rogers, J.A., and Lewis, J.A. (2009). Omnidirectional printing of flexible, stretchable, and spanning silver microelectrodes. *Science* 323: 1590–1593.

23 Hui, Z., Liu, Y., Guo, W., Li, L., Mu, N., Jin, C., Zhu, Y., and Peng, P. (2017). Chemical sintering of direct-written silver nanowire flexible electrodes under room temperature. *Nanotechnology* 28: 285703 (12pp).

24 Abbel, R., Van Lammeren, T., Hendriks, R., Ploegmakers, J., Rubingh, E.J., Meinders, E.R., and Groen, W.A. (2012). Photonic flash sintering of silver nanoparticle inks: A fast and convenient method for the preparation of highly conductive structures on foil. *MRS Commun.* 2: 145–150.

25 Farraj, S., Smooha, A., Kamyshny, A., and Magdassi, S. (2017). Plasma-induced decomposition of copper complex ink for the formation of highly conductive copper tracks on heat-sensitive substrates. *Appl. Mater. Interf.* 9: 8766–8773.

26 Hong, S., Yeo, J., Kim, G., Kim, D., Lee, H., Kwon, J., Lee, H., Lee, P., and Ko, S.H. (2013). Nonvacuum, maskless fabrication of a flexible metal grid transparent conductor by low-temperature selective laser sintering of nanoparticle ink. *ACS Nano* 7: 5024–5031.

27 Allen, M.L., Aronniemi, M., Mattila, T., Alastalo, A., Ojenpera, K., Suhonen, M., and Seppa, H. (2008). Electrical sintering of nanoparticle structures. *Nanotechnology* 19: 175201 (4 pp).

28 Perelaer, J., De Gans, B.-J., and Schubert, U.S. (2006). Ink-jet printing and microwave sintering of conductive silver tracks. *Adv. Mater.* 18: 2101–2104.

29 Kong, Y.L., Tamargo, I.A., Kim, H., Johnson, B.N., Gupta, M.K., Koh, T.-W., Chin, H.-A., Steingart, D.A., Rand, B.P., and McAlpine, M.C. (2014). 3D printed quantum dot light-emitting diodes. *Nano Lett.* 14: 7017–7023.

30 Liu, D., Ren, J., Wang, J., Xing, W., Qian, Q., Chen, H., and Zhou, N. (2020). Customizable and stretchable fibre-shaped electroluminescent devices via multicore-shell direct ink writing. *J. Mater. Chem. C.* doi: 10.1039/d0tc03078c.

31 Li, R.-Z., Hu, A., Zhang, T., and Oakes, K.D. (2014). Direct writing on paper of foldable capacitive touch pads with silver nanowire inks *ACS. Appl. Mater. Interf.* 6: 21721–21729.

32 Nair, N.M., Daniel, K., Vadali, S.C., Ray, D., and Swaminathan, P. (2019). Direct writing of silver nanowire-based ink for flexible transparent capacitive touch pad. *Flex. Print. Electron.* 4: 045001.

19

Flexible Displays for Medical Applications

Uwadiae Obahiagbon[1], Karen S. Anderson[2,3], and Jennifer M. Blain Christen[4]

[1] School of Electrical, Computer and Energy Engineering at Arizona State University, Tempe, AZ
[2] Virginia G. Piper Center for Personalized Diagnostics, The Biodesign Institute at Arizona State University, Tempe, AZ
[3] Corresponding author
[4] School of Electrical, Computer and Energy Engineering at Arizona State University, Tempe, AZ

19.1 Introduction

19.1.1 Flexible Displays in Medicine

The flexible display market has grown rapidly over the past decade. Particularly, active-matrix organic light-emitting diode (AMOLED) revenues alone increased from $3.5 billion in 2016 to $12 billion in 2017 [1]. Although the emergence of flexible displays has been largely driven by applications in military, mobile devices, wearables, TV screens, and lighting panels, several other industries (healthcare, gaming, paper) are projected to become major players in the next decade [2]. There is an ever-increasing trend toward higher resolution/quality images for clinical diagnosis by radiologists, robust display systems for surgeries, miniaturization and parallelization in analytical instrumentation, wearable monitors, and low-cost disposable sensor platforms. Flexible, foldable, and rollable displays could potentially be made using liquid crystal display (LCD), organic light-emitting diodes (OLED), and electronic paper display (EPD) technologies. Of these three technologies, OLED displays are particularly attractive for medical applications partly because they are emissive. LCDs work on the principle of polarization using a backlight (from a cold cathode fluorescent lamp or an inorganic light-emitting diode, ILED), whereas EPDs are reflective displays. The potential for revolutionizing healthcare and personal health monitoring using OLEDs has been a subject of great interest to researchers and industry [3–5]. In particular, wearable and low-cost disposable electronics are on the path toward defining a paradigm shift in healthcare and medicine. This is mostly due to the demand for sensors capable of noninvasive monitoring of bioprocesses. The utility of flexible displays, especially organic-based optoelectronics, toward the development of integrated and miniaturized biochemical sensors is the focus of this chapter.

19.1.2 A Brief Historical Perspective

Since the invention of the OLED at Kodak in 1987 [6] (~ 20 years after the LCD [7]), there has been significant investment, developments, and improvements in its performance. Today, flat panel display technology has become a part of our everyday lives, continuing to play a prominent role in consumer products like smartphones and TVs. The high-volume production of flat panel displays creates an avenue for developing low-cost devices, supported by

Flexible Flat Panel Displays, Second Edition. Edited by Darran R. Cairns, Dirk J. Broer, and Gregory P. Crawford.
© 2023 John Wiley & Sons Ltd. Published 2023 by John Wiley & Sons Ltd.

already existing infrastructure. Researchers have therefore continued to explore alternate applications for flat panel display technology. Reports about the use of organic display technology as excitation sources for bioanalytical applications only began to show up about 10 years after the invention [8]. Although LCD and OLED displays dominate the flat panel display market [9], at the time of this writing, flexible LCDs have not yet been commercialized. Flexible and particularly bendable OLEDs, however, have found their way into today's consumer products. The continued research and development toward the application of flexible OLEDs has opened up promising pathways toward complete integration (with microfluidics) and miniaturization, especially for disposable and wearable sensors for point-of-care (POC) applications. Researchers have therefore continued to explore the potential use of OLEDs in sensor applications [10].

19.1.3 Application of Flexible Displays for Biochemical Analysis

The basic unit of organic flat panel/flexible displays is the OLED, a single pixel. Compared to ILED, OLEDs have physical and optoelectronic characteristics that make them suitable for sensor design. Typically, ILEDs are made out of III-V semiconductor materials, which are not easily deposited over large areas. Notably, the feasibility of fabricating OLEDs on different substrates (glass, plastic, polyimide, metal, etc.) has led to several applications. Particularly, the fabrication of flexible displays is enabled by the unique properties of the organic materials used, in that their flexibility allows for deposition on flexible substrates over a large area. Flexible display technology has opened up a lot of avenues for miniaturization and integration of technologies for biochemical analysis. Microarrays, which are on the order of the size of a pixel, could deliver high throughput analysis for biological and chemical applications. The attractive characteristic of OLEDs, including lower production cost, also make them suitable for POC applications especially for resource-poor settings. Owing to their flexibility, self-emission properties (no backlighting required), ultra-thin profile (< 1 μm excluding substrate), conformal, light weight, transparency, full color capabilities, and cost-effective fabrication (large area, solution processed [11, 12], or roll-to-roll manufacturing on polyethylene terephthalate [PET] films) [13], OLEDs have proven to be the material of choice for flexible flat panel display technology and sensor development [14, 15].

19.1.4 OLEDs and Organic Photodiodes as Optical Excitation Sources and Detectors

Optical transduction systems typically require an excitation light source that probes a sample in order to measure a concentration-dependent parameter, e.g. absorbance, reflection, transmission, or light emissions from a fluorescent dye. A photodetector (PD) is used to convert the concentration-dependent optical parameter to an electric current, which can be amplified, processed, analyzed, and displayed for analysis. Standard light sources (including xenon, mercury, lasers) are not easily integrated with miniaturized sensors (see Sections 19.1.5 and 19.5.1) or at best lead to bulky analytical instruments. Although ILEDs and laser diodes are gradually becoming the dominant light source for many analytical applications [16, 17], OLEDs have been shown to be viable excitation sources for optical detection systems [3–5]. With advances in technology, OLEDs have sufficient brightness and stability required for them to function as excitation sources especially in disposable POC sensors. From an operational standpoint, OLEDs could be operated at low voltages (5 V) or at higher voltages in pulsed mode to obtain higher brightness while maintaining stability. Furthermore, OLEDs have a lower power consumption than their LCD and light-emitting diode (LED) counterparts. OLEDs also have a wide viewing angle (suitable for bending and folding applications) with devices fabricated across the visible range (red, green, and blue emitters) combined with active matrix (thin-film transistor [TFT] technology) making a crispy full color display. Figure 19.1 (left) shows normalized emission spectra for OLEDs across the visible and near infrared spectra. These can be used to excite fluorophores or measure absorption across the broad spectrum for different applications.

One of the challenges associated with miniaturizing central laboratory optical detection devices is the form factor and geometries of various components used in their design, including standard PDs such as photomultiplier tubes (PMT), avalanche photodiodes (APD), charge-coupled device (CCD), or complementary metal oxide semiconductor (CMOS) detectors. Inorganic photodiodes based on Si have been used extensively as PDs in many applications. However, organic photodiodes (OPD) are relatively easily fabricated with processes that are compatible with organic flat panel display fabrication, enhancing the ease with which miniaturized and compact devices could be fabricated [18].

Figure 19.1 (right) shows several integrated sensor configurations using OLEDs and OPDs. It is easier to implement the "front detection" (transmission) configuration, where the bio-affinity compound/sensing element/microfluidic channel is sandwiched between the OLED excitation source and a PD. In the "back-detection" (reflection/coplanar) configuration, the OLED and the PD are next to each other on the same plane. For example, in fluorescence applications, the OLED is patterned to allow detection of the emitted fluorescence through the spaces or over the entire backside in the case of transparent OLEDs. There are a few applications that are better suited to reflection and absorption configurations [19, 21] (see Section 19.7.1). The configuration choice depends on design. An added advantage of using OLEDs and OPDs in place of traditional/standard sources and detectors is that the emission wavelength and responsivity of OLEDs and OPDs can be relatively easily tuned (mostly based on material selection), to maximize sensitivity at a particular wavelength. Furthermore, OLEDs and OPDs are ideal candidates for disposable applications as they are not only fabricated by simple planar polymer microfabrication but also inexpensive and biodegradable. Additionally, in sensing applications, OLEDs are either operated in dc mode or in pulsed mode. With advances in signal processing and miniaturized integrated

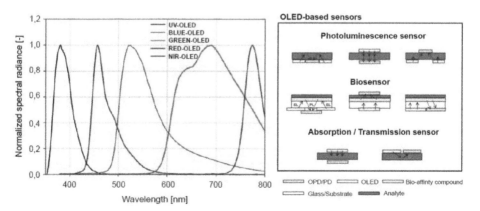

Figure 19.1 Normalized emission spectra of monochromatic OLEDs (left). Orientations of OLED-based photo-organic sensor systems (right). Adapted from [20]. Copyright © [20], with permission from Krujatz F. et al.

circuits, there is a realistic potential to take advantage of pulsed-mode excitation for noise reduction, minimizing recalibration, and enhancing dye stability in compact devices. A major challenge with optical detection systems in general is the need to suppress or minimize the excitation light intensity reaching the detector. This intensity increases (inverse square law) as the distance between the excitation source and detector decreases. In fluorescence applications, usually the excitation light reaching the detector can be several orders of magnitude higher than the emitted fluorescence of interest [22]. Technologies and techniques to minimize this effect are discussed elsewhere [23, 24]. The broad electroluminescence (EL) spectrum and tails of OLED sources is detrimental to sensitivity (spectral crosstalk). For applications requiring high sensitivity, a potent way to minimize this effect is the use of interference filters. Realizing miniaturized optoelectronic diagnostic devices for biological, environmental, drug, and food safety applications has generated a lot of interest over the years. This trend has been enabled by the vast research, development, and significant progress in the area of flexible OLEDs and integrated lab-on-a-chip (LOC) devices. Particularly, the detection of biologically relevant species has been the focus of several research groups [5, 12, 21, 25–27].

19.1.5 Device Integration

The integration of organic optoelectronic devices for detecting chemical or biological species has received attention in the past two decades [8, 28]. Integration of optical transduction systems with standard light sources and detectors is either not possible due to geometric constraints or logistically not feasible. Although ILEDs could be integrated with LOC or microfluidic channels, there are significant limitations in the array density and complexity of design. For example, Cho et al. [29], physically machined' ~ 500 μm deep wells in the face of ILEDs to create a reaction well into which recognition elements were immobilized for oxygen detection. This negates the benefit of large area surface emission, which is an attractive quality of flexible display technology. Perhaps for direct immobilization on an emitter surface, flat panel technology combined with microarray printing will not only yield a simpler process but also a more robust and flexible configuration (see Section 19.5). Flexible display technology not only provides a path for direct integration of the sensing element as part of the fabrication process, the planar and relatively simple fabrication procedure allows for the integration of optical components (e.g. interference filters) and other technologies (LOC, microfluidics and flexible electronics, e.g. organic TFTs [OTFTs]). Standard light sources are not easily integrated with sensing platforms due to form factor constraints or cost. Furthermore, the heat generated from most of these sources could damage the sensor or analyte of interest. Obviously, device integration of flexible display technology with LOC and microfluidic protein printing technologies does not have these limitations. Negligible joule heating of the sensing element and components occur when OLEDs are used as an excitation source, even when the sensing element is in contact with the OLEDs4. Also, advantageous is the fact that the planar and flexible form factor allows for patterning and assembly. Monolithic [30], structural, or modular [31] integration of organic optoelectronics – OLEDs and OPD – with microfluidics, planar optics, and LOC is simple because of the planar and flexible configuration. The ease of fabrication and processing (low temperature, sequential/layer-by-layer deposition, solution processing, screen printing) makes it possible to develop organic optoelectronics on several substrates. The processing steps involved are suitable for mass production. Ultimately this leads to compact and miniaturized devices, making these technologies attractive for integrated systems. Integrated optical excitation sources, detectors, and sensing elements are useful for the development of miniaturized high-throughput sensor arrays and for wearable applications.

19.1.6 Fluorescence, Photoluminescence Intensity, and Decay-time Sensing

In the sections that follow, applications of flexible display technology for biomedical sensing are presented. In this section, we present a brief introduction to the optical sensing modalities most often used in flexible display-based sensors. The emissive characteristics of organic flexible display technology makes it suitable for developing sensors that require an excitation/incident light (optical sensors). In fluorescence applications, energetic photons from OLEDs could be used to excite fluorophores (typically used as a label, e.g. in immunofluorescence-based assays, to determine the concentration of species in an analyte/sample). The fluorescence emission intensity is converted to an electric current by a PD. For a given system using optical filters, the emitted fluorescence intensity measured by the PD is expressed as shown in Equation 19.1

$$I_{PD} = KI_0 \left[1 - \left(10^{-\epsilon lc} \right) \right] \tag{19.1}$$

where, ϵ is the extinction coefficient or molar absorptivity of the fluorophore; c, is the molar concentration; l is the optical path length through the sample; I_0, is the excitation source intensity; K, is a system dependent (lumped) constant that depends on the photodiode responsivity, $R(\lambda)$, collection efficiency of the system, η_{LCE}, excitation (T_X) and emission (T_M) filter transmission, and fluorophore quantum yield (QY_{fl}). Equation 19.1 is based on Beer–Lambert law, where the absorption $A = \epsilon cl$.

Another very common sensing mode explored extensively and adopted by several groups (developing flexible display-based sensors) is photoluminescence (PL) quenching, as described by the Stern–Volmer (SV) relation [32]. It relates broadly to variations in the quantum yield (or decay time) of a dye due to a photochemical process and specifies the relationship between the concentration of the target species (quencher), the intensity, and the decay time in sensor applications. For example, the PL intensity from certain oxygen-sensitive dyes (e.g. Ru-based, Pt octaethylporphyrin [PtOEP]), or the Pd analog PdOEP) are collision quenched by oxygen. In its basic form, a linear relationship between the concentration, PL intensity, and the decay time is observed. The SV relationship [3, 33], as shown in Equation 2

$$\frac{I_0}{I} = \frac{\tau_0}{\tau} = 1 + K_{SV}[Q] \tag{19.2}$$

where, I_0 and τ_0 are the unquenched intensity and decay time values, respectively; I and τ are the measured or quenched intensity and decay time, respectively; whereas K_{sv} is the Stern-Volmer constant, which relates to how sensitive the dye is to the quenching species, and Q is the concentration of the quenching species, e.g. O_2. The concentration of target species is correlated to changes in intensity and decay time of a dye.

The PL quenching effect has been applied extensively toward the detection of oxygen using flexible display technology (see Sections 19.2 and 19.3). Sensing could be done in the intensity or the decay time/lifetime mode. A major advantage to being able to sense in the decay time mode is that there is no need for frequent recalibration as with the intensity mode where calibration and ratiometric measurements are required to account for variations in OLED intensity and fluorophore stability. However, lifetime mode sensing requires the use of fast response time PDs,

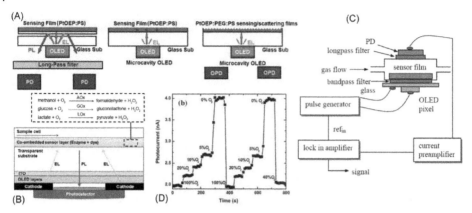

Figure 19.2 (A) Evolution of the back-detection organic-based O_2 sensor: (left to right) original setup; sensing film with more permeable higher molecular weight PS matrix directly dropcast on the back of the OLED substrate excited by a microcavity(μC) OLED with a significantly narrower emission band at the absorption band of PtOEP; modified scattering matrix of blended PS:PEG for higher PL. (B)Illustration of the working principle of bio-catalytic-based biosensors (C) Schematic (not to scale) of the experimental setup in the *I* mode (D)The sensing signal excited by the μC OLED detected by CuPc/C70 OPD at various O_2 concentrations. Figures (A) and (D)adapted from [41], Copyright © [41], with permission from Elsevier (B)Adapted from [20]. Copyright © [20], with permission from Krujatz F, et al.; (C) Adapted from [35]. Copyright © [35], with permission from Society of Photo-Optical Instrumentation Engineers (SPIE).

hence the use of PMTs in many laboratory setups. A typical laboratory setup for PL sensing is shown in Figure 19.2 (C).

19.2 Flexible OLEDs for Oxygen Sensors

Dissolved oxygen (DO) or gaseous phase oxygen sensors find numerous applications in research, biological, environmental, industrial, and medical application. For medical use, oxygen sensors are deployed in ventilators, anesthetic machines, and oxygen analyzers and monitors. There are several types of oxygen sensors in use today. The sensing mechanisms deployed include electrochemical (Clark electrode), resistive, thermoelectric, paramagnetic, and photoluminescence ref [34]. Most of the listed oxygen sensor technologies are more sensitive to higher oxygen concentrations or are designed to work at high temperatures. The standard Clark oxygen sensor uses membranes and electrolytes, requiring maintenance. They are also bulky, making them difficult to deploy for many applications. For most of the electro-galvanic fuel cell oxygen sensors used in medical equipment, the electrolytes are used up even if the sensor is not in use limiting their shelf life, even when refrigerated as recommended by many manufacturers.

A promising approach to more accurate and flexible oxygen monitoring is based on PL (see Section 19.1.6). The technique involves detecting and tracking the quenching of the PL intensity, *I* or the decrease in the lifetime, τ of an excited oxygen-sensitive dyes, e.g. ruthenium tris 4,7-diphenyl-1,10-phenanthroline chloride (Ru(dpp)), Pt octaethylporphyrin (PtOEP), or Pd octaethylporphryn (PdOEP). OLED-based PL oxygen sensors provides a path for miniaturized, sensitive, flexible, and field-deployable oxygen sensors that are particularly sensitive to low oxygen concentrations and work great at low temperatures [35]. They require minimal maintenance and are disposable at extremely low cost. For most optical detection schemes a common problem is the deployment of compact light sources and detectors (see Section 19.1.4). Early reports on this technique include those by Savvate'ev et al., in which the authors demonstrated the development of sub-second-fast fluorescence-based oxygen sensor, integrating an OLED with a chemical sensor platform in a very simple design [36], similar to Figure 19.2 (A) and (B). The design simply involves fabricating the OLEDs on

one side of a substrate (glass) and the sensing element on the other side of the same substrate. The fluorescence emission could be measured from the spaces between the pixels (which are aligned with a PD) [30] on the transparent indium tin oxide (ITO) side in the so called "back-detection mode" as shown in Figure 19.2 (A) and (B). Different configurations are possible as shown in the figure, which indicates some improvements made in oxygen sensor configuration for improved performance over the years. Choudhury at al., demonstrated oxygen sensors in the front- and back-detection configuration and oxygen detection in the gas and solution phase (DO) [37]. It is worth noting that a PMT was used as the PD in this study to enable lifetime sensing using a laboratory instrumentation setup similar to Figure 19.2.

Ghosh et al., integrated a tris-8-hydroxy-quinolinato aluminum (Alq_3) OLED, PdOEP-doped polystyrene (PS) oxygen-sensitive film and used a thin-film nc-Si-based photodiode as a detector. This is an improvement over previous work [36, 37] reported by the same group in which a less stable Ru(dpp) dye (less resistant to photobleaching and shorter lifetime) was embedded in sol gel films. Encapsulating the dye in PS has been shown to perform better than when the dye was encapsulated in sol gels [38]. Although the predicted linear SV dependence of I and τ was confirmed, these sensors are mostly deployable for PL intensity sensing. A major problem of sensing in the τ mode is the response time of the photodiodes ($\sim$250 μs and 2 ms for the (nc)-Si film-based PD and a-(Si,Ge):H-based PD) investigated by Ghosh et al.; compared to the nanosecond scale time response typical of PMTs and present-day Si pin photodiodes. As in the case of the glucose sensors (see Section 19.3), the response time for the photodiodes used in this work is too slow to monitor the decay of the dyes, which usually varies in the range of $\sim$ 5 μs to 1 ms for fluorophores with relatively long lifetimes. At the time of this writing, typical silicon pin photodiodes have response times ranging from 100 μs to 10 ns. With improvements in research and development into new materials, device geometries and operating conditions, organic photodiodes may soon be commercially available and capable of handling such speed requirements [18]. Peumans et al. reported the development of a multilayer OPD that could possiblly have a response time shorter than 1 ns [39]. Other researchers have shown OPDs with responses times in the 100–500 ns range [33, 40].

Nalwa et al. investigated an all-organic platform to demonstrate oxygen and glucose sensing in lifetime mode, comparing the response obtained from and LED to an OLED excitation, while using tuned P3HT:PCBM OPD (640 nm) as the detector [42]. The configuration is similar to that shown in Figure 19.2 (B) and a typical response of the sensor to O_2 is shown in Figure 19.2 (D). Figure 19.3 shows the response of the integrated sensing platform to oxygen and glucose, indicating a decrease in PL intensity and lifetime with increasing O_2 concentration; and a corresponding decrease in PL intensity with decreasing glucose concentration due to enzymatic reaction mechanisms (see Section 19.3). The authors also compared the response of their sensor when an ILED was used versus an OLED. The sensing film was a PtOEP encapsulated in PS with a peak emission at 640 nm. The fast response of the OPD enabled measurements in lifetime mode. Similar results were obtained by Mayr et al. [43, 44], where OPDs and the sensing element

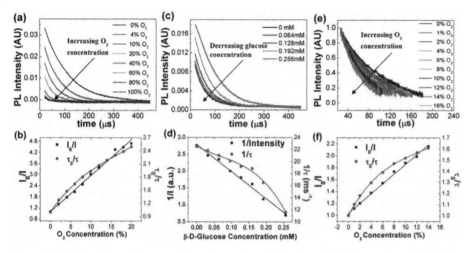

Figure 19.3 The effect of concentration of gas-phase O2 (a, e), and of glucose (c) on the OPD's temporal photocurrent response. The excitation sources were LED (for (a–d)) and OLED (for (e–f)). Adapted from [42]. Copyright © [42], with permissions from John Wiley & Sons, Inc.

were fabricated on the same substrate or on walls of a capillary tube, the walls of the tube acted as a waveguide for PL intensity and lifetime measurements.

Even as researchers continue to strive for improvements in various aspects of flexible organic oxygen sensors, focus will be on (but not limited to) the following: enhancing the light outcoupling from OLEDs [3] (e.g. using microlenses and microcavity OLEDs), eliminating the OLED's post-pulse (transient) EL tail, developing compatible fast response, low noise, and high external quantum efficiency (EQE) OPDs, for fully functional and standalone organic flexible sensing platforms. Oxygen sensing is very important for sensing other species including alcohol, lactate, and glucose as depicted in Figure 19.2 (B), using the appropriate enzymatic reactions (see Sections 19.3 and 19.5.2 for more details).

19.3 Glucose Sensing Using Flexible Display Technology

There is a vast array of glucose sensors available on the market today, most of which deploy electrochemical detection methods (amperometric or coulometric) [45]. Aside from the well-established use of glucose monitoring for managing diabetes, glucose sensing has significance in biochemical research and for industrial applications. Most glucometers monitor glucose (in blood samples obtained from a finger prick) by monitoring the release of electrons from the oxidation of glucose (amperometric) in the presence of glucose-selective enzyme, e.g. glucose oxidase (GOx) or by measuring the amount of charge required to oxidize glucose (coulometric). Continuous glucose monitors (CGM) on the other hand could be invasive (implantable) [46, 47] or noninvasive and measure glucose from urine, sweat, tears, saliva, or exhaled air [48]. Although many of these well-established glucose-sensing platforms have great sensing characteristics (accuracy, sensitivity, selectivity, and response time), they show significant deficiencies in terms of their short-term stability and shelf life, often requiring very frequent recalibration with a standard glucose meter [45]. Other drawbacks of available glucose monitors include the bulkiness of the device, warm-up period, and the use of battery-driven potentiostat circuits [45, 47]. These are significant considerations, especially for self-monitoring of blood glucose applications. Significant research has been ongoing in this area, with significant attempts at improving and optimizing glucose sensors. For example, research into functional semipermeable membranes, enzymes, self-powered units, and the development of non-enzymatic sensors employing precious metal catalysts or carbon nanotubes or nanowires [49, 50]. Glucometers, CGM, and multi-analyte sensing, (see Section 19.5.2), and industrial applications could benefit from applying flexible display technology toward glucose sensing. Such benefits could include miniaturization, reducing recalibration, and simplifying the sensor design by monitoring glucose concentration using PL intensity or lifetime. The use of OLED display technology enables integration and miniaturization, which could play a significant role in the development of next-generation glucose monitors. Choudhury et al., reported the structural integration of glucose sensors with individually addressable OLED pixels [51]. The authors developed a blue-emitting 4,48-bis-(2,28-diphenylviny1)-1,18-biphenyl rutheniu tris 4,7- dipheny1-1,10-phenanthroline chloride blue OLED-oxygen-sensitive dye (DPVBi/Ru(dpp)) and a green A1q3/PtOEP OLED-dye co-embedded with GOx in a thin film on a glass or plastic substrate or as a liquid in a microchannel for DO measurements. The sensing element was attached to the OLEDs back-to-back yielding a compact structure similar to that shown in Figure 19.2 (A) and (B). Standard excitation sources (Xenon, lasers, mercury) may render such applications cost prohibitive or too bulky for decentralized or point-of-need testing. Although the PL response (intensity, I and lifetime, τ) was monitored with a photomultiplier (PMT), which has an internal gain and enables a high signal-to-noise ratio (SNR), other researchers have since replaced the use of expensive/bulky detectors (PMT, CCD, and CMOS cameras) with organic photodiodes [18]. This has led to a more compact, compatible, and flexible point-of-use design, with a form factor suitable for one-time-use, disposable designs, wearable and high-throughput implementations or for in vivo glucose monitoring. The authors identify fluctuations in light intensity, instability of oxygen-sensitive dye, leaching and local vanations of oxygen as significant issues associated with PL-based sensing. These variations in the sensor could be combated by using ratiometric measurements or by simultaneous monitoring of glucose and oxygen. Personal glucose sensors based on flexible display technology are capable of measuring glucose in the normal range (65–105 mg/dL), hypoglycemic (< 65 mg/dL) and hyperglycemic glucose (> 105 mg/dL).

19.4 POC Disease Diagnosis and Pathogen Detection Using Flexible Display Optoelectronics

Flexible OLED POC immunosensors are devices that incorporate a biological sensing element usually an antigen (Ag) or antibody(Ab), a detector agent (usually colorimetric or fluorescent), an OLED excitation source, a PD (photodiode), and readout electronics, converting the biological parameters of a sample to an interpretable output. Central laboratory techniques and biosensors that operate based on optical transduction mechanism have become common place in disease diagnosis, vaccine screening, and drug treatment monitoring applications. Although there are other transduction mechanisms (electrochemical, mechanical, thermal, and electromagnetic), optical transduction offers a number of advantages, which include: a wide electromagnetic spectral range, measurement could be made in different modes (intensity, frequency, polarization, phase) [52], multi-analyte arrays could be easily fabricated (see Section 19.5), and finally, some of optical transduction techniques have high sensitivity and selectivity, e.g. fluorescence. Particularly, combining optical detection schemes with immunometric assays has been very successful at central laboratories yielding a high sensitivity and low limit-of-detection (LOD) for most analytical procedures (e.g. the gold standard enzyme linked immunosorbent assay, ELISA). Immunosensors take advantage of the high affinity and specificity of Ag–Ab interactions toward the detection of species in an analyte. Antibodies and Ags are also used as biomarkers required to make a clinical diagnosis. In some fluorescence applications, an Ab-conjugated fluorophore is used as the detector label, such that the fluorescence emissions (due to bound labels) are directly correlated to the concentration of the target species in a sample/analyte. Fluorescence has been widely used for several biological and chemical analytical procedures including proteomic analysis, DNA sequencing, cell studies, environmental monitoring, and food and water analysis. Some challenges associated with deploying central laboratory analytical tools at the POC include the high infrastructure and logistic cost, as well as the bulky form factor due to the expensive and complicated optoelectronic and automated optomechanical components required for high performance. Furthermore, very often these systems will require frequent calibration and alignment of the optics. Miniaturization of laboratory scale platforms is a potent approach to improve patient access and reduce diagnostic cost. Flexible display emitters are particularly suited for deployment in systems that rely on optical detection. Many groups have reported attempts to integrate optical detection schemes with paper-based, and polydimethylsiloxane (PDMS)/poly (methyl methacrylate) (PMMA)/glass microfluidics, using OLEDs as the excitatio source and inorganic or organic photodiodes (OPD) as detectors [12, 20, 21, 25, 26, 53–59]

Early reports on integrated OLEDs for fluorescence sensing include the work reported by Hofmann et al. in 2005 [54]. The authors reported the development of a disposable PDMS microchannel, integrated with a yellow OLED excitation source for the detection of urinary human serum albumin (HSA) as a marker for renal disease (microa-luminuria, MAU). The authors adopted a filter-less discrimination by using an orthogonal detection geometry, where the fluorescence emissions were captured using a fiber optic spectrometer, oriented orthogonally to the OLED excitation. Although this system demonstrated proof-of-concept toward a fully integrated device, the authors report an LOD of 10 mg/L and a linear range between 10 and 100 mg/L HSA. The linear range for the detection of HSA was clinically relevant (dinical cut-off levels: 15–10 mg/L). Later work by the same group [58] reported an injection molded (PS) two-channel microfluidic LOC fluorescence detection system for cardiac markers myoglobin and creatine kinase-muscle/brain (CK-MB). An LOD of 1.5 ng/mL for both myoglobin and CK-MB was reported using a combination of absorption filters, linear and reflective polarizers, an ILED (InGaN) as excitation source, and an OPD and Si PD as the detector.

Pais and Banergee et al. detailed the development of a disposable LOC device with integrated thin-film green OLED (Figure 19.4 (A)–(B)) as excitation source and an OPD as detector for fluorescence detection [26]. The system was designed with crossed polarizers (Figure 19.4 (E)). The excitation source was polarized with the first polarizer (polarizer 1) and the emitted fluorescence was collected through a second polarizer (polarizer 2) oriented perpendicularly to the first. The goal was to suppress the excitation source reaching the detector. The authors reported that the crossed polarizer scheme reduced the excitation light leakage by 25 dB while reducing the fluorescence emission by approximately only 3 dB. Rhodamine 6G and fluorescein were used as model dyes to test the system and an LOD of 100 nM and 10 nM for rhodamine 6G and fluorescein, respectively. This represents one of the first reports that demonstrated the potential toward fully integrated monolithic LOC system (Figure 19.4 (D)) comprising polarization filters, an

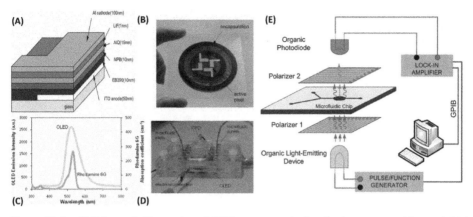

Figure 19.4 (A) Schematic illustration of OLED structures, indicating layer composition and thickness. (B) Photograph of the fabricated OLED. (C) Emission spectrum of the OLED and the absorption spectra of rhodamine 6 G dye. (D) Photograph of the integrated device. (E) Organic excitation/detection system with microfluidic device for POC fluorescence-based assays. Figure adapted from [60]. Copyright © [60], with permission from IEEE.

OLED excitation source, and a detector using an organic photodiode and not any of the standard detectors (PMT, CCD, CMOS). By improving the design of the OPD (10x responsivity increase), the authors demonstrated an improvement from 100 nM to 1 nM for rhodamine 6G (although with the use of an AC lock-in amplifier detection setup to reduce the noise and therefore increase the SNR) [60].

Sensor applications for early detection, however, require higher sensitivity and a lower LOD, especially for detecting markers that are minimally expressed. Hence, it is desirable that integrated devices perform at the level of clinical or central laboratory analytical devices to be useful for early detection at the POC. Additionally, many platforms under development still rely on external sources of pressure (e.g. syringe pumps, pipettes) to drive the reagents/reactants involved in the assay.

Recently we reported the integration of a nitrocellulose-based microfluidic platform with a high sensitivity fluorescence detection system for infectious disease [61]. We have also reported on the design, fabrication, and characterization of an OLED-based fluorescence detection platform using a silicon photodiode and interference filters capable of detecting 100 fM (~10 pg/mL) of Dylight549. The OLED multilayer stack, deposited by thermal evaporation on a flexible 125 μm thick DuPont Teijin film (Teonex polyethylene naphthalate, PEN), an ITO patterned flexible plastic substrate, temporarily bonded to a rigid alumina carrier. The layer composition and thicknesses are shown in Figure 19.5 (A). The high performance of this system is directly linked to the use of bright OLEDs, high-quality interference filters, and charge-integration readout electronics [4, 25, 62]. Here we substitute the typical bulky light sources with a thin-film flexible OLED excitation; transfer optics by adopting a "face-to-face" transmission architecture (Figure 19.5 (C)); and PMTs with a photodiode. Compared to the work reported by Pais et al., we substitute the complicated bulky readout electronics (preamplifiers and lock-in amplifiers) with a simple and efficient charge-integration readout. The constant current generated (due to fluorescence emissions reaching the photodiode) is stored as charge on a capacitor (current or charge-to-voltage converter). This architecture averages out the noise and trades time for accuracy, such that the system yields a concentration-dependent output voltage monitored by a

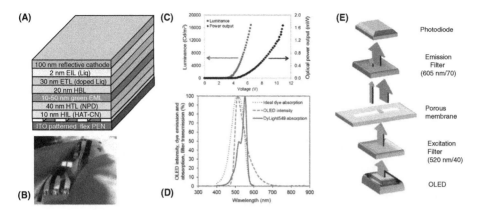

Figure 19.5 (A) Flexible bottom emitting ASU green OLED device test structure; illustration indicating layer composition and thickness. (B) Photograph of the fabricated green OLED on a flexible PEN substrate. (C) Luminance and optical power output of OLED as a function of voltage (D) Emission spectrum of the OLED, ideal (matching) dye absorption (Nile red) and the absorption spectra of Dylight549. (E) Schematic of assembled fluorescence detection platform using interference filters with paper-based microfluidics for POC fluorescence-based assays.

simple and inexpensive microcontroller platform. A significant problem with most LOC systems is the difficulty in deploying a standalone system for field application (as such, many bright ideas eventually end up on the shelves or in cabinets in laboratories, or at most restricted to laboratory use only). The architecture reported by Katchmann et al. enables a field-deployable platform for disease diagnosis, vaccine screening campaigns, and drug treatment monitoring. Combining our flexible display and protein microarray technology, we have demonstrated the utility of this platform for the detection of IgG antibodies to multiple viral antigens in patient sera [25]. We envision a low-cost multiplexed disposable diagnostic platform, enabled by the monolithic integration of our filter design with flexible display technology platform, protein microarray printing, and microfluidic technologies (for more details on multiplexed detection using OLEDs, see Section 19.5).

Recently, Shu et al. reported the first integrated fluorescence sensing system fabricated fully by (vacuum free) solution processing [12]. A blue OLED, OPD, and sensor integrated on a glass chip (with a glass channel) was developed, for potential applications in fluorescence sensing. The system was tested with a model dye (fluorescein amidite, FAM) using linear (crossed) polarization filters to suppress the excitation light reaching the detector and isolate fluorescence emissions. The advantage of using crossed polarizers is that such a system could be used for detection of multiple fluorophores at different excitation and emission wavelengths. Thus, compared to interference or absorption filter-based systems, the crossed polarizer-based detection system is not limited to use with fluorophores that match with a specific portion of the electromagnetic spectrum, (as with interference filter-based systems) dictated by the bandpass filters used. However, with the crossed polarizer-based system reported by Shu et al., a 1 μM (~0.33 mg/mL) LOD was reported.

From a practical design standpoint, minimizing the separation between the emitter, detector, and sensing element or microfluidic chamber, is essential to maximizing the in-coupling of the excitation light toward the sample and the out coupling of the emitted fluorescence reaching the detector. Effective design steps should be targeted toward directing the isotropic fluorescence emissions toward the detector [28], while minimizing the excitation light reaching the detector (see Section 19.1.4) to increase the SNR. This approach ensures that the use of complicated transfer optics is minimized, enabling the use of small pixel size OLED arrays not only for high-throughput applications but also for multispectral analysis.

Venkatraman et al. reported the development of an integrated OLED-lateral flow immuno-assay (OLED-LFIA) with a nitrocellulose membrane [21]. The authors used color plastic filters (absorption filter) adhesively bonded to the OLED (on PET) and an emission filter positioned on top of the LFIA strips. The authors investigated the use of quantum dots (QD) for improved sensitivity in a model immunoreaction assay; mouse (anti-flu) and (donkey) anti-mouse conjugate. The performance (visual sensitivity) of the QD-LFIA was compared to the conventional gold nanoparticles (AuNP). However, both assays were designed for visual observation-based qualitative LFIA, upon excitation with an OLED. Ambient light photographs of the test the development of the test line takes

Image J software. An LOD of 3 nM and 21 nM using the QD-LFIA and Au-LFIA labels, respectively, was reported, representing a 7X improvement using the QD over AuNP. The reported LOD could be improved by incorporating effective design strategies, e.g. those detailed by Obahiagbon et al. [24]. The same group [59] reported improvements to their previous work by integrating OLEDs with a LFIA and an OPD (Figure 19.6 (A)). However, this time the application was the development of a quantitative colorimetry (AuNP) transmission mode detection configuration (through the paper-based membrane). The intensity of the test line is measured over time as the analyte reaches and binds to the test line. In other words, the development of the test line takes ~20–30 minutes and the signal intensity is monitored during this time as shown in Figure 19.6 (B) and (C). With the transmission mode configuration, the intensity decreased over time as expected.

The current reaching the OPD decreases, since the concentration of the AuNP bound on the test line increased over time, scattering more light away from the detector. Although not stand-

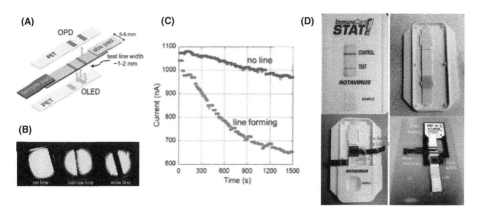

Figure 19.6 (A) Schematic diagram of integration approach of organic optoelectronics (light emission and detection) with LFIA (5–6 mm wide and 60–100 mm long). (B) Optical density of test line under green LED illumination. (C) OPD photocurrent of LFIA test line over time. (D) Test and control lines formation, integration of organic optoelectronic devices (OLED and OPD) with LFIA strip inside the cassette, and concept for fully integrated opto/microelectronics LFIA package. The OLED and OPD are on separate substrates, but have the same dimensions and are carefully superimposed. Adapted from [59]. Copyright © with permission from IEEE.

alone-operational, the authors also presented a concept in which the OLED and OPD (fabricated on a flexible substrate) would be integrated with commercially available rotavirus test kits (LFIA test for rotavirus in a cassette) as shown in Figure 19.6 (D). The required signal control, processing, and readout electronics issues need to be resolved with the rotavirus detection system using OLEDs and OPDs and enhanced sensitivity will be required for clinical level performance.

The integration of flexible organic optoelectronics with microfluidic LOC technologies provides a path toward miniaturization and deployment of inexpensive and ultimately disposable POC devices. We envision that within the next decade several products applying flexible display technology for disease diagnosis will become commercially available.

19.5 Flexible Display Technology for Multi-analyte Sensor Array Platforms

19.5.1 Integrated LOC and Flexible Display Devices

Central to the concept of LOC systems is the development of integrated micro-analytical platforms. The goal is to miniaturize central laboratory processes, reduce sample volume and processing time, at low manufacturing cost. This concept ultimately yields disposable inexpensive tests especially for POC diagnostic applications. Organic electronics are particularly suited for miniaturization since they are deposited on planar/flexible substrates at low temperature. Additionally, they can be patterned, and fabricated by processes that are compatible with mass manufacturing. The low-temperature processing of flexible displays enables fabrication of emissive devices on plastic, glass, fabrics, polyimide, and a host of other flexible substrates. Hence, organic optoelectronic devices could be integrated with microfluidics for LOC applications, especially for optical detection [26, 54, 64]. Flexible organic optoelectronics can be deposited directly on polymers used in microfluidics (e.g. PDMS) or on substrates that can easily be bonded directly to microfluidics and LOC devices [5]. The overall flexibility and robustness of organic display technology makes it suitable for integration with LOC devices for a wide array of applications in healthcare and medicine. Integration is enabled by advancement in research and commercialization of these respective technologies. For example, the pixel size in flexible display technology is on the order of the size of a typical microfluidic channel. Hence, there are significant benefits in applying the concepts and designs in microfluidics, LOC, and flexible display technology, toward the development of integrated devices. One such benefit is the development of multi-analyte sensors and multiplexed high-throughput analytical platforms for applications in personal health monitoring, infectious disease detection, and vaccine screening. Such is the focus of Section 19.5.2.

19.5.2 Multiplexed Sensor Platforms

The inherent design and characteristics of flexible display technology (including OLEDs and TFTs) are suitable and meet the requirements for multiplex biomarker detection. A two-dimensional array of pixels can be patterned and individually addressed, targeting/probing different detection sites or a group of sites, for low sample volume and high throughput applications. We have explored the potential of combining high density protein microarray printing, developed at the Virginia G. Piper Center for Personalized Diagnostics, with flexible display technology developed at the Flexible Electronics and Display Center (FEDC) (both at Arizona State University). Each element in the array consists of pitch-matched OLED-photodiode pairs in a sandwiched transmission-mode configuration as conceptualized in Figure 19.7 (A), yielding a compact and highly sensitive device. This approach eliminates the need for transfer/focusing optics (lenses, fiber, beam spitters), motorized stages, or expensive low-light detectors and emitters [4, 24]. High sensitivity and low LOD could be achieved using OLEDs as excitation source to probe fluorescent markers as discussed in Section 19.4. An added benefit of a fluorescence-based assay is a clear path to developing a quantitative readout, which aids clinicians in making a diagnosis. Miniaturized fluorescence detection platforms using OLEDs for multiplexed detection of infectious diseases, is not yet commercially available. With heightened interest in this area by academia and industry, it is not hard to imagine that such devices may become available in a couple of years.

There have been several efforts to develop multi-analyte sensors using OLEDs over the years. Notably, Shinar et al. reported extensively on the detection of oxygen and used the same principles in the development of an enzyme-based multi-analyte array [27, 38] for detection of glucose, lactate, ethanol, and oxygen by PL intensity quenching or decay time decrease of an oxygen-sensitive dye (see Sections 19.1.6, 19.2, and 19.3). Detection of DO plays an important role for the detection of glucose, lactate, and ethanol, since their concentration could be determined by sensing the concentration of DO after the completion of the oxidation reaction by glucose oxidase (GOx), lactate oxidase (LOx), and ethanol oxidase (AOx), respectively (see Figure 19.2 (B)). The authors developed 2 x 2 mm OLED pixels and integrated the sensing film (dye and enzyme in PS sealed cells) attached to a glass slide. The slide was then attached to the OLED array, back-to-back, eliminating the need for any optical couplers or focusing optics. The performance was comparable to other detection schemes that use standard light sources. The ease of fabricating the OLEDs enables a compact sensing array for multiple analytes and presents a viable alternative to multiplexed sensors designed using microelectrodes where specificity could be a significant challenge. The ability to fabricate OLEDs of different colors on

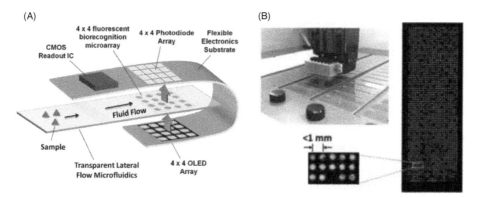

Figure 19.7 (A) LOC flexible hybrid electronics concept configuration combining an integrated 4 x 4 OLED array for optical excitation and a 4 x 4 photodiode detector array with protein microarray technology for multiplexed fluorescence analysis and biorecognition of 16 different disease biomarkers. (B) Micro-scale printing of high sensitivity fluorescent biorecognition protein microarrays on microscope slides for multiple biomarker detection at the Biodesign Institute, Arizona State University. Figure (B) reused from [4]. Copyright © (2016), with permission from IEEE.

the same substrate, forming an array, introduces a wide range of opportunity in enhancing the robustness of OLED array platforms, increasing accuracy via redundancy and multispectral analysis. With proper isolation, which can be integrated into the array fabrication process, it is possible to individually address the OLEDs in the array (sequentially) or probe all the elements of the array consecutively (in parallel). Later, the same group reported an improvement to their oxygen sensor using directionally emitting, narrow band microcavity OLEDs for simultaneous monitoring of pH and oxygen [3].

Other groups have reported the development of OLED-based multi-analyte arrays for sensing O_2, CO, temperature, pH, and ammonia [22, 44].

19.6 Medical Diagnostic Displays

Displays play a vital role in modern-day medicine. Some application modes of medical-grade displays include mammography, surgical/intervention, dentistry, digital pathology, general radiology, and multimodal applications. Some specific application areas include intravascular ultrasound (IVUS), cross-sectional MRI/CT, fMRI, and surgical displays, each with different monitor specifications/recommendations for different imaging modalities [65]. Presently, the dominant technologies for monitors/displays are LED/cold cathode fluorescent lamps backlit LCD displays and OLED displays. The high contrast, color reproduction, brightness, light weight, thinness, high refresh rates, and wide viewing angles are desirable properties for OLED-based surgical and radiology monitors. Sony unveiled the world's first medical grade OLED monitor in 2012, other companies including LG have OLED-based products commercially available. High resolution monitors (3, 5, 8, and 10 MP) monitors are now available to radiologists and surgeons for making diagnosis and performing surgical interventions. The demand for thinner, larger, lighter, higher brightness, contrast, color, lower power, and lower cost continues to drive the development of alternate display technologies. Advances in resolution (pixel pitch), screen size, color, lighting remote calibration, and management will continue. Cathode ray tube (CRT) display is a mature and now extinct technology (replaced largely by LCD technology) despite several decades of development and significant investment. It is interesting that the (previous gold) standard CRT technology has now disappeared commercially. The quest for the "ideal medical-grade display" will continue for a long time [66]. OLED technologies are still being plagued by cost and yield challenges.

The race is on at the moment between LCD and OLED technologies [7], the future will tell!

19.7 Wearable Health Monitoring Devices Based on Flexible Displays

In the near future wearable electronics will become an integral part of patient care. These devices have the potential to play an important role in the relationship between healthcare providers and patients, by providing noninvasive continuous physiological monitoring. Today, patients who desire to take proactive steps toward monitoring their health are able to do so with any of the commercially available wearable trackers and smartphones. In 2013, Samsung and LG started producing flexible AMOLED displays on polyimide substrates for mobile phones and wearable devices. This revolutionary advancement was used in devices like the Galaxy S7 edge, LG G Flex

2, and more recently in Apple watch series 3 and the iPhone X. AMOLEDs are now being mass produced for applications in the mobile phone, portable tablets, and wearables space, with Samsung producing about 9 million flexible OLEDs per month. Creativity is endless with companies like Royole reporting the development of a 10-μm-thick flexible OLED full-color display and touch-sensor platform with a 1-mm bend radius. Conformal characteristics of OLEDs make them suitable for wearable applications. In the sections that follow, we explore the applications of flexible display technology for monitoring vital signs, light treatment, and wearable devices.

19.7.1 Monitoring Vital Signs Using Flexible Display Technology

Personal health monitoring has received a lot of attention and investment over the years. This is partly due to the fact that individuals continue to seek higher involvement and play a more proactive role in their healthcare. Physical sensors have been around for a long time and have been used to monitor parameters like temperature, heart rate, blood pressure, and oxygenation. Other flexible platforms being investigated and developed to monitor physical parameters are presented elsewhere [67]. In healthcare, it is vital to be able to accurately and conveniently measure or monitor physical signatures using sensors that are in contact or in close proximity to the skin. The use of photoplethysmography (PPG) or pulse oximetry is a well-established technique for monitoring heart rate and arterial blood oxygenation. Nevertheless, research in developing more accurate, reliable and easy-to-use sensors has continued to be the center of focus for researchers and industry. The PPG technique measures heart rate and capillary oxygen saturation (SPO_2) by measuring blood volume changes or taking advantage of the change in absorption coefficient of oxygenated (absorbs in green and blue, reflects red) and deoxygenated blood. These sensors are usually configured in two modes: transmission and reflection mode oximetry (see Section 19.1.4). Both techniques use light at two wavelengths for a good reference measurement. Typically, in transmission mode, photodiode(s), red and near IR LED pairs on opposite sides of tissue are used because of higher penetration at longer wavelengths. Reflection mode typically uses a photodiode, green, and red LED pairs on the same side of the tissue. Currently available trackers use the latter configuration. The performance (reliability and reproducibility) of PPG sensors is greatly impacted by artifacts. Posture, movement, skin type, and location of the sensor on the body could significantly impact the result [68]. The reflection mode is known to be more susceptible to noise and absorption in tissue. The mechanical flexibility and ease of implementing large arrays with flexible display technology could help mitigate many of the challenges with PPG platforms. The flexibility of OLED technology helps establish conformal contact with the skin and other soft human tissue improving the SNR for most sensors. Application of flexible display technology enables high-volume production, improved performance, and reduced cost.

Smith et al. explored the possibility of reducing manufacturing cost and improving the diagnostic functionality of wearable biomedical devices using flexible flat panel display technology [69]. This effort included the development of a prototype smart bandage-style PPG using flexible OLEDs and a thin-film pin photodiode for heart rate monitoring (Figure 19.8 (A)–(C)) using green and red OLEDs paired with (a-Si)-flexible pin photodiodes in a two-dimensional array. This configuration has the potential to overcome some of the limitations of PPG sensors, especially artifacts due to alignment and positional dependence, and motion artifacts, improving the SNR.

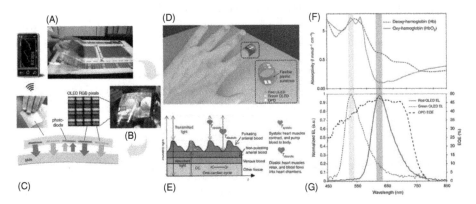

Figure 19.8 (A) Flexible displays on Gen2 370 x 470 mm plastic substrate being debonded from temporary rigid alumina carrier. (B) 7.4 inch diagonal full-color flexible OLED display manufactured at Arizona State University FEDC. (C) Concept for conformal Bluetooth® connected disposable optical heart rate monitor smart bandage manufactured using flexible OLED display technology with alternating red and green OLED and a-Si pin photodiode pixels arranged in two-dimensional PPG sensor array. (D) Pulse oximetry sensor composed of two OLED arrays and two OPDs. (E) A schematic illustration of a model for the pulse oximeter's light transmission path through pulsating arterial blood, non-pulsating arterial blood, venous blood and other tissues over several cardiac cycles. The a.c. and d.c. components of the blood and tissue are designated, as well as the peak and trough of transmitted light during diastole ($T_{diastolic}$) and systole ($T_{systolic}$), respectively. (F) Absorptivity of oxygenated (orange solid line) and deoxygenated (blue dashed line) hemoglobin in arterial blood as a function of wavelength. The wavelengths corresponding to the peak OLED electroluminescence (EL) spectra are highlighted to show that there is a difference in deoxy- and oxy-hemoglobin absorptivity at the wavelengths of interest. (G) OPD EQE (black dashed line) at short circuit, and EL spectra of red (red solid line) and green (green dashed line) OLEDs. Figures (A) to (C) adapted from [69], Copyright © [69], with permission from Electronic Letters. (D) to (F) adapted from [70]. Copyright © [70], with permission from Springer Nature.

Arias et al. developed an all-organic solution processed optoelectronic finger sensor for pulse oximetry [70]. A green (532 nm), red (626 nm) OLED, and OPD fabricated on flexible substrates was used to develop a transmission-mode pulse oximeter to determine the ratio of oxygenated to deoxygenated hemoglobin in blood (Figure 19.8 (D)–(G)).

The authors reported a higher performance than the standard NIR-LED-based pulse oximeters with 1% and 2% error for pulse rate and oxygenation status, respectively. The application of flexible display technology for physical sensors continues to generate interest in the research community, leading to several publications [71, 72]. We envision that some of these sensors will soon begin to find their way into commercial products in the near future, perhaps integrated with other stretchable and flexible platforms [73]. These sensors will potentially monitor human vitals, activity, and give feedback to healthcare providers. Cutting-edge advances like the development of a 3-μm thick tattoo-like ultraflexible organic photonic skin will continue to emerge. Yokota et al. reported the development of a green and red polymer light-emitting diode (PLED) [74] and integrated the PLEDs with an OPD to develop an ultraflexible reflective mode pulse oximeter. Generally, PPG sensors require calibration and measurements are susceptible to artifacts. For example, most measurement setups do not account for light scattering in tissue. Advances in signal processing for pulse wave analysis and developments in the flexible and wearable space, wireless technologies, internet of things (IoT), and smart connected health will continue to drive the efforts toward the development of more compact, portable, low-cost, integrated, and accurate PPG sensors. Again, designers could capitalize on the tunability of OLED and OPD response to maximize the sensors' SNR (compared to their inorganic counterparts – LEDs and inorganic PD – which are not easily tuned) and thus improve the overall accuracy. These flexible sensors will not only be limited to peripheral sensing (fingers and ears) but also have the potential to be placed on any part of the body.

19.7.2 Flexible Display Technology for Phototherapy

Phototherapy has been demonstrated to be an effective method of treating certain medical conditions. Some long wavelength light (630–1000 nm) application areas include stroke, myocardial infarction, wound healing, pain relief, prevention of oral mucositis in cancer patients, hatching rate and survival of chickens, nasal symptoms of allergic rhinitis [75], skin cancer and shorter wavelength therapy (~390–470 nm) for neonatal jaundice (to help babies get rid of excess bilirubin) [76]. The application of flexible display technology for phototherapy greatly simplifies the integration of light sources into wearable items for various light therapy applications. For example, the simplicity of treating hyperbilirubinemia using simple light therapy could be greatly improved with added functionality by integrating flexible display technology in today's "bili-light machines or bili-blankets." Since the development of the first phototherapy machine by Cremer et al. [77], using 40-W fluorescent bulbs, researchers have used several light sources (halogen bulbs, fluorescent light, and ILEDs) in an effort to increase the effectiveness of the treatment by increasing the intensity.

Flexible display technology holds great promise for applications in this area, as OLEDs have the required intensity, are flexible and could easily be integrated and fabricated on large areas, eliminate unnecessary heating of the patient, are inexpensive especially for poor-resource settings, and have tunable wavelength across the visible spectrum. A "multiparametric bili-blanket" for maximum area and intensity exposure, that also measures other physical parameters (temperature, heart rate, and hydration levels) is easily conceivable, including additional functionality like skin bilirubin measurements [78].

Previously, we reported the potential application of flexible display technology for optogenetic neurostimulation of different branches of the auricular vagus nerve (via the outer ear) for a noninvasive prescription-drug-free treatment of chronic inflammatory disease and metal disorders [79]. The system is enabled by a flexible and addressable two-dimensional blue (455-nm) OLED array required to stimulate specific afferent vagus nerve fibers in the outer ear. Standard electrical transcutaneous and implantable therapies are either too large to match the size of specific nerves for individual stimulation or too invasive a surgery for many patients except as a last resort. A conceptual configuration of our approach is shown in Figure 19.9 (A). This technology relies on optogenetics, which is a relatively new biological technique used to control genetically modified-light-sensitive ion channels in neurons, essentially selectively turning neurons on or off. Anatomically, the vagus nerve provides a path to stimulating desired portions of the brain. As proof-of-concept, we demonstrated an in vitro optical stimulation of cultured neurons using cells expressing channelrhodopsin-2 (ChR2-YFP), a light-switched cation-selective channel that opens upon absorption of photons. Figure 19.9 (B) shows an increase in spike rate when the tissue is illuminated with a 1-mW/mm^2 blue OLED. It is expected that a red activated opsin would give a higher response because of the higher penetration depth of red light in tissue and the higher intensity/EQE obtainable from the red OLEDs (Figure 19.9 (C)).

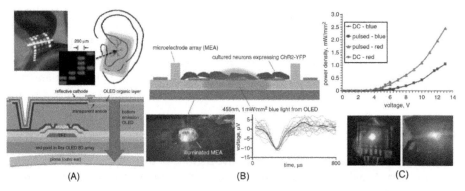

(A) (B) (C)

Figure 19.9 (A) Concept for transcutaneous auricular vagus nerve optogenetic neurostimulator using flexible red OLED display technology to treat mental health disorders and chronic inflammatory disease using disposable biophotonic smart bandage applied to patient's outer ear. Reconfigurable, two-dimensional, optical stimulus pattern of activated red OLED pixels targets specific auricular branches of vagus nerve while not stimulating neighboring off-target afferent nerve fibers to minimize side effects. (B) Experimental test configuration with 32-channel microelectrode array (MEA) connected to Intan neural recording system used to demonstrate 1 mW/mm^2 of pulsed 455-nm light from blue OLED test structure can optically stimulate cultured neurons in vitro using cells expressing ChR2-YFP. Also, shown (graph) is the time aligned neural spike activity under blue OLED optical stimulation. Green and black waveforms are the mean values of all recordings captured before (green) and after (black) optical stimulation. (C) Red and green OLED light optical output (intensity) against forward bias voltage for our 5-mm^2 flexible OLED test structures. Adapted from [4]. Copyright © 2015, with permission from Electronic Letters.

19.7.3 Smart Clothing Using Flexible Display Technology

Wearable electronics in the form of textiles [80] provides a means for large area mobile sensor networks with numerous implications for monitoring physiological parameters and for IoT. The future of clothing will perhaps no longer be limited to the basic functions such as covering, fashion, and protection. An ever-growing trend is the desire to integrate advanced function into clothing (functional clothing/E-textiles), such as communication capability, power generation, full body sensors, electrode stimulation systems, transdermal therapy, and augmented reality. Classical electronic circuit design using silicon and other inorganic semiconductors poses significant challenges including stiffness, difficulty in producing large-area coverage, and high cost. To overcome this problem, researchers have started investigating the integration of organic electronics in clothing to provide more flexible and low-cost unobtrusive wearable textiles. Integration of OLEDs with other flexible conducting polymers and displays (electrochromic) [81] will most likely revolutionize clothing as we know it [82]. This provides an opportunity for medical applications beyond monitoring and control of physiological parameters into the diagnostics space. For example, functional clothing with integrated sensors capable of monitoring or probing the contents of sweat for disease biomarkers, wearable body sensors for health monitoring in sports and fitness applications, as well as functional clothing for the incontinent, kids, and for older individuals [83, 84].

Early reports integrating OLED display technology in textiles include the work by Jamietz et al., in which the authors detailed the integration of encapsulated OLEDs into textiles [85]. The authors explored the use of solution processing for this purpose, which provides a path for integrating the OLED and textile manufacturing processes ("filamentary OLED") – fabricated on a glass fiber. In their work, the authors also integrated OLED using flexible foil taps embroidery and spacer warp

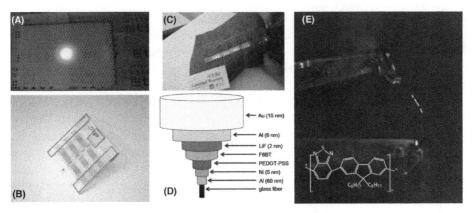

Figure 19.10 (A) Integrated solid encapsulated OLED. (B) Laminated, encapsulated OLED substrate on the PEN foil with stripped contact pads. (C) Integration of the OLED stripe on textiles and the corresponding testing of the OLEDs. (D) Scheme of the cylindrical/filamentary OLED-setup. (E) Photograph of a working OLED using a glass fiber as substrate. Adapted from [85]. Copyright © [85], with permission from Trans Tech Publications Ltd.

knitting (see Figure 19.10 (A)–(C)). Depositing OLEDs on a cylindrical substrate by dip-coating, the authors overcame the challenges with fabricating OLEDs on a small diameter cylindrical substrate, demonstrating the fabrication of a glass-fiber OLED (see Figure 19.10 (D)–(E)).

More recently, Choi and co-workers at the Korea Advanced Institute of Science and Technology [63] developed a very highly flexible OLED integrated-clothing-shaped technology for wearable displays. Using a combination of thermal evaporation, spin coating, and ALD, the authors fabricated flexible fabric-based OLEDs with a bending radius of 2 mm, high luminous intensity, and efficiency. PEN fibers at a pitch of 11 μm were woven into a soft fabric with 520-μm thick fibers. The mechanical simulations performed estimated the Young's modulus and Poisson's ratio to be ~3.5 GPa and 0.3, respectively.

Integration of optical and electrochemical transduction-mode fabric-based sensors will enable diagnostic capabilities in clothing. Functional wearable clothing platforms have a vast array of applications, improving quality of life. Sensing and monitoring technologies could be integrated into clothing by simply using coated fabrics as substrates, attaching fabricating devices directly to clothing, or better still, by integrating them into the fabric manufacturing process [86]. Finally, issues with flexibility, integration into fabric/textile manufacturing (knitting, weaving), reliability, and autonomy need to be resolved.

19.8 Competing Technologies, Challenges, and Future Trends

ILEDs have been used in numerous bioanalytical applications and are now found more often in central laboratory analytical instruments and numerous applications [16, 75, 17]. A lot of development and investment has gone into their development. While OLEDs have significant advantages over ILEDs (including flexibility, ease of integration, and fabrication), ILEDs have a longer lifetime and a narrower electroluminescence spectrum. Hybrid platforms may take advantage of both technologies. The invention of the quantum dot LCDs represents a major competition on the medical displays front as well.

Flexible, curved, rollable, and foldable displays could theoretically be made using LCD, OLED, and EPD technologies; however, flexible LCDs and EPDs have not been commercialized yet [1]. There has been an aggressive investment in OLED flexible displays particularly AMOLEDs. The very thin and flexible platform provides a path for other application areas, especially in the design of biomedical sensors.

Self-powered sensors and IoT applications will most likely take the center stage in the near future. The ease of fabricating and integrating organic solar cells with organic flexible display technology will play a pivotal role in the development of self-powered sensors. The development of flexible organic TFT technology (OTFT) will enable faster and truly flexible TFT, opening more avenues for truly wearable technologies. Applications in virtual/augmented reality gaming will see a lot of growth even as other areas like emotional lighting continues to gather momentum. For example, the development of self-powered flexible OLEDs with integrated organic solar cells coated on windows and public spaces could be used in the development of behavioral sensors, to control mood [87]. As researchers continue to explore simpler manufacturing processes like screen printing, inkjet printing and solution processing, sensor development stands to benefit a great deal from these techniques. Easily printed flexible OLEDs may become a reality in the future. As technology advances, the use of smartphones, modular designs, and attachments will likely increase in their use. Capitalizing on the OLED displays, LEDs and cameras on smartphones for diagnostics and medical applications will likely see an increase as well. Coupled with integration for multi-analyte sensing, multimodal sensors on wearable devices will enhance the robustness and accuracy of OLED-based sensor platforms. There may be great value in integrating electrochemical and optical sensors on the same platform to take advantage of the benefits of both technologies. For example, fabricating flexible ion-sensitive field-effect transistor-based sensors [88] using TFT technology, alongside optical sensors for glucose monitoring.

19.9 Conclusion

Flexible display technology has become a part of our everyday lives, from smartphones and TVs to fitness trackers and other biomedical sensors. The potential of flexible display technologies (especially those based on organic electronics) and the utility in biomedical sensor design has been explored. With numerous applications in medicine and healthcare, organic optoelectronics could play a pivotal role in biomedical sensor design and implementation. Ongoing research into organic optoelectronics sensors for medicine, healthcare, and diagnostics [89] will continue to define new paradigms in the years to come. The desirable properties of organic optoelectronic material and the relative simplicity of its fabrication processes make it more likely that researchers will continue to develop more flexible platforms for medical applications based on organic-type displays rather than LCD technology (liquid crystals and inorganic LEDs). A versatile tool kit is being developed as researchers continue to forge a path toward the improvement of all aspects of OLED sensing platforms. So far, there have been several advances in the potential use of flexible display technology for biomedical sensors including monolithic integration of OLEDs with sensors and OPDs, OLED-based spectrometers in the visible spectrum, development of microcavity OLEDs, a better understanding of encapsulation, wavelength tuning, optical waveguiding potentially replacing filters, enhanced light extraction (higher external quantum efficiency using corrugated substrates), and evanescent wave schemes using OLEDs [18, 19, 28, 90–92]. These milestones represent significant steps toward achieving robust, accurate, sensitive, integrated, and field-deployable sensors based on flexible display technology.

Acknowledgment

Funding: This work was supported in part by the National Science Foundation Smart and Connected Health [grant number ITS-1521904] SCH: TNT: "Disposable high sensitivity point-of-care immunosensor for multiple disease and pathogen detection" and National Cancer Institute Cancer Detection, Diagnosis, and Treatment Technologies for Global Health [grant number CA211415]: "Rapid point-of-care detection of hpv-associated malignancies."

Conflicts of Interest

Dr. Anderson is a consultant and has stock options with Provista Diagnostics and is on the institutional patent submissions for HPV serologic testing. Other authors report no conflicts.

References

1 Jerry, K. (2016). Flexible display market tracker. Techreport, IHS Technology. http://news.ihsmarkit.com/press-release/technology/flexible-amoled-market-more-tripled-12-billion-2017-ihs-markit-says.

2 Technavio. (2017). Global flexible displays market 2017–2021. Technical report, Technavio. https://www.technavio.com/report/global-displays-global-flexible-displays-market-2017-2021.

3 Liu, R., Cai, Y., Mok, P.J., Ming, H.K., Shinar, J., and Shinar, R. (2011). Organic light emitting diode sensing platform: Challenges and solutions. *Adv. Funct. Mater.* 21(24): 4744–4753.

4 Smith, J.T., Katchman, B.A., Kullman, D.E., Obahiagbon, U., Lee, Y.-K., O'Brien, B.P., Raupp, G.B., Anderson, K.S., and Blain Christen, J. (2016). Application of flexible OLED display technology to point-of-care medical diagnostic testing. *J. Diag. Test.* 12(3): 273–280.

5 Williams, G., Backhouse, C., and Aziz, H. (2014). Integration of organic light emitting diodes and organic photodetectors for lab-on-a-chip bio-detection systems. *Electronics* 3(1): 43–75.

6 Tang, C.W. and VanSlyke, S.A. (1987). Organic electroluminescent diodes. *Appl. Phys. Lett.* 51(12): 913–915.

7 Chen, H.-W., Lee, J.-H., Lin, B.-Y., Chen, S., and Wu, S.-T. (2018). Liquid crystal display and organic light-emitting diode display: Present status and future perspectives. *Light Sci. Appl.* 7: 17168. EP Review Article.

8 Camou, S., Kitamura, M., Gouy, J.-P., Fujita, H., Arakawa, Y., and Fujii, T. (2003). Organic light emitting device as a fluorescence spectroscopy's light source: One step towards the lab-on-a-chip device. *Proceedings of the 5th International Conference on Applications of Photonic Technology*, 4833: 1–8.

9 Kalyani, N.T., Swart, H., and Dhoble, S. (2017). History of organic light-emitting diode displays. In: *Principles and Applications of Organic Light-Emitting Diodes (OLEDS)* (ed. N.T. Kalyani, H. Swart, and S. Dhoble), 205–225. Woodhead Publishing Series in Electronic and Optical Materials, Woodhead Publishing.

10 Buckley, A., Underwood, I., and Yates, C. (2013). TT The technology and manufacturing of polymer {OLED} on complementary metal oxide semiconductor (CMOS) microdisplays. In: *Organic Light-Emitting Diodes (OLEDS)* (ed. A. Buckley), 459–511. Woodhead Publishing Series in Electronic and Optical Materials, Woodhead Publishing.

11 Pierre, A. and Arias, A.C. (2016). Solution-processed image sensors on flexible substrates. *Flex. Print. Electron.* 1(4): 043001.

12 Shu, Z., Kemper, F., Beckert, E., Eberhardt, R., and Tunnermann, A. (2017). Highly sensitive on-chip fluorescence sensor with integrated fully solution processed organic light sources and detectors. *RSC Adv.* 7: 26384–26391.

13 Crawford, G.P. (2005). *Flexible Flat Panel Display Technology*, 1–9. Wiley-Blackwell.

14 Deshpande, R., Pawar, O., and Kute, A. (2017). Advancement in the technology of organic light emitting diodes. International Conference on Innovations in Information, Embedded and Communication Systems (ICIIECS), 1–5.

15 Nikam, M., Singh, R., and Bhise, S. (2012). Organic light emitting diodes: Future of displays. *Int. J. Clin. Audit* 3(3): 31–34.

16 Bui, D.A. and Hauser, P.C. (2015). Analytical devices based on light-emitting diodes — A review of the state-of-the-art. *Anal. Chim. Acta.* 853: 46–58.

17 Yeh, P., Yeh, N., Lee, C.-H., and Ding, T.-J. (2017). Applications of LEDs in optical sensors and chemical sensing device for detection of biochemicals, heavy metals, and environmental nutrients. *Renew. Sustain. Energy Rev.* 75(461): 468.

18 Manna, E., Xiao, T., Shinar, J., and Shinar, R. (2015). Organic photodetectors in analytical applications. *Electronics* 4(3): 688–722.

19 Prabowo, B.A., Chang, Y.-F., Lee, -Y.-Y., Su, L.-C., Yu, C.-J., Lin, Y.-H., Chou, C., Chiu, N.-F., Lai, H.-C., and Liu, K.-C. (2014). Application of an OLED integrated with BEF and giant birefringent optical (GBO) film in a SPR biosensor. *Sens. Actuators B Chem.* 198: 424–430.

20 Krujatz, F., Hild, o., Fehse, K., Jahnel, M., Werner, A., and Bley, T. (2016). Exploiting the potential of oled-based photo-organic sensors for biotechnological applications. *Chem. Sci. J.* 7: 134.

21 Venkatraman, V. and Steckl, A.J. (2015). Integrated OLED as excitation light source in fluorescent lateral flow immunoassays. *Biosens. Bioelectron.* 74: 150–155.

22 Kraker, E., Haase, A., Lamprecht, B., Jakopic, G., Konrad, C., and Kostler, S. (2008). Integrated organic electronic based optochemical sensors using polarization filters. *Appl. Phys. Lett.* 92(3): 033302.

23 Dandin, M., Abshire, P., and Smela, E. (2007). Optical filtering technologies for integrated fluorescence sensors. *Lab. Chip.* 7: 955–977

24 Obahiagbon, U., Smith, J.T., Zhu, M., Katchman, B.A., Arafa, H., Anderson, K.S., and Christenasser, J.M.B. (2018). A compact, low-cost, quantitative and multiplexed fluorescence detection platform for point-of-care applications. *Biosens. Bioelectron.* 117: 153–160.

25 Katchman, B.A., Smith, J.T., Obahiagbon, U., Kesiraju, S., Lee, Y.-K., O'Brien, B., Kaftanoglu, K., Blain Christen, J., and Anderson, K.S. (2016). Application of flat panel OLED display technology for the point-of-care detection of circulating cancer biomarkers. *Sci. Rep.* 6: 29057.

26 Pais, A., Banerjee, A., Klotzkin, D., and Papautsky, I. (2008). High-sensitivity, disposable lab-on-a-chip with thin-film organic electronics for fluorescence detection. *Lab. Chip.* 8: 794–800.

27 Shinar, J. and Shinar, R. (2008). Organic light-emitting devices (OLEDs) and OLED-based chemical and biological sensors: An overview. *J. Phys. D: Appl. Phys.* 41(13): 133001.

28 Eeshita, M., Fadzai, F., Rana, B., Joseph, S., and Ruth, S. (2015). Tunable near UV microcavity OLED arrays: Characterization and analytical applications. *Adv. Funct. Mater.* 25(8): 1226–1232.

29 Cho, E.J. and Bright, F.V. (2001). Optical sensor array and integrated light source. *Anal. Chem.* 73(14): 3289–3293.

30 Tam, H.L., Choi, W.H., and Zhu, F. (2015). Organic optical sensor based on monolithic integration of organic electronic devices. *Electronics* 4(3): 623–632.

31 Koetse, M., Rensing, P., Van Heck, G., Meulendijks, N., Kruijt, P., Enting, E., Wieringa, F., and Schoo, H. (2008). Optical sensor platforms by modular assembly of organic electronic devices. *Conference of the SPIE 7054, Organic Field-Effect Transistors VII and Organic Semiconductors in Sensors and Bioelectronics*, 70541I: 1–3.

32 McNaught, A.D. and Wilkinson, A. (2006). *IUPAC. Compendium of Chemical Terminology* (2nd ed.). Blackwell Scientific Publications. . XML on-line corrected version: http://goldbook.iupac.org (2006–) created by M. Nic, J. Jirat, and B. Kosata; updates compiled by A. Jenkins.

33 Sagmeister, M., Tschepp, A., Kraker, E., Abel, T., Lamprecht, B., Mayr, T., and Kostler, S. (2013). Enabling luminescence decay time-based sensing using integrated organic photodiodes. *Anal. Bioanal. Chem.* 405: 5975–5982.

34 Ghosh, D. (2008). Structurally integrated luminescence based oxygen sensors with organic LED/ oxygen sensitive dye and PECVD grown thin film photodetectors. Ph.D. thesis, Iowa State University. Available at http://lib.dr.iastate.edu/rtd/15698.

35 Ghosh, D., Shinar, R., Cai, Y., Zhou, Z., Dalal, V.L., and Shinar, J. (2007). Advances in oled-based oxygen sensors with structurally integrated OLED, sensor film, and thin-film Si photodetector. *Proc. SPIE*, 6659: 7.

36 Savvate'ev, V., Chen-Esterlit, Z., Aylott, J.W., Choudhury, B., Kim, C.-H., Zou, L., Friedl, J.H., Shinar, R., Shinar, J., and Kopelman, R. (2002). Integrated organic light-emitting device/ fluorescence-based chemical sensors. *Appl. Phys. Lett.* 81(24): 4652–4654.

37 Choudhury, B.J., Shinar, R., and Shinar, J. (2004), Luminescent chemical and biological sensors based on the structural integration of an OLED excitation source with a sensing component. *Proceedings of the SPIE*, 5214: 64–72.

38 Cai, Y., Shinar, R., Zhou, Z., and Shinar, J. (2008). Multianalyte sensor array based on an organic light emitting diode platform. *Sens. Actuators B Chem.* 134(2): 727–735.

39 Peumans, P., Bulovie, V., and Forrest, S.R. (2000). Efficient, high-bandwidth organic multilayer photodetectors. *Appl. Phys. Lett.* 76(26): 3855–3857.

40 Clifford, J.P., Konstantatos, G., Johnston, K.W., Hoogland, S., Levina, L., and Sargent, E.H. (2008). Fast, sensitive and spectrally tuneable colloidal-quantum-dot photodetectors. *Nat. Nanotechnol.* 4(1).

41 Liu, R., Xiao, T., Cui, W., Shinar, J., and Shinar, R. (2013). Multiple approaches for enhancing all-organic electronics photoluminescent sensors: Simultaneous oxygen and PH monitoring. *Anal. Chim. Acta.* 778: 70–78.

42 Nalwa, K.S., Cai, Y., Thoeming, A.L., Shinar, J., Shinar, R., and Chaudhary, S. (2010). Polythiophene-fullerene based photodetectors: Tuning of spectral response and application in photoluminescence based (bio)chemical sensors'. *Adv. Mater.* 22(37): 4157–4161.

43 Lamprecht, B., Tschepp, A., Cajlakovic, M., Sagmeister, M., Ribitsch, V., and Kostler, S. (2013). A luminescence lifetime- based capillary oxygen sensor utilizing monolithically integrated organic photodiodes. *Analyst* 138: 5875–5878.

44 Mayr, T., Abel, T., Kraker, E., Kostler, S., Haase, A., Konrad, C., Tscherner, M., and Lamprecht, B. (2010). An optical sensor array on a flexible substrate with integrated organic opto-electric devices. Eurosensor XXIV Conference, *Procedia Eng.* 5: 1005–1008.

45 Kulkarni, T. and Slaughter, G. (2016). Application of semipermeable membranes in glucose biosensing. *Membranes* 6(4): 55.

46 Nichols, S.P., Koh, A., Storm, W.L., Shin, J.H., and Schoenfisch, M.H. (2013). Biocompatible materials for continuous glucose monitoring devices. *Chem. Rev.* 113 (4): 2528–2549. PMID: 23387395.

47 Witkowska Nery, E., Kundys, M., Jelerl, P.S., and Jonsson-Niedziolka, M. (2016). Electrochemical glucose sensing: Is there still room for improvement? *Anal. Chem.* 88(23): 11271–11282. PMID: 27779381.

48 Makaram, P., Owens, D., and Aceros, J. (2014). Trends in nanomaterial-based non-invasive diabetes sensing technologies. *Diagnostics* 4(2): 27–46.

49 Li, Z., Chen, Y., Xin, Y., and Zhang, Z. (2015). Sensitive electrochemical nonenzymatic glucose sensing based on anodized cuo nanowires on three-dimensional porous copper foam. *Sci. Rep.* 5: 16115. PMID: 26522446.

50 Park, S., Chung, T.D., and Kim, H.C. (2003). Nonenzymatic glucose detection using mesoporous platinum. *Anal. Chem.* 75(13): 3046–3049.

51 Choudhury, B., Shinar, R., and Shinar, J. (2004b). Glucose biosensors based on organic light-emitting devices structurally integrated with a luminescent sensing element. *J. Appl. Phys.* 96(5): 2949–2954.

52 Cunningham, A.J. (1998). *Introduction to Bioanalytical Sensors, Techniques in Analytical Chemistry*. Wiley Interscience.

53 Banerjee, A., Shuai, Y., Dixit, R., Papautsky, I., and Klotzkin, D. (2010). Concentration dependence of fluorescence signal in a microfluidic fluorescence detecto. *J. Lumin.* 130(6): 1095–1100.

54 Hofmann, o., Wang, X., deMello, J.C., Bradley, D.D.C., and deMello, A.J. (2005). Towards microalbuminuria determintionon a disposable diagnostic microchip with integrated fluorescence detection based on thin-film organic light emitting diodes. *Lab. Chip.* 5(8): 863–868.

55 Lefevre, F., Juneau, P., and Izquierdo, R. (2015). Integration of fluorescence sensors using organic optoelectronic components for microfluidic platform. *Sens. Actuators B Chem.* 221: 1314–1320.

56 Manzano, M., Cecchini, F., Fontanot, M., Iacumin, L., Comi, G., and Melpignano, P. (2015). Oled-based dna biochip for campylobacter spp. detection in poultry meat samples. *Biosens. Bioelectron.* 66: 271–276.

57 Marcello, A., Sblattero, D., Cioarec, C., Maiuri, P., and Melpignano, P. (2013). A deep-blue oled-based biochip for protein microarray fluorescence detection. *Biosens. Bioelectron.* 46: 44–47.

58 Ryu, G., Huang, J., Hofmann, O., Walshe, C.A., Sze, J.Y.Y., McClean, G.D., Mosley, A., Rattle, S.J., deMello, J.C., deMello, A.J., and Bradley, D.D.C. (2011). Highly sensitive fluorescence detection system for microfluidic lab-on-a-chip. *Lab. Chip.* 11: 1664–1670.

59 Venkatraman, V. and Steckl, A.J. (2017). Quantitative detection in lateral flow immunoassay using integrated organic optoelectronics. *IEEE Sens. J.* 17(24): 8343–8349.

60 Yun Shuai, A., Banerjee, D., Klotzkin, I., and Papautsky, I. (2008). On-chip fluorescence detection using organic thin film devices for a disposable lab-on-a-chip. IEEE, 169–172.

61 Zhu, M., Obahiagbon, U., Anderson, K.S., and Christen, J.B. (2017). Highly sensitive fluorescence-based lateral flow platform for point-of-care detection of biomarkers in plasma. 2017 IEEE Healthcare Innovations and Point of Care Technologies (HI-POCT), 249–252.

62 Obahiagbon, U., Kullman, D., Smith, J.T., Katchman, B.A., Arafa, H., Anderson, K.S., and Christen, J.B. (2016), Characterization of a compact and highly sensitive fluorescence-based detection system for point-of-care applications. IEEE Healthcare Innovation Point-Of-Care Technologies Conference 2016 (HI-POCT), 117–120.

63 Choi, S., Kwon, S., Kim, H., Kim, W., Kwon, J.H., Lim, M.S., Lee, H.S., and Choi, K.C. (2017). Highly flexible and efficient fabric-based organic light-emitting devices for clothing-shaped wearable displays. *Sci. Rep.* 7(1): 6424.

64 Hofmann, O., Bradley, D.D.C., deMello, A.J., and deMello, J.C. (2008). *Lab-on-a-Chip Devices with Organic Semiconductor-Based Optical Detection*, 97–140. Springer Berlin Heidelberg.

65 Ehsan, S., Aldo, B., Dev, C., Ken, C., Craig, C., Kevin, C., Flynn, M.J., Bradley, H., Nick, H., Jeffrey, J., Moxley-Stevens, D.M., William, P., Hans, R., Lois, R., Ehsan, S., Jeffrey, S., Robert, A.U., Jihong, W., and Charles, E.W. (2005). Assessment of display performance for medical imaging systems: Executive summary of aapm tg18 report. *Med. Phys.* 32(4): 1205–1225.

66 Indrajit, I.K. and Verma, B.S. (2009). Monitor displays in radiology: Part 2. *Indian J. Radiol. Imaging* 19(2): 94–98. IJRI-19-94 [PIT].

67 Khan, Y., Ostfeld, A.E., Lochner, C.M., Pierre, A., and Arias, A.C. (2016). Monitoring of vital signs with flexible and wearable medical devices. *Adv. Mater.* 28(22): 4373–4395.

68 Allen, J. (2007). Photoplethysmography and its application in clinical physiological measurement. *Physiol. Meas.* 28(3): R1–R39.

69 Smith, J., Bawolek, E., Lee, Y.K., O'Brien, B., Marrs, M., Howard, E., Strnad, M., Christen, J.B., and Gory, M. (2015). Application of flexible flat panel display technology to wearable biomedical devices. *Electron. Lett.* 51(17): 1312–1314.

70 Lochner, C.M., Khan, Y., Pierre, A., and Arias, A.C. (2014). All-organic optoelectronic sensor for pulse oximetry. *Nat. Commun.* 5: 5745.

71 Koetse, M., Rensing, P., Van Heck, G., Sharpe, R., Allard, B., Wieringa, F., Kruijt, P., Meulendijks, N., Jansen, H., and An Schoo, H. (2008). *In plane optical sensor based on organic electronic devices. Proc. of the SPIE*, 7054: 1–8.

72 Quang, T.T. and Nae-Eung, L. (2016). Flexible and stretchable physical sensor integrated platforms for wearable human- activity monitoringand personal healthcare. *Adv. Mater.* 28(22): 4338–4372.

73 Xu, H., Liu, J., Zhang, J., Zhou, G., Luo, N., and Zhao, N. (2017). Flexible organic/inorganic hybrid near infrared photo- plethysmogram sensor for cardiovascular monitoring. *Adv. Mater.* 29(31): n/a–n/a.

74 Yokota, T., Zalar, P., Kaltenbrunner, M., Jinno, H., Matsuhisa, N., Kitanosako, H., Tachibana, Y., Yukita, W., Koizumi, M., and Someya, T. (2016). Ultraflexible organic photonic skin. *Sci. Adv.* 2(4): e1501856. PMID: 27152354 .

75 Yeh, N., Ding, T.J., and Yeh, P. (2015). light-emitting diodes light qualities and their corresponding scientific applications'. *Renew. Sustain. Energy Rev.* 51: 55–61.

76 Ennever, J., McDonagh, A., and Speck, W. (1983). Phototherapy for neonatal jaundice: Optimal wavelengths of light. *J. Pediatr.* 103(2): 295–299.

77 Cremer, R., Perryman, P., and Richards, D. (1958). Influence of light on the hyperbilirubinaemia of infants. *The Lancet* 271(7030): 1094–1097. Originally published as 1, 7030.

78 Zecca, E., Barone, G., Luca, D.D., Marra, R., Tiberi, E., and Romagnoli, C. (2009). Skin bilirubin measurement during phototherapy in preterm and term newborn infants. *Early Hum. Dev.* 85(8): 537–540.

79 Smith, J., Shah, A., Lee, Y., O'Brien, B., Kullman, D., Sridharan, A., Muthuswamy, J., and Blain Christen, J. (2016a). Optogenetic neurostimulation of auricular vagus using flexible oled display technology to treat chronic inflammatory disease and mental health disorders. *Electron. Lett.* 52(11): 900–902.

80 Stoppa, M. and Chiolerio, A. (2014). Wearable electronics and smart textiles: A critical review. *Sensors* 14(7): 11957–11992.

81 Meunier, L., Kelly, F.M., Cochrane, C., and Koncar, V. (2011). Flexible displays for smart clothing: Part ii electrochromic displays. *Indian J. Fibre Text. Res.* 36(4): 429–435.

82 Cochrane, C., Hertleer, C., and Schwarz-Pfeiffer, A. (2016). 2 – smart textiles in health: An overview. In: *Smart Textiles and Their Applications* (ed. V. Koncar), 9–32. Woodhead Publishing Series in Textiles, Woodhead Publishing.

83 Kim, Y., Wang, H., and Mahmud, M. (2016). WWW Wearable body sensor network for health care applications. In: *Smart Textiles and Their Applications* (ed. V. Koncar), 161–184. Woodhead Publishing Series in Textiles, Woodhead Publishing.

84 McCann, J. (2013). Smart protective textiles for older people. In: *Smart Textiles for Protection* (ed. R. Chapman), 244–275. Woodhead Publishing Series in Textiles, Woodhead Publishing.

85 Janietz, S., Gruber, B., Schattauer, S., and Schulze, K. (2013). Integration of OLEDs in textiles. In: *Smart and Interactive Textiles (eds. P. Vincenzini and C.),* 14–21. *Advances in Science and Technology*, volume 80. Trans Tech Publications.

86 Cochrane, C., Meunier, L., Kelly, F.M., and Koncar, V. (2011). Flexible displays for smart clothing: Part I overview. *Indian J. Fibre Text. Res.* 36(4): 422–428.

87 Park, J.W. (2013). LL arge-area {OLED} lighting panels and their applications. In: *Organic Light-Emitting Diodes (OLEDs)* (ed. A. Buckley), 572–600. Woodhead Publishing Series in Electronic and Optical Materials, Woodhead Publishing.

88 Smith, J.T., Shah, S.S., Gory, M., Stowell, J.R., and Allee, D.R. (2014). Flexible isfet biosensor using igzo metal oxide tfts and an ito sensing layer. *IEEE Sens. J.* 14(4): 937–938.

89 Bansal, A.K., Hou, S., Kulyk, O., Bowman, E.M., and Samuel, I.D.W. (2014). Wearable organic optoelectronic sensors for medicine. *Adv. Mater.* 27(46): 7638–7644.

90 Abel, T., Sagmeister, M., Lamprecht, B., Kraker, E., Kostler, S., Ungerbock, B., and Mayr, T. (2012). Filter-free integrated sensor array based on luminescence and absorbance measurements using ring-shaped organic photodiodes. *Anal. Bioanal. Chem.* 404(10): 2841–2849.

91 Bernhard, L., Elke, K., Martin, S., Stefan, K., Nicole, G., Harald, D., Birgit, U., Tobias, A., and Torsten, M. (2011). Integrated waveguide sensor utilizing organic photodiodes. *Phys. Status Solidi Rapid Res. Lett.* 5(9): 344–346.

92 Chamika, H., Rajiv, K., Eeshita, M., Teng, X., Akshit, P., Rana, B., Dennis, S.W., Tom, T., Joseph, S., and Ruth, S. (2018). Enhanced light extraction from oleds fabricated on patterned plastic substrates. *Adv. Opt. Mater.* 6(4): 1701244.

Index

Note: Page numbers followed by "*f*" and "*t*" indicate figures and tables, respectively.

Flexible Flat Panel Displays, Second Edition. Edited by Darran R. Cairns, Dirk J. Broer, and Gregory P. Crawford.
© 2023 John Wiley & Sons Ltd. Published 2023 by John Wiley & Sons Ltd.